International Conference on 'Quantum Control, Exact or Perturbative, Linear or Nonlinear' to celebrate 50 years of the scientific career of Professor Bogdan Mielnik

(Mielnik50)

Journal of Physics: Conference Series Volume 624

Mexico City, Mexico
22 – 24 October 2014

Editors:

Nora Bretón
David Fernández
Piotr Kielanowski

ISBN: 978-1-5108-0757-0
ISSN: 1742-6588

Printed from e-media with permission by:

Curran Associates, Inc.
57 Morehouse Lane
Red Hook, NY 12571

Some format issues inherent in the e-media version may also appear in this print version.

Printed by Curran Associates, Inc. (2015)

For permission requests, please contact the Institute of Physics
at the address below.

Institute of Physics
Dirac House, Temple Back
Bristol BS1 6BE UK

Phone: 44 1 17 929 7481
Fax: 44 1 17 920 0979

techtracking@iop.org

Additional copies of this publication are available from:

Curran Associates, Inc.
57 Morehouse Lane
Red Hook, NY 12571 USA
Phone: 845-758-0400
Fax: 845-758-2634
Email: curran@proceedings.com
Web: www.proceedings.com

TABLE OF CONTENTS

INTERDISCIPLINARY PROBLEMS IN QUANTUM MECHANICS

Preface

The International Conference on 'Quantum Control, Exact or Perturbative, Linear or Nonlinear', took place in Mexico City on 22-24 October 2014. It was held with the aim of celebrating the first fifty years of scientific career of Bogdan Mielnik, an outstanding scientist whose professional trajectory spans over Poland and Mexico and who is currently Professor Emeritus in the Physics Department of Centro de Investigación y de Estudios Avanzados del IPN (Cinvestav) in Mexico.

Bogdan Mielnik was born on May 6th, 1936 in Warsaw, Poland. He studied elementary and high school until 1953. In the autumn of 1953 he started the studies in the Faculty of Mathematics and Physics at the University of Warsaw, and at the end of 1957 he did his master work under the direction of Professor Jerzy Plebański. In 1962 he was invited to the newly opened Research Center of IPN (Cinvestav), in Mexico, as an assistant and PhD student of Jerzy Plebański. On October 22nd, 1964, he submitted to Cinvestav his PhD Thesis entitled *Analytic functions of the displacement operator*, marking the official beginning of his scientific career. It is worth mentioning that Bogdan Mielnik is the first PhD graduate of the Physics Department of Cinvestav, so with this Conference our Department was also celebrating an important date on its calendar. A more detailed information can be found in the website `http://www.fis.cinvestav.mx/mielnik50/`.

It was our great pleasure to see that many collaborators and former students of Bogdan Mielnik attended this Conference. The articles collected in this volume are the written contributions of the majority of talks presented at the conference. They have been organized according to the research subjects that Bogdan Mielnik has been involved in. Thus, the articles of JG Hirsch, L Hughston, G Morales-Luna, O Rosas-Ortiz and G Torres-Vega deal with *Fundamental Problems in Quantum Mechanics*. On the other hand, the papers by F Delgado, H Hernández-Coronado, G Herrera Corral, F Rojas, KB Wolf and M Znojil belong to the subject of *Quantum Control and Dynamical Manipulation*, while the articles of D Bermudez, A Contreras-Astorga, E Díaz-Bautista, JC González, V Hussin and VS Morales-Salgado are related with *Factorization Method, Supersymmetric Quantum Mechanics and Coherent States.* Finally, the papers of S Cruz y Cruz, M Enríquez, A Jaimes-Nájera and R Kerner address some *Interdisciplinary Problems in Quantum Mechanics.*

Participants of the Conference.

We would like to conclude by thanking for the support of the Physics Department of Cinvestav, the Academic Affairs Office of Cinvestav, and the Mexican National Council of Science and Technology (Conacyt, projects 152574 and 166581). Without their support, neither the Conference would have been held nor this Conference Volume would have been published.

The Editors
Nora Bretón, David Fernández, Piotr Kielanowski
Physics Department, Cinvestav

Mexico City, April 2015

Journal of Physics: Conference Series **624** (2015) 011001 doi:10.1088/1742-6596/624/1/011001

Bogdan Mielnik's Inauguration Talk

Esteemed Colleagues:

In my vagabond life, I had not many prizes or recognitions. I was thus surprised when one day I was notified that there was a silver sculpture of some prehistorical figure (Juchiman) waiting for me as a scientific prize offered by the University of Tabasco. Hmm... but, who was Juchiman? Everything indicates that he indeed existed (with some advantage over Saint Juan Diego). He was not a God, nor Saint but a soldier, perhaps a guardian living more than 3000 years ago in the kingdom of Olmecas (still before Mayas). His enormous, pensive figure, sculptured in stone was discovered by archeologists in the jungle of Tabasco. There are a lot of unsolved questions. Obviously, Juchiman could not predict the present day science, and its troubles. Did his figure need to wait until so distant future (more than 3000 years) to participate in our scientific discussions?

And suddenly, against any logic, I had a feeling that I have received some message from the past. What could it mean? Was I supposed to do something? I was thinking intensely, but could not guess.

Then, one night, I had a curious dream. I found myself in the Tabasco jungle about 3000 years ago and I have seen an enormous figure of Juchiman, who seemed submerged in some deep meditation. I felt, I had some occasion to ask... I approached carefully:

"Esteemed Juchiman!" I said, "I came from a distant epoch, which is in an enormous trouble. Pulverized by wars, our world is now a victim of a horrible crisis. Perhaps, you could have some advice, from your unpolluted, primeval world. What should we do?"

Uneasy conversation (picture by F. Vergara).

Juchiman for a moment seemed indifferent. Then he looked at me with a strange grin:

"In the first place", he told, "We are here in much worse crisis! So please do not distract me. If I commit an error, a catastrophe can happen to all our civilization. This is not only our problem, for it may happen that your extremely artificial future will never materialize. So, you will have nowhere to return, in fact, you will neither be born... So, please, don't distract me. For your own good, just vanish as fast as you can!..."

I suddenly felt a horror... I made an effort to return, but I couldn't... Once more, and nothing: I was still facing Juchiman in the Tabasco jungle! I made almost a superhuman effort, and suddenly... Bam! I was back. Uff!... I looked over my room: it seemed more or less the same. I dressed fast and went to the Physics Department. It seemed more or less the same... It looks that the catastrophe was averted!

So, be welcome in our artificial future! Maybe, it will be an occasion to discuss our artificial theories? Let's enjoy the occasion! Best regards!

Bogdan Mielnik

Mexico City, October 22, 2014

The organizers of the conference would like to thank for the support from the Physics Department and the Office of Academic Affairs of the Centro de Investigación y de Estudios Avanzados and from the Consejo Nacional de Ciencia y Tecnología (Conacyt), projects 152574 and 166581.

Journal of Physics: Conference Series **624** (2015) 011003 doi:10.1088/1742-6596/624/1/011003

Peer review statement

All papers published in this volume of *Journal of Physics: Conference Series* have been peer reviewed through processes administered by the proceedings Editors. Reviews were conducted by expert referees to the professional and scientific standards expected of a proceedings journal published by IOP Publishing.

Mielnik50 IOP Publishing
Journal of Physics: Conference Series **624** (2015) 012001 doi:10.1088/1742-6596/624/1/012001

Quantumness, Randomness and Computability

Aldo Solis and Jorge G. Hirsch

Instituto de Ciencias Nucleares, Universidad Nacional Autónoma de México, Apdo. Postal 70-543, México 04510 D.F.

E-mail: hirsch@nucleares.unam.mx

Abstract. Randomness plays a central role in the quantum mechanical description of our interactions. We review the relationship between the violation of Bell inequalities, non signaling and randomness. We discuss the challenge in defining a random string, and show that algorithmic information theory provides a necessary condition for randomness using Borel normality. We close with a view on incomputablity and its implications in physics.

1. Introduction

The empirical success of quantum mechanics in the description of microscopic phenomena is indeed impressive. We visualize matter as made of molecules, formed by atoms with electrons orbiting around a nucleus with protons and neutrons, which are built with quarks, bound together by interaction carriers, like gluons, pions and photons. The mathematical formalism allows the description of element formation inside the stars, the structure of materials, the energetic collisions which take place in many scenarios including high-energy accelerators. A plethora of new technologies, devices, materials, has born out of it.

At the same time, quantum physics has put at the reach of our hands the limitations of our imagination, of our capacity to describe employing words this microscopic world. The very notion of reality is waning, despite many efforts on the opposite direction. Particles and waves are two central notions of the classical description of the physical world. They are associated with the objects, formed by particles, and the interactions between them, whose perturbations propagate as waves. Particles are localized, waves have extended fronts. A century of wave-particle duality has messed it all. The violation of the Bell inequalities opened a Pandora's box. If we insist in having well defined objects in the quantum domain, they must have very strange properties: a measurement of an observable property in one part of a quantum system is correlated with the outcome of a different measurement realized in other part of the system in a way which can be described as non-local, contextual and at random. This property, called entanglement, is found in most composed quantum systems. Good overviews on these subjects were published associated with the 50 years of Bell's theorem [1, 2].

By non-local it is implied that the correlations in the outcomes of the measurements are obtained in space-time regions which are causally disconnected, i.e., no signal traveling at most at the velocity of light in vacuum can be sent from one region to the other. Contextual refers to a connection between the measurement device and the property to be measured. The quantum formalism associates observables with expectation values of non-commuting operators, implying the uncertainty principle. The better the value of one observable is known, the lesser we can say about the other. More than that, if these complementary properties are assumed to have definite but unknown values when they are not measured, flat contradictions appear in their description. The simplest way to avoid these contradictions is to assume that, if these properties exist at all, they manifest themselves in accordance to the measurement device, the context in which the experiment is performed.

If the quantum correlations are associated with quantum objects, they must have non-local and contextual properties. What would prevent us to employ contextuality, measuring in a given way one part of quantum system to generate through the non-local correlation a specific outcome of the measurement in the other part, far away, sending in this way a signal faster than light? We cannot do that, because we can select the property to be measured, but cannot influence the outcome: it is either trivial, when we measure a property which was already prepared to have a specific value, or random, when we measure a complementary property.

We can see that randomness is an imprescindible companion of non-locality and contextuality, if no-signaling (the short name of the impossibility to send information faster than light in vacuum) would be enforced. And we really want it, because special relativity tells us that being able to send information between causally disconnected space-time regions is equivalent to send information back in time is some reference frames. And this would cause logical paradoxes which would demolish the basics of science as we know them.

When instead of thinking about objects and their interactions, the quantum system is analyzed from the information perspective, interesting alternative views arise. Reality and information can be considered as two sides of the same coin. Randomness, complementarity and entanglement emerge from the fact that from individual measurements there is a finite amount of information available. One way to formalize these ideas is to postulate that an elementary system can only give a definite result in one specific measurement. It follows that the outcomes of other independent measurements must be irreducibly random [3].

When the intrinsic quality of quantum randomness is accepted as a postulate, a world of applications emerges. "Private randomness" is defined by the presence of correlations that cannot be reproduced with local variables. It is quantified by the violation of Bell inequalities, and is associated with the impossibility to predict a given string employing a classical computer and classical information [4–8].

The concept of probability is associated with quantum predictions through the Born rule: the square of the wave function provides the probability density of observing a given event. It is associated with an intrinsic quantum randomness. Employing composite systems, like pairs of entangled photons, the violation of Bell inequalities guarantees their quantum origin. It is assumed to imply that the information generated with their measurements did not exist before them, certifying their privacy, and that they are "as random as they can be".

Can we prove that a sequence of measurements is already random? This is a challenging question, whose answer is explored in the following section.

2. Randomness

How can be known that a given series of numbers is random? Suppose we give you the sequence "1,5,7,0,7". It looks pretty random, and maybe you accept it as a good candidate for a sequence of random numbers: But if we give you the numbers "1,2,3,4,5,6" instead; you can say: "Well, that's not random, you are just counting". We offer you the sequence "2,4,6,8,10,12"; and you answer: "you are saying the same numbers times two!". With our best effort we come with the sequence "1,1,2,3,5,8" and irritated you say: "That's the beginning of the Fibonacci sequence!". Even if you accepted the first sequence as random, we may say it is not, a closer look reveals that it contains the first digits of $\pi/2$.

It looks like we can always find a way to argue that a sequence is not random. The sequences above are "easy", but take "8,3,2,0,9"; it does not seem to have a pattern and if we say something like: take the first digit and add -5, then add -1, then -2 and finally add 9, it may look like we just made that up. So what descriptions can be considered as patterns?

The notion of randomness in this context is "the absence of a pattern". The idea of pattern refers to a procedure that describes the string. Such a procedure doesn't need any kind of decision or creativity, is just a set of instructions that need to be followed to reproduce the string in an "algorithmic" way [9].

2.1. Turing Machines

One of the greatest achievements in the past century was the formalization of the algorithmic procedures, this was done in many different ways and all of them turn out to be equivalent. The most intuitive one is due to Alan M. Turing and is called "Turing Machines" [10]. An excellent exposure of this subject can be find in [11].

A Turing machine can be seen as a man following a set of rules, this man can read and write in a boundless (we don't want to limit the amount of paper he can use) tape divided in squares, every square can have only one symbol at a time. He can read only one symbol at a time and, depending on that symbol and the current state of mind ("memory and thoughts") he will write another symbol, change his state of mind and move to the left or right.[1]

This definition may seem abstract. However, nowadays we are so used to manipulate computers that another way to make this a comprehensible definition is to associate a Turing machine with a general purpose computer which has an infinite amount of memory (because the tape is boundless), running a specific program. This brings an interesting question: Is there a Turing machine that can be identified with a general purpose computer running any program? Yes, this is a really important class of machines called Universal Turing Machines (UTM). Every UTM can "run" every possible program, i.e. can mimic the behavior of every other Turing machine. This is done with the proper input, in the same way a general purpose computer can "run" every algorithm, which must be written in an appropriate code. Therefore, for every machine M with input i must exist an input x such that a UTM $U(x)$ can imitate $M(i)$ when is fed with x, and therefore $M(i) = U(x)$.

2.2. Information, Randomness and Turing machines.

The existence of a UTM is very important for our purposes. We cannot take a Turing machine as a description of a pattern, because for every finite sequence we can make a Turing machine whose output is the sequence. Just take a program that says something like

print "0111001..."

(putting the sequence as argument of 'print'). In this way we can always find an algorithm that prints the sequence. The trick is simple: the sequence is contained in the program. But then, the number of symbols needed to print the sequence is by necessity larger than the sequence itself.

We can make contact with the intuition suggested above: when a pattern gets really complicated we think that it should not be considered as a pattern. In the realm of algorithms we can say: if the program is longer than the sequence itself that's not a pattern. This gives us the opportunity to define a pattern in a sequence as a program, whose output is the sequence, with fewer symbols than the sequence. The sequences without patterns as those we want to call random.

[1] The way to formalize memory and thoughts is just by defining a finite set of states $\{q_i\}$. Every time the machine makes an operation the state of the machine changes to another state q_j. The concept of state of mind can be applied to machines too. In the case of a calculator made with gears this states describe the physical state of the gears.

Our relation with computers provides us with a familiar example, where the key concept is compression. We employ many compression algorithms, like zip, to save space in the hard drive or send information faster. We can identify the strings that do have patterns with the compressible ones, because there is an algorithm whose output is the string. It offers a very good intuition about random strings: A random string cannot be compressed.

It may seem hard to find a random string but that's not the case. As an example take all the binary strings with 3 symbols, there are 8 of them. And there are only 6 strings shorter: 2 binary strings with one symbol (0 and 1), and 4 with two symbols (00, 01, 10 and 11). If we try to represent the binary strings with 3 symbols with shorter strings, at least two of them cannot be included, therefore there are at least 2 random strings. Algorithms are programs in a given language, which are represented by binary strings also. This argument can be applied to any number of symbols, implying that random strings are quite common.

2.3. Algorithmic information theory

The algorithmic information theory [9, 12, 13] was developed in the past century by Andrey Kolmogorov, Ray Solomonoff, Gregory Chaitin and others. Within this theoretical framework we can, instead of just talking about random and non-random strings, ask: what is the smallest number of symbols necessary to print a given sequence? For example, take the sequence made of '1' a million times. This sequence have a million symbols but it will be very simple to write a program to print it, and such a program will easily have less than a million symbols.

The least number of symbols is called the algorithmic information content of the sequence. Random sequences have, by definition, an information content close to their length. Other mathematical objects, like real numbers, can be identified with a binary sequences using their binary expansion. In this way we can define the algorithmic information in a real number. It is worth to mention that the paper in which Turing introduce Turing machines is dedicated to computable numbers, showing the most real numbers are incomputable.

With this formalism we can investigate randomness, in the way we have just defined it. Can we use it to construct an algorithm able to classify strings in random and non-random? Unfortunately, this is not possible. The demonstration is simple.

Suppose we have a method to distinguish between random and non-random strings, and this method is algorithmic, so can be written as a program and feed to a UTM. To complement it, we write another piece of code whose output contains all the possible strings, one by one, in order, i.e. 0,1,00,01,10,00,... . With these programs we can make another program following these steps:

(i) produce a sequence,
(ii) check if the sequence is random,
(iii) if it is random print it,
(iv) go to (i).

The outcome of these instructions is a sequence of random strings, all of them in order of length. Suppose the length of this new program is L symbols, this program will eventually print a random string of $L+1$ symbols. But then we have just printed a random sequence of $L+1$ symbols with less than $L+1$ symbols! This contradicts the definition of algorithmic randomness, therefore there cannot exist an algorithm that can decide the randomness of a string.

2.4. Prefix free strings and the Omega number

The algorithmic information theory has changed over time and today is really common to find it as prefix-free algorithmic information theory, where prefix-free is referred to the possible programs accepted by the UTM. For instance, if the UTM accepts "1000" as a program, then the string "10001", or any other starting with "1000", can't be accepted. In Figure 1 we represent all the possible binary strings, showing explicitly the first three lines, containing the sequences with 1, 2 and 3 symbols, respectively. Below every string we find all the possible extensions of this string. If a string is accepted as a program, all the possible strings below it can't be accepted as programs.

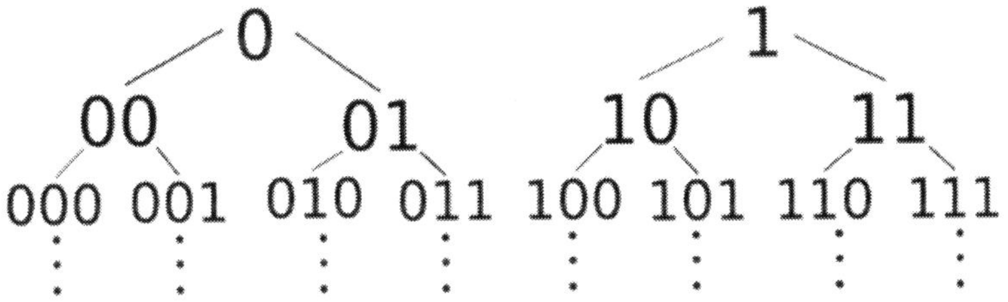

Figure 1. All possible binary string ordered in a tree structure.

A schematic procedure to obtain the accepted programs could be this: we have a coin and want to produce a program by flipping it several times, following these steps:

(i) Flip a coin.
(ii) Incorporate the output (0 or 1) at the end of the string you already have.
(iii) Check if the string of results is an accepted program.
(iv) if it is accepted, halt; if not go to (i).

Note that we are not missing any possible program, because every time we obtain an accepted program, it closes a branch of the tree in Fig. 1. We are not allowed to continue adding new symbols to this string (because of the prefix-free condition), and must start building a new string from its first symbol. For those of us who are familiar with programming, this step is like compiling the code in a given programming language, and checking if it compiles, i.e., is an accepted program, or not. Each different string of length n has a probability $P_i = \frac{1}{2^n}$ of being generated. The prefix-free condition guarantees that $\sum_{i=\text{possible programs}} P(i) = 1$.

Once we have an accepted program, which in this view includes all the input data files, the next question is: will it generate an output? Which is equivalent to the question: will this program stop, or will it be running forever? Running a program we can learn if it halts, in the event it halts. The problem is to decide when to give up on a program that does not halt. While many special cases can be answered, Turing demonstrated that a general solution is impossible. No algorithm, no mathematical theory, can ever tell us which programs will halt. The demonstration goes in the following direction: Suppose there is an algorithm that takes the description of a Turing machine and some input x, this algorithm does not halt if the machine halt when is feed with the input x and halt if the machine will be running forever. Because this is an algorithm there must be a Turing machine that formalize this, and we can feed this algorithm with its own description, and because of the way we define it this machine does not halt if it halts, which is an obvious contradiction. Therefore there cannot exist an algorithm able to decide if a Turing machine will halt. This is known as the halting problem, and is closely related with Gödel's theorem, which states that there is an infinite number of true statements in Mathematics which cannot be formally proved [9].

Chaitin considered the ensemble of all possible programs, and asked the question: What is the probability of getting a halting program chosen at random from the ensemble, as outlined in the example given above? If we can associate to each accepted program i a function defined as $H(i) = 1$ is the program halts, and $H(i) = 0$ if not, we can evaluate the probability of obtaining a halting program for a given Universal Turing Machine U as

$$\Omega_U = \sum_{i=\text{possible programs}} H(i)P(i) = \sum_{i=\text{possible programs}} H(i)\frac{1}{2^{n(i)}}$$

where $n(i)$ is the number of bits (symbols in the general case) in the program i. Ω_U is called the Chaitin's Omega number for U.

Omega is perfectly well defined. It is a specific number which can have a value between zero and one (is a probability), but it is impossible to compute in its entirety. Some of its first digits can be calculated. For example, if the computer programs 1, 00 and 010 halt in a given universal Turing Machine, the first digits of this Ω_U would be 0.111. But the relevant aspect is that the first N digits of Ω_U cannot be computed using a program significantly shorter than N bits long. In this sense, any Ω_U has an infinite number of bits which are well and uniquely defined, but cannot be calculated by any finite program [9]. It does not matter how long is the program, the number of bits which are impossible to calculate is infinite. It is worth to mention that Ω_U is completely different from some irrational numbers, like π or $\sqrt{2}$, for whom billions of digits can be computed with quite compact programs [14].

The consequences of the impossibility to compute Ω_U go far beyond computational sciences. Together with the findings of Gödel and Turing, they demonstrate that, given any finite set of axioms, there is an infinite number of truths that are unprovable in that system. It implies directly that an infinite number of mathematical true statements are impossible to prove, they must be assumed as new axioms, without any justification. Philosophically, we are touching the limits of the principle of sufficient reason, which states that everything happens for a reason or, in other words, if something is true, there must be a reason for that. In mathematics most of what is true has no reason at all [9].

3. Randomness in physics and Borel normality

The close connection between the previous section and the foundation of physics can be seen in the following quote "The discovery that individual events are irreducibly random is probably one of the most significant findings of the twentieth century ... for the individual event in quantum physics, not only do we not know the cause, there is no cause. The instant when a radioactive atom decays, or the path taken by a photon behind a half-silvered beamsplitter are objectively random" [15].

In physics we employ the concept of random in different contexts. Sometimes we say something is random when we ignore the state of the system, sometimes when the behavior of the system is really complicated (that's really close to the previous definitions), and finally we have intrinsic randomness, applied to individual quantum systems where we can't, in general, predict the result of individual measurement outcomes.

What we mean when we say something is unpredictable? In the case of quantum systems we say that because the result of individual measurements are not, in general, in a one to one relation with the conditions of the experiment. One could say that there are other parameters (known as hidden variables) that determine the outcome, but thanks to the so called no-go theorems, e.g. Bell's theorem [16], Kochen-Specker [17], etc., this is unlikely to be the case. The most accepted interpretation, at least in the area of quantum information, is that the result of any individual measurement is "created" in the act of measurement. This assumption provides the basis of quantum random number generation, certified using Bell-type inequalities [18, 19].

We are confronted with two different concepts of randomness, the first one related with algorithmic information theory that define a random sequence as a incompressible one, and the second one related with unpredictability of measurements. On one side they are similar because, if a sequence is incompressible, it has no patterns at all and if we can't predict something is because there are no regularities or patterns that we can use to predict the result of the measurement. On the other side the word random is used, in the first case, to describe a property of sequences of numbers while the second one is applied to single experiments where each measurement gives us a number.

Can we relate both concepts? Imagine we make a series of measurements of polarization of photons, assigning a 1 if we get vertical polarization and a 0 in the other case, and put all the results together forming a string. Is the final sequence compressible? It will depend on the initial state the photons were prepared. If we prepare all of them in a similar vertical polarization $|1\rangle$, we can predict that we will measure a sequence like "11111. . . 1" that is very compressible. If we prepare the photons in the state $\frac{1}{10}|0\rangle + \frac{\sqrt{99}}{10}|1\rangle$, even though we don't know the result of any measurement in advance, we can anticipate that the outputs will form a compressible sequence. Suppose we get something like:

"111111101111111111110111111110111111111111111111111111101111111010..."

where the ratio between zeroes and ones is $\frac{1}{99}$. We can compress this sequence by writing the number of 1 instead of 1 themselves, like this:

"70110802407010. . . "

We can reconstruct the original sequence just by replacing a number between zeroes by its equivalent in ones, and to perform this translation doesn't need too much code.

Due to the fact we can compress sequences with more ones than zeroes (or more zeroes than ones), both must appear the same number of times. Therefore the most incompressible sequence must have the form:

$$\frac{|0\rangle + e^{i\theta}|1\rangle}{\sqrt{2}}.$$

In this case there is no obvious way to compress the sequence. While there is no algorithm able to test the compressibility, we can use another condition related to compressibility. From the examples above it is clear that we need equal number of zeroes and ones; but suppose you split the sequences every two symbols:

"11 10 11 01 11 11 11 01 1101 . . . "

Employing the same arguments of compressibility, we must ask for this number to be random that all two-symbol subsequences appear the same number of times. Let $N(x, y)$ be the number of occurrences of the substring y in the string x then we need:

$$N(x, \text{"0"}) \approx N(x, \text{"1"}) \approx L\frac{1}{2},$$

where L is the length in bits of the sequence; and

$$N(x, \text{"00"}) \approx N(x, \text{"01"}) \approx N(x, \text{"10"}) \approx N(x, \text{"11"}) \approx \frac{L}{2}\frac{1}{4},$$

as we have $L/2$ subsequences of two symbols, each one with probability $\frac{1}{4}$. The general expression is:

$$N(x, y) \approx \frac{L}{n}\frac{1}{2^n},$$

where n is the length in bits of "y". And therefore:

$$\left|\frac{N(x, y)}{L/n} - \frac{1}{2^n}\right| < \epsilon$$

where ϵ is a "small" number.

C. Calude showed in [20] the relation between the previous ideas and incompressibility. The proposal is to select

$$\epsilon = \sqrt{\frac{\log L}{L}},$$

where again L is the length of the sequence, and check the above inequality to all subsequences shorter, in bits, than $\log(\log(L))$. If for all subsequences the inequality is satisfied we will say the sequence is Borel normal[2]. In [20] is proved that almost all incompressible sequences are Borel normal.

Borel normality is a necessary, but not sufficient condition, to incompressibility. An example of a Borel normal sequence that can be compressed is:

$$0100011011000001010011100101110111\ldots$$

which is called the Champernowne's constant, built with all binary numbers put in order and forming a single string (0, 1, 00, 01, 10, 11, ...). It was the first example of a Borel normal number, later is was realized that almost all numbers are normal numbers.

Even though we cannot compare effectively the relation between unpredictability and incompressibility or randomness, we can compare Borel normality and unpredictability in a sequence of individual measurement outcomes.

The concepts of Borel normality and incomputablity inspired a comparison between quantum and computer generated random sequences. While quantum randomness can be proven incomputable; that is, it is not exactly reproducible by any algorithm, software-generated random numbers, known as pseudo-random, can be reproduced if the computer code and the seed are known. Calude et al. [21] performed finite tests of randomness inspired by algorithmic information theory. They report that all tests produced evidence – with different degrees of statistical significance – of differences between quantum and non-quantum sources. But in a recent investigation of randomness in ten sequences of bits generated from the differences between detection times of photon pairs generated by spontaneous parametric downconversion, they are found fulfilling the randomness criteria without difficulties [22]. A deeper study is being carried out to clarify this point.

4. Computability and undecidability

We have explored the close relationship between quantum physics and randomness, the definition of random as the absence of a pattern, the association of a pattern with an algorithm, and, through algorithmic information theory, with Borel normality, arriving at a necessary condition for a string to be random which has been employed to analyze the randomness of sequences of photon detections.

We can now go a step forward, and explore the relation between physical laws, algorithms and their computability. As algorithms can be mapped to Turing machines, programs built with finite alphabets, they are infinitely many but numerables. The set of rational numbers is also infinite numerable, but the real numbers are non-numerable infinite. It follows that the absolute majority of irrational numbers are incomputable in principle: it is impossible to write a computer code to evaluate any of them. It is also impossible to name them, even trying to write a full encyclopedia devoted to each of them is worthless. This concept is thrilling for some of us. It invites to question the meaning of the physical theories, which associate physical observables with real numbers. But, has it any practical implication, or is just a curious but irrelevant information?

[2] This definition is related with the normal numbers E. Borel defined.

A physical law can be seen as a pattern, a regularity which can be described with a mathematical equation relating measurable quantities. It is assumed that the equation can be solved, given a set of initial or boundary conditions, and the solutions provide predictions for the future behavior of the system. Any practitioner has found practical difficulties in doing so. In many cases the dimensions of the basis in which the solution must be expressed are huge, exceeding the ever growing capacities of the supercomputers. In other cases the dynamical equations are chaotic, i.e., display extreme sensitivity to the initial conditions, which puts a limit in the predictive power independent of the computers. We are used to these limitations and have found ways to deal with them, estimating the uncertainties in the predictions and looking for more powerful methods to reduce them.

But there are some mathematical equations which do not have any algorithm which can be used to solve them. A beautiful example is given by the following problem in number theory: Does the equation $\sum a_{ijk...} x^i y^j z^k ... = 0, x_i, a_i \in \mathbb{Z}$, which is a polynomial equation with integer coefficients, have roots in the integers? It has been proved in [11] that there is no algorithm to answer this question. It is incomputable. No computer can answer it.

An interesting connection of this problem with quantum physics is offered in [23]. It is suggested to evaluate the square of the left hand side of the previous equation, which will then be always equal or greater than zero. It can be interpreted as a Hamiltonian, where the integer numbers are associated with photon numbers. If a physical realization of this Hamiltonian can be constructed, measuring the ground state energy of the system would provide an answer, if it is zero, to the integer values of $x, y, z, ..$ which are the solutions, proving their existence. There are many subtle points in this proposal, which can be explored in [23].

Are there relevant problems in physics belonging to this category? Examples of undecidability and incompleteness results in classical analysis within a sufficiently rich axiom system can be extended to undecidability and incompleteness results in axiomatized formulations of physics. Many undecidability and incompleteness results deal with the integrability of Hamilton-Jacobi equations, and with the possibility of proving that a given dynamical system has a chaotic behavior [24].

Consider the following question: Given a quantum many-body Hamiltonian, is the system it describes gapped or gapless? It is perhaps the simplest and most relevant question in condensed matter, and can be rephrased into the mass problem in quantum field theory. For a finite number of components (like spins in a Heisenberg chain), the solution can be numerically found. The challenge is to find the thermodynamic limit, when the number of components tends to infinity. It has recently been shown that "this problem is undecidable, in the same sense as the Halting Problem was proven to be undecidable by Turing. A consequence of this is that the spectral gap of certain quantum many-body Hamiltonians is not determined by the axioms of mathematics, much as Godels incompleteness theorem implies that certain theorems are mathematically unprovable." [25]

This is a really interesting situation. One is invited to consider the possibility to solve problems which are demonstrated to be incomputable, for whom it has been proved the impossibility to compute even the existence of a solution, by employing physical systems which exactly map the problem into an experimental setting. Measuring the output would provide the answer. In a way it would be a return to analogous computation. In fact, the digital computation we employ daily also employs programs which are written in bits, each having a physical realization in the computer. The difference is that in the usual case we could, in principle, perform by hand the same computation. To solve incomputable problems using devices which are not equivalent to Universal Turing Machines is the proposal of a research area called hypercomputation [26]. The possibility of effectively employing it in solving problems seems at the present very far away.

We can go further and apply this kind of ideas to measurement itself. The "physical variables" that we use are always considered as results of procedures that we can perform to measure them. Are this procedures algorithmic? We have algorithmic procedures to measure many physical quantities like distance or voltage (with some limitations). The creation or invention of the procedure is not algorithmic but, once it is written, every trained person having the proper equipment could perform it following a list of instructions. Can we imagine a "physical variable" whose process of measurement is not algorithmic? If we have a variable whose process is not algorithmic there should be no fixed list of instructions to measure it. This may seem unreal but take a Turing machine and suppose we want to measure if this machine will halt or not, there is no algorithm to do that, the only option to know the answer is to sit and wait.

5. Conclusion

The relation between physics and randomness is indeed rich, being randomness quite complicated by itself. Being considered in the past a curiosity or a metaphysical question outside of physics, the developments in the last decades in quantum optics an quantum information, in particular the applications of individual measurements of quantum systems to random number generation and security, have brought them to a central stage.

There are very interesting questions about the meaning in physics of concepts like law, pattern, and their computability. These questions are not only related with the results of measurements but with some central elements of physics like "physical variables".

Acknowledgements

We are very grateful for the contributions of Bogdan Mielnik to illuminate the dark side of quantum mechanics. This work was supported in part by Conacyt, Mexico. AS is a fellow of Conacyt, Mexico.

References

[1] Special issue on 50 years of Bell's theorem 2014 *Journal of Physics A: Mathematical and Theoretical* **47** number 42
[2] Wiseman H 2014 *Nature.* **510** 467–469
[3] Zeilinger A 1999 *Foundations of Physics,* **29** 631–643
[4] Scarani V 2010 *Nature* **464** 988–989
[5] Pironio et al S 2010 *Nature* **464** 1021–1024
[6] Christensen B G, McCusker K T, Altepeter J B, Calkins B, Gerrits T, Lita A E, Miller A, Shalm L K, Zhang Y, Nam S W, Brunner N, Lim C C W, Gisin N and Kwiat P G 2013 *Phys. Rev. Lett.* **111** 130406
[7] Bancal J D, Sheridan L and Scarani V 2014 *New J. Phys.* **16** 033011
[8] Nieto-Silleras O, Pironio S and Silman J 2014 *New J. Phys.* **16** 013035
[9] Chaitin G 2006 *Scientific American* **March** 74–81
[10] Turing A 1937 *Proceedings of the London Mathematical Society* **42** 230–265
[11] Davis M 1953 *Computability and Unsolvability* (Dover)
[12] Calude C 2002 *Information and Randomness: An Algorithmic Perspective* 2nd ed (Secaucus, NJ, USA: Springer-Verlag New York, Inc.) ISBN 3540434666
[13] Li M and Vitnyi P M 2008 *An Introduction to Kolmogorov Complexity and Its Applications* 3rd ed (Springer Publishing Company, Incorporated) ISBN 0387339981, 9780387339986
[14] Lopez-Ortiz A 1998 How to compute digits of pi? URL `https://cs.uwaterloo.ca/~alopez-o/math-faq/mathtext/node12.html`
[15] Zeilinger A 2005 *Nature* **438** 743
[16] Bell J S and Aspect A 2004 La nouvelle cuisine *Speakable and Unspeakable in Quantum Mechanics* (Cambridge University Press) pp 232–248 2nd ed ISBN 9780511815676 cambridge Books Online URL `http://dx.doi.org/10.1017/CBO9780511815676.026`
[17] Kochen S and Specker E 1975 The problem of hidden variables in quantum mechanics *The Logico-Algebraic Approach to Quantum Mechanics* (*The University of Western Ontario Series in Philosophy of Science* vol 5a) ed Hooker C (Springer Netherlands) pp 293–328 ISBN 978-90-277-0613-3 URL `http://dx.doi.org/10.1007/978-94-010-1795-4_17`

[18] Nieto-Silleras O, Pironio S and Silman J 2014 *New Journal of Physics* **16** 013035 URL http://stacks.iop.org/1367-2630/16/i=1/a=013035
[19] Bancal J D, Sheridan L and Scarani V 2014 *New Journal of Physics* **16** 033011 URL http://stacks.iop.org/1367-2630/16/i=3/a=033011
[20] Calude C 1994 *in G. Rozenberg, A. Salomaa (eds.) Developments in Language Theory* 113–129
[21] Calude C S, Dinneen M J, Dumitrescu M and Svozil K 2010 *Phys. Rev. A* **82**(2) 022102 URL http://link.aps.org/doi/10.1103/PhysRevA.82.022102
[22] Solis A, Angulo Martínez A M, Ramírez-Alarcón R, Cruz-Ramírez H, U'Ren A B and Hirsch J G 2015 *Physica Scripta* in press
[23] Kieu T D 2003 *Int.J.Theor.Phys.* **42** 1461–1478
[24] da Costa N C and Doria F A 1991 *International Journal of Theoretical Physics* **30** 1041–1073
[25] Cubitt T, Perez-Garcia D and Wolf M M 2015 *arXiv* **[quant-ph]** 1502.04135
[26] Copeland B J and Proudfoot D 1999 *Scientific American* **April** 98–103

Universal Quantum Measurements

Dorje C. Brody[1,2] **and Lane P. Hughston**[1,2]

[1]Department of Mathematics, Brunel University London, Uxbridge UB8 3PH, UK
[2]Department of Optical Physics and Modern Natural Science, St Petersburg National Research University of Information Technologies, Mechanics and Optics, 49 Kronverksky Avenue, St Petersburg 197101, Russia

E-mail: Dorje.Brody@brunel.ac.uk, Lane.Hughston@brunel.ac.uk

Abstract. We introduce a family of operations in quantum mechanics that one can regard as "universal quantum measurements" (UQMs). These measurements are applicable to all finite-dimensional quantum systems and entail the specification of only a minimal amount of structure. The first class of UQM that we consider involves the specification of the initial state of the system—no further structure is brought into play. We call operations of this type "tomographic measurements", since given the statistics of the outcomes one can deduce the original state of the system. Next, we construct a disentangling operation, the outcome of which, when the procedure is applied to a general mixed state of an entangled composite system, is a disentangled product of pure constituent states. This operation exists whenever the dimension of the Hilbert space is not a prime, and can be used to model the decay of a composite system. As another example, we show how one can make a measurement of the direction along which the spin of a particle of spin s is oriented ($s = \frac{1}{2}, 1, \ldots$). The required additional structure in this case involves the embedding of $\mathbb{CP}^1$ as a rational curve of degree $2s$ in $\mathbb{CP}^{2s}$.

1. Introduction

As we enter into what may be the dawning of an age of quantum engineering, the widespread interest in quantum information, quantum communication, quantum cryptography, and quantum computation thus entailed has had the effect of reawakening research in finite dimensional quantum systems. Indeed, from a mathematical point of view an n-dimensional quantum system can be given a very satisfactory treatment, bringing elements of algebraic geometry, symplectic geometry, Riemannian geometry, and convex analysis into play that are perhaps less obviously central in the infinite-dimensional situation. Foundational issues are being revisited as well, in part with a view to underpinning applications, but also on account of the fact that when the fog of technical issues associated with the infinite dimensional case clears one can in the finite-dimensional case—in its own way no less rich—identify more distinctly some of the problematic aspects of the general theory.

As a way of honouring the scientific career of Professor Bogdan Mielnik we propose in the present paper to construct a variety of generalized measurement operations arising in the finite-dimensional case that only involve a minimal amount of structure on the Hilbert space. To the extent that the structure involved is indeed minimal, the class of measurement operations associated with it is maximal—or "universal"—that is to say, applicable to any quantum system. We shall look at a number of explicit examples of such universal quantum measurements (UQMs) and comment on their possible applications.

We begin in Section 2 with a brief résumé of quantum mechanics from a modern perspective, emphasizing geometrical and probabilistic aspects of the theory. The idea is to model simple quantum systems in such a way that they can be regarded as elements of more complicated systems. To model a quantum system we need to specify a collection of suitably inter-related mathematical objects representing different aspects of the system. In the spirit of von Neumann (1955, IV.1) we can thus say that when a quantum system is given it is "characterized for the experimenter by the enumeration of all the effectively measurable quantities in it and their functional relations with one another." We keep the point of view pragmatic, with a view to modelling various systems of the type that might be involved in applications to information technology. No attempt will be made to model the universe as a whole, or to address the "measurement problem", and the models we look at will be mainly nonrelativistic—or, more precisely, pre-relativistic, since we do not bring the geometry of space and time into play.

We find it convenient to use an index notation for Hilbert space operations in our development of the theory of quantum state transformations. In Proposition 1 we present a useful characterization of positive maps, and in Proposition 2 we present a simplified proof of Choi's theorem on completely positive maps. In Section 3 we use the language of σ-algebras to present a rather general approach to modelling experiments on quantum systems and the state transformations that occur when an experiment is performed. Then in Section 4 we turn to the construction of universal quantum measurements that do not involve any structure other than that which is implicit in the geometry of the Hilbert space of the system. In particular, no preferred "observables" are selected. Such a measurement can be regarded as a determination of the state of the system. The input state can be any mixed state, and the output is a pure state. If measurements are performed on a large number of independent identical copies of a quantum system, then by gathering the data of the resulting pure output states one can determine the input state. Thus, such measurements are "tomographic" or "informationally complete" .

Then we consider examples involving some symmetry breaking, but again without the specification of any observables. The first of these is a "disentangling" operation, introduced in Section 5. These operations exist whenever the dimension of the Hilbert space is not a prime, and rely on the fact that if the dimension of the Hilbert space is a composite then the Hilbert space can be regarded as the tensor product of two or more Hilbert spaces of lower dimension, which we regard as "constituent" spaces. Disentangling measurements take the form of basic UQMs (of the type described above) but operating at the level of the constituents. The input state is a general mixed state, and the output state takes the form of a disentangled composite of pure states. The required additional structure involves a canonical Segre embedding of form

$$\mathbb{CP}^{p_1-1} \times \mathbb{CP}^{p_2-1} \times \mathbb{CP}^{p_3-1} \cdots \to \mathbb{CP}^{n-1}, \tag{1}$$

where the prime factorization of n is given by

$$n = \prod_i p_i \, . \tag{2}$$

Finally, in Section 6 we consider a class of "coherent" measurements that generalize the idea of determining the axis of spin in the case of a particle with spin. The additional structure involved in this class of generalized measurements takes the form of an embedding of $\mathbb{CP}^{n-1}$ as a rational variety of degree d in $\mathbb{CP}^{N-1}$, where

$$N = \frac{(n+d-1)!}{(n-1)!\,(d)!}. \tag{3}$$

For example, if $n = 3$, the additional structure is given by the embedding of $\mathbb{CP}^2$ as a rational surface of degree d in $\mathbb{CP}^{N-1}$ for $N = \frac{1}{2}(d+1)(d+2)$. In the case of degree two we obtain the classical Veronese embedding of $\mathbb{CP}^2$ as a rational surface in $\mathbb{CP}^5$.

2. States and operations

In the standard approach to quantum theory taught to undergraduate physicists, the usual line of attack is to introduce states as elements of a Hilbert space, then observables as Hermitian operators; a Hamiltonian is introduced as a special observable, and the evolution of the state vector, in the absence of external interventions, is given by the Schrödinger equation. When a measurement of an observable is made, the outcome is one of the eigenvalues of the associated Hermitian operator, and the probability of any particular outcome can be worked out using the Born rule; the associated change in the state of the system is then determined by the so-called projection postulate of von Neumann in the more general form introduced by Lüders (1951) to take into account degeneracies (see Adler *et al.* 2001, sections 2-3, for a discussion of the status of the Lüders postulate). We shall assume that the reader is familiar with the standard approach, and how it is conventionally applied, and we take for granted its many strengths.

Nevertheless, from a modern perspective we can to some extent dispense with the foundational notions of observables as Hermitian operators and states as state vectors. Even if for convenience, or out of habit, we still use the associated "old-fashioned" language from time to time, these traditional notions are strictly speaking no longer required.

As mentioned, in what follows we use an index notation in Hilbert space calculations. We model a typical quantum system with the introduction of a finite dimensional complex vector space $\mathcal{H}^\alpha$ ($\alpha = 1, 2, \ldots, n$) of some specified dimension $n \geq 2$, equipped with a complex conjugation operation $\mathbf{C}$, which maps elements of $\mathcal{H}^\alpha$ to the associated dual space $\mathcal{H}_\alpha$. Thus if X^α is a typical element of $\mathcal{H}^\alpha$, then under the complex conjugation map X^α gets mapped by $\mathbf{C}$ to the dual vector $\bar{X}_\alpha$ in $\mathcal{H}_\alpha$. For any two elements X^α and Y^α in $\mathcal{H}^\alpha$, if we form the sum $aX^\alpha + bY^\alpha$, where a and b are complex numbers, then under complex conjugation this gets mapped in an antilinear way to $\bar{a}\bar{X}_\alpha + \bar{b}\bar{Y}_\alpha$. The complex conjugation map acts in such a way that it likewise sends elements of $\mathcal{H}_\alpha$ to $\mathcal{H}^\alpha$, with $\mathbf{C}^2 = \mathbf{1}$. This structure is to be regarded as in place when we state that $\mathcal{H}^\alpha$ is a finite-dimensional complex Hilbert space.

A compelling case for use of the index notation is made in Geroch (1974). One could in principle use the notation of bras and kets, or an abstract notation; but for our purpose the index notation is more effective, since tensorial objects arise frequently in the discussion of quantum systems. The use of an index label on the Hilbert space itself acts as a reminder of the structure of the space, a convention that is handy when complicated tensor products of such spaces are brought into play. Thus we write $\mathcal{H}^\alpha$ for the Hilbert space, $\mathcal{H}_\alpha$ for its dual, and

$$\mathcal{H}^\alpha_\beta = \mathcal{H}^\alpha \otimes \mathcal{H}_\beta \tag{4}$$

for the tensor product of $\mathcal{H}^\alpha$ and $\mathcal{H}_\beta$. Similarly we write

$$\mathcal{H}^{\alpha\beta} = \mathcal{H}^\alpha \otimes \mathcal{H}^\beta, \quad \mathcal{H}_{\alpha\beta} = \mathcal{H}_\alpha \otimes \mathcal{H}_\beta, \quad \mathcal{H}^{\alpha\gamma}_{\beta\delta} = \mathcal{H}^\alpha \otimes \mathcal{H}_\beta \otimes \mathcal{H}^\gamma \otimes \mathcal{H}_\delta, \tag{5}$$

and so on. We use the summation convention to write $X^\alpha \bar{X}_\alpha$ for the inner product between the ket X^α and the bra $\bar{X}_\alpha$, and we say that the vector X^α is normalized if $X^\alpha \bar{X}_\alpha = 1$. The scheme that we have in mind for the manipulation of multi-index tensorial objects in the present setting is essentially the same as the general "abstract index" setup outlined, for example, in section 3 of Penrose (1968), and sections 2.1-2.2 of Penrose & Rindler (1984). Following the conventional terminology we often refer to the elements of $\mathcal{H}^\alpha$ as "state vectors", and the associated rays as "states". This way of speaking is a convenient relic of the old-fashioned approach.

Now we are in a position to introduce the ideas of states as positive operators and operations as maps from states to states. There is a substantial body of literature concerned with operational approaches to quantum mechanics, much of which is relevant to the "modern" approach to the subject. See, for example, Davies (1976), Davies & Lewis (1970), Haag & Kastler (1964), Holevo (1982), Krauss (1971, 1983), Mielnik (1968, 1969, 1974), Segal (1947), and references cited therein. We begin with the notion of a state. The elements of $\mathcal{H}^{\alpha}_{\beta}$ act as linear operators on the Hilbert space and its dual. Under complex conjugation, any element $F^{\alpha}_{\beta} \in \mathcal{H}^{\alpha}_{\beta}$ is mapped to an element $\bar{F}^{\beta}_{\alpha} \in \mathcal{H}^{\beta}_{\alpha}$. Since the vector spaces $\mathcal{H}^{\alpha}_{\beta}$ and $\mathcal{H}^{\beta}_{\alpha}$ are naturally isomorphic, we can say that $F^{\alpha}_{\beta} \in \mathcal{H}^{\alpha}_{\beta}$ is mapped under complex conjugation to another element $\bar{F}^{\alpha}_{\beta} \in \mathcal{H}^{\alpha}_{\beta}$. Then we say that F^{α}_{β} is Hermitian if $F^{\alpha}_{\beta} = \bar{F}^{\alpha}_{\beta}$. We shall write $\mathcal{O}^{\alpha}_{\beta}$ for the space of Hermitian operators. An element $F^{\alpha}_{\beta} \in \mathcal{O}^{\alpha}_{\beta}$ is said to be positive (nonnegative) if $F^{\alpha}_{\beta}\xi^{\beta}\bar{\xi}_{\alpha} \geq 0$ for all $\xi^{\alpha} \in \mathcal{H}^{\alpha}$, and strictly positive if $F^{\alpha}_{\beta}\xi^{\beta}\bar{\xi}_{\alpha} > 0$ for all $\xi^{\alpha} \in \mathcal{H}^{\alpha}$. A Hermitian operator is positive if and only if there exists a collection of one or more linearly independent vectors Z^{α}_{i}, with complex conjugates $\bar{Z}_{i\alpha}$, each normalised to unity, such that F^{α}_{β} is of the form

$$F^{\alpha}_{\beta} = \sum_{i} f_i Z^{\alpha}_{i} \bar{Z}_{i\beta} \tag{6}$$

where the f_i are positive constants. If a Hermitian operator is strictly positive, then one can find a set of n mutually orthogonal vectors, each normalised to unity, such that F^{α}_{β} can be written in the form (1) and where the f_i are strictly positive.

We shall write ${}^{+}\mathcal{O}^{\alpha}_{\beta}$ for the cone of positive Hermitian operators. By a "state" we mean any nontrivial element of ${}^{+}\mathcal{O}^{\alpha}_{\beta}$. If A^{α}_{β} and B^{α}_{β} are elements of ${}^{+}\mathcal{O}^{\alpha}_{\beta}$, and if a and b are positive constants, not both zero, then $aA^{\alpha}_{\beta}+bB^{\alpha}_{\beta}$ is also an element of ${}^{+}\mathcal{O}^{\alpha}_{\beta}$. By a "pure" state, we mean a state of the form $Z^{\alpha}\bar{Z}_{\beta}$ for some (nontrivial) vector Z^{α}. A state F^{α}_{β} is said to be normalized if $F^{\alpha}_{\alpha} = 1$. In what follows, we do not require that states are necessarily normalized. The terms "state" and "density matrix" are used more or less interchangeably in the modern literature; we shall usually reserve the term "density matrix" for a normalized state. It should be clear that with each state vector X^{α} one can associate a corresponding pure state $X^{\alpha}\bar{X}_{\beta}$. For this reason, it is indeed consistent, even if somewhat misleading, to regard state vectors as representing a class of "states". In fact, the physics literature seems to be divided on the issue of what constitutes a state. Some physicists take the view, in effect, that an individual quantum system is always in a pure state, and that so-called mixed states, represented by density matrices, correspond to "ensembles" of pure states (see Hughston *et al.* 1993 on this point). Other physicists seem to be happy with the idea that an individual quantum system can be in a mixed state, but that this represents a state of ignorance concerning the "true" state of the system, which is pure.

In our scheme the state of an individual system is represented by a density matrix, which may or may not be pure, and the question of how we use this density matrix, and what calculations we perform with it, depends on the particular model we are constructing. In some models, for example, it can be fruitful to introduce the idea of an ensemble in the form of a probability measure on the space of pure states, which in turn can be represented by a probability measure on the complex projective space $\mathbb{CP}^{n-1}$ associated with the given n-dimensional Hilbert space.

By an "operation" on a quantum system we mean a positive linear map from ${}^{+}\mathcal{O}^{\alpha}_{\beta}$ to itself. Thus we need to consider elements of the space $\mathcal{O}^{\alpha\beta'}_{\alpha'\beta}$ and characterize those elements that constitute positive maps, i.e. maps from states to states. Here we write $\mathcal{O}^{\alpha\beta'}_{\alpha'\beta}$ for the space of Hermitian elements of the tensor product space $\mathcal{H}^{\alpha\beta'}_{\alpha'\beta}$. The primed letters α', β', and so on, are regarded as extra letters of the alphabet. The action of $\phi^{\alpha\beta'}_{\alpha'\beta} \in \mathcal{O}^{\alpha\beta'}_{\alpha'\beta}$ on a state $F^{\alpha}_{\alpha'}$ is given by

$$F^{\alpha}_{\alpha'} \to \phi^{\alpha\beta'}_{\alpha'\beta} F^{\beta}_{\beta'} \, . \tag{7}$$

We say that $\phi^{\alpha\beta'}_{\alpha'\beta}$ is a positive map if under the action indicated above it maps any positive operator to another positive operator.

Proposition 1. *The map* $\phi^{\alpha\,\beta'}_{\alpha'\beta} \in \mathcal{O}^{\alpha\,\beta'}_{\alpha'\beta}$ *is positive if and only if for all* X^α, $Y^\alpha \in \mathcal{H}^\alpha$ *we have*

$$X^{\alpha'}\bar{X}_\alpha \phi^{\alpha\,\beta'}_{\alpha'\beta} Y^\beta \bar{Y}_{\beta'} \geq 0. \tag{8}$$

Proof. We require that for any positive operator $F^\beta_{\beta'}$ the transformed operator $\phi^{\alpha\,\beta'}_{\alpha'\beta} F^\beta_{\beta'}$ should be positive. Thus for all X^α we require that

$$X^{\alpha'}\bar{X}_\alpha \phi^{\alpha\,\beta'}_{\alpha'\beta} F^\beta_{\beta'} \geq 0. \tag{9}$$

In particular, if $F^\beta_{\beta'}$ is a pure state $Y^\beta \bar{Y}_{\beta'}$ then we obtain (8). Conversely, suppose that (8) holds. Now, any state can be represented as a positively weighted sum of pure states. If we fix X^α and consider the inequality (8) for various choices of Y^α, we deduce that (9) holds for any positive $F^\beta_{\beta'}$. Since this is true for any choice of X^α, we deduce that $\phi^{\alpha\,\beta'}_{\alpha'\beta}$ is a positive map. □

Let us write ${}^+\mathcal{O}^{\alpha\,\beta'}_{\alpha'\beta}$ for the space of such positive maps. For applications we frequently require a stronger condition that limits the class of admissible maps to a subspace of the space of positive maps consisting of so-called "completely positive" maps. The condition of complete positivity ensures that if $\phi^{\alpha\,\beta'}_{\alpha'\beta}$ acts "locally" on any state of the larger composite system obtained by forming the tensor product of the Hilbert space $\mathcal{H}^\alpha$ and any "ancilla" Hilbert space $\mathcal{H}^j$ (not necessarily of the same dimension), then the result is a positive operator. More precisely, suppose that a typical Hilbert space vector of such a composite system is given by $X^{\alpha j} \in \mathcal{H}^{\alpha j}$. One can think of the index "clump" αj as constituting an index for the Hilbert space vector of the composite system. A general state of the composite system is given by a positive operator of the form $F^{\alpha j}_{\beta k}$. By saying that $F^{\alpha j}_{\beta k}$ is a positive operator in this situation we mean that for any composite state vector $X^{\beta k}$ we have

$$X^{\beta k}\bar{X}_{\alpha j} F^{\alpha j}_{\beta k} \geq 0. \tag{10}$$

Now clearly if we let $\phi^{\alpha\,\beta'}_{\alpha'\beta}$ operate "only on the first element of the composite" through the transformation

$$F^{\alpha\, j}_{\alpha' j'} \rightarrow \phi^{\alpha\,\beta'}_{\alpha'\beta} F^{\beta\, j}_{\beta' j'}, \tag{11}$$

then this gives us a linear map from Hermitian operators to Hermitian operators on the composite system, and we can call $\phi^{\alpha\,\beta'}_{\alpha'\beta}$ a "local" operation. We say that $\phi^{\alpha\,\beta'}_{\alpha'\beta}$ is "completely positive" if for any such composite system the given transformation takes states to states. Note that we do not require here that such an operation should preserve the normalization of a state.

Proposition 2. *The map* $\phi^{\alpha\,\beta'}_{\alpha'\beta}$ *is completely positive if and only if for some* $N \geq 1$ *there exists a family of operators* $K^\alpha_\beta(i)$ *for* $i = 1, \ldots, N$ *such that*

$$\phi^{\alpha\,\beta'}_{\alpha'\beta} = \sum_{i=1}^{N} K^\alpha_\beta(i) \bar{K}^{\beta'}_{\alpha'}(i). \tag{12}$$

Proof. We require that the transformed state (11) is positive for any initial composite state. This must be true in particular for a pure state of the form $F^{\beta\, j}_{\beta' j'} = \xi^{\beta\, j}\bar{\xi}_{\beta' j'}$. The transformed state in that case is $\phi^{\alpha\,\beta'}_{\alpha'\beta}\xi^{\beta\, j}\bar{\xi}_{\beta' j'}$, and to ensure that it is positive we require that for any composite state vector $X^{\alpha i}$ we have

$$X^{\alpha' j'}\bar{X}_{\alpha j}\, \phi^{\alpha\,\beta'}_{\alpha'\beta}\, \xi^{\beta\, j}\bar{\xi}_{\beta' j'} \geq 0. \tag{13}$$

If we set

$$Z^{\alpha'}_{\beta'} = X^{\alpha' j'} \bar{\xi}_{\beta' j'}, \quad \bar{Z}^{\beta}_{\alpha} = \bar{X}_{\alpha j} \xi^{\beta j}, \tag{14}$$

then the inequality (8) takes the simple form

$$\phi^{\alpha\,\beta'}_{\alpha'\beta} Z^{\alpha'}_{\beta'} \bar{Z}^{\beta}_{\alpha} \geq 0. \tag{15}$$

This is the condition that the Hermitian form obtained by clumping the indices on $Z^{\alpha'}_{\beta'}$ and $\bar{Z}^{\beta}_{\alpha}$ should be positive. As a consequence one sees by the theory of positive Hermitian forms that $\phi^{\alpha\,\beta'}_{\alpha'\beta}$ admits an expansion of the form (12). Conversely, if $\phi^{\alpha\,\beta'}_{\alpha'\beta}$ takes the form (12), then it is straightforward to check that the resulting map is completely positive. One needs to verify that for any composite state $F^{\beta\,j}_{\beta' j'}$ and any composite vector $X^{\alpha' j'}$ it holds that

$$X^{\alpha' j'} \bar{X}_{\alpha j} \sum_{i=1}^{N} K^{\alpha}_{\beta}(i)\, \bar{K}^{\beta'}_{\alpha'}(i)\, F^{\beta\,j}_{\beta' j'} \geq 0. \tag{16}$$

But this follows at once since

$$X^{\alpha' j'} \bar{X}_{\alpha j} \sum_{i=1}^{N} K^{\alpha}_{\beta}(i)\, \bar{K}^{\beta'}_{\alpha'}(i)\, F^{\beta\,j}_{\beta' j'} = \sum_{i=1}^{N} W^{\beta' j'}(i) \bar{W}_{\beta j}(i)\, F^{\beta\,j}_{\beta' j'}, \tag{17}$$

where for each value of i we define

$$W^{\beta' j'}(i) = X^{\alpha' j'} \bar{K}^{\beta'}_{\alpha'}(i), \quad \bar{W}_{\beta j}(i) = \bar{X}_{\alpha j} K^{\alpha}_{\beta}(i), \tag{18}$$

and we thus observe that each term on the right hand side of (12) is positive by virtue of the assumed positivity of the state $F^{\beta\,j}_{\beta' j'}$. □

The result of Proposition 2 is the theorem of Choi (1975), with a proof that is on account of the use of the index notation perhaps more transparent than the original.

3. Experiments and state transformations

Rather than taking a "one size fits all" approach to quantum theory, the idea is to construct a number of different models. Each model involves the specification of a quantum system, the experiments that can be made on it, the possible outcomes, and the resulting state transformations. The quantum system is represented by a Hilbert space $\mathcal{H}^{\alpha}$ along with a state w^{α}_{β}. Each experiment that can be performed is described by a measurable space $(\Omega, \mathcal{F})$ endowed with some structure that relates it to the Hilbert space $\mathcal{H}^{\alpha}$ and the state w^{α}_{β}. Here the set Ω represents all of the possible outcomes of chance when the experiment is performed, and $\mathcal{F}$ is a collection of subsets of Ω forming a σ-algebra.

More precisely, we require that Ω itself should belong to $\mathcal{F}$, that the empty set $\varnothing$ should belong to $\mathcal{F}$, and that the union of any countable collection of elements of $\mathcal{F}$ should belong to $\mathcal{F}$. Distinct σ-algebras correspond to distinct experiments. If $\mathcal{E}$ and $\mathcal{F}$ are σ-algebras on Ω, and if $\mathcal{E}$ is a sub-σ-algebra of $\mathcal{F}$, then we can say that the experiment $\mathcal{F}$ is a refinement of the experiment $\mathcal{E}$. Conversely, we can say that the experiment $\mathcal{E}$ is a "course-grained" version of the experiment $\mathcal{F}$. In this way one obtains hierarchies of experiments.

In a given experiment $(\Omega, \mathcal{F})$, if $\omega \in \Omega$ is the outcome of chance, then the result of the experiment is the smallest element $A \in \mathcal{F}$ such that $\omega \in A$. We say that A is the smallest element of $\mathcal{F}$ containing ω if ω belongs to no proper subset of A which is also an element of $\mathcal{F}$. In general, distinct outcomes of chance can give rise to the same result for an experiment. It may be that for each $\omega \in \Omega$ the subset $\{\omega\} \subset \Omega$ that only contains ω belongs to $\mathcal{F}$. This happens for example in "refined" experiments where the result of the experiment is sufficient to determine the outcome of chance. We distinguish between the outcomes of chance (which are elements of the set Ω) and the results of experiments (which are minimal elements of $\mathcal{F}$). Each outcome of chance belongs to a unique minimal element of $\mathcal{F}$. One might ask why we introduce the entire σ-algebra $\mathcal{F}$ if only the "minimal" elements of $\mathcal{F}$ count as possible results of the experiment represented by $\mathcal{F}$. The reason is that when we consider an experiment, we generally like to consider alongside it the various course-grained versions of the experiment. Thus given $\mathcal{F}$ we wish to consider as well the various σ-subalgebras associated with it, the minimal elements of which are not necessarily minimal elements of $\mathcal{F}$.

The next ingredient that we require for the specification of an experiment is a system of state transformations $\mathbb{T} = \{T(A),\ A \in \mathcal{F}\}$. Thus $\mathbb{T}$ takes the form of a transformation-valued measure on $(\Omega, \mathcal{F})$, satisfying the following conditions:

(i) For each $A \in \mathcal{F}$, the associated state transformation $T(A)$ is given by a completely positive map of the form

$$T(A) : w^{\alpha}_{\beta} \to T^{\alpha\beta'}_{\beta\alpha'}(A)\, w^{\alpha'}_{\beta'} . \tag{19}$$

(ii) The system $\mathbb{T} = \{T(A),\ A \in \mathcal{F}\}$ is countably additive. Thus, $T^{\alpha\beta'}_{\beta\alpha'}(\varnothing) = 0$, and if the sets $\{A_n : n \in \mathbb{N}\}$ are disjoint, and such that $A = \cup_n A_n$, then

$$T^{\alpha\beta'}_{\beta\alpha'}(A) = \sum_{n\in\mathbb{N}} T^{\alpha\beta'}_{\beta\alpha'}(A_n) . \tag{20}$$

(iii) For each $A \in \mathcal{F}$, $T(A)$ is trace-reducing. Thus, we have

$$\frac{T^{\gamma\beta'}_{\gamma\alpha'}(A)\, w^{\alpha'}_{\beta'}}{w^{\gamma}_{\gamma}} \leq 1 . \tag{21}$$

(iv) $T(\Omega)$ satisfies the law of total probability. Thus, we have

$$\frac{T^{\gamma\beta'}_{\gamma\alpha'}(\Omega)\, w^{\alpha'}_{\beta'}}{w^{\gamma}_{\gamma}} = 1 . \tag{22}$$

Again, the transformations are specified for each element of $\mathcal{F}$, not merely for the minimal elements. In this way, we also determine the relevant transformations for each course-grained version of the experiment.

Once we have specified the system of transformations on $\mathcal{F}$, then in the experiment $(\Omega, \mathcal{F}, \mathbb{T})$, or coursed-grained version thereof, the probability that the outcome of chance ω lies in the set $A \in \mathcal{F}$ is given by

$$\mathbb{P}(\omega \in A) = \frac{T^{\gamma\beta'}_{\gamma\alpha'}(A)\, w^{\alpha'}_{\beta'}}{w^{\gamma}_{\gamma}} . \tag{23}$$

If A_ω denotes the smallest element of $\mathcal{F}$ containing ω, and if $\mathbb{P}(\omega \in A_\omega) \neq 0$, then the normalized state transformation associated with ω is

$$w^{\alpha}_{\beta} \to \frac{T^{\alpha\beta'}_{\beta\alpha'}(A_\omega)\, w^{\alpha'}_{\beta'}}{T^{\gamma\beta'}_{\gamma\alpha'}(A_\omega)\, w^{\alpha'}_{\beta'}} . \tag{24}$$

The state transformations associated with the usual projective measurements (with or without selection) in quantum mechanics take this form, and so do the transformations associated with more general discrete positive operator-valued measures. For example, in the maximally course-grained experiment corresponding to a non-selective projective measurement we have $\mathcal{F} = \{\Omega, \varnothing\}$. In that case, the outcome of the measurement is trivial in the sense that we have $\mathbb{P}(\omega \in \Omega) = 1$ and $\mathbb{P}(\omega \in \varnothing) = 0$. Nevertheless, the state transformation will in general be nontrivial. For example, in the case of a non-selective projective measurement of the energy of a finite-dimensional system with a nondegenerate Hamiltonian, no "outcome" is recorded other than the fact the experiment was done ("the result lies in the admissible set of possible outcomes"), yet the state transforms from a general state to a state which is diagonal in the energy basis.

In the continuous case, the probability of any particular outcome of chance is zero. In that situation we model the state transformations as follows. We suppose that there exists a measure $\mu(\mathrm{d}\omega)$ on Ω and a transformation density $t^{\alpha\beta'}_{\beta\alpha'}(\omega)$ with the property that for any $A \in \mathcal{F}$ we have

$$T^{\alpha\beta'}_{\beta\alpha'}(A) = \int_\Omega \mathbb{1}\{\omega \in A\} t^{\alpha\beta'}_{\beta\alpha'}(\omega)\, \mu(\mathrm{d}\omega)\,. \tag{25}$$

Then the probability distribution for the outcome of chance is

$$\mathbb{P}(\omega \in \mathrm{d}\omega) = t^{\gamma\beta'}_{\gamma\alpha'}(\omega)\; w^{\alpha'}_{\beta'}\; \mu(\mathrm{d}\omega)\,, \tag{26}$$

and the normalized state transformation is given by

$$w^\alpha_\beta \to \frac{t^{\alpha\beta'}_{\beta\alpha'}(\omega)\; w^{\alpha'}_{\beta'}}{t^{\gamma\beta'}_{\gamma\alpha'}(\omega)\; w^{\alpha'}_{\beta'}}\,. \tag{27}$$

In what follows, we shall be concerned mainly with the continuous situation where we have a finite dimensional Hilbert space and the outcome space Ω has the structure of a manifold on which a natural candidate for the measure $\mu(\mathrm{d}\omega)$ on Ω is available.

4. Tomographic measurements

We consider the case where the measurable space Ω representing the possible outcomes of chance is the manifold $\mathbb{CP}^{n-1}$, the space of pure states associated with the given Hilbert space $\mathcal{H}^\alpha$. This is a rather natural choice to look at first since it does not involve the introduction of any additional structure on the quantum system. Thus all finite dimensional quantum systems admit a version of the following measurement operation. Write $\Omega = \mathbb{CP}^{n-1}$, let x denote a typical point in Ω, and let $Z^\alpha(x)$ denote a representative vector in $\mathcal{H}^\alpha\backslash\{0\}$ lying on the fibre above the point $x \in \Omega$. Then we can construct a system of transformations $\mathbb{T}$ by setting

$$T^{\alpha\beta'}_{\beta\alpha'}(A) = n \int_\Omega \mathbb{1}\{x \in A\} \frac{Z^\alpha(x) Z^{\beta'}(x) \bar{Z}_\beta(x) \bar{Z}_{\alpha'}(x)}{(Z^\gamma(x)\bar{Z}_\gamma(x))^2} \mu(\mathrm{d}x) \tag{28}$$

for any element A of the Borel σ-algebra on Ω. Here

$$\mu(\mathrm{d}x) = \frac{\mathcal{D}Z(x)}{\int_\Omega \mathcal{D}Z(x)} \tag{29}$$

defines the uniform probability measure on $\mathbb{CP}^{n-1}$, where

$$\mathcal{D}Z = \frac{\epsilon_{\alpha\beta\cdots\gamma} Z^\alpha \mathrm{d}Z^\beta \cdots \mathrm{d}Z^\gamma\, \epsilon^{\alpha\beta\cdots\gamma} \bar{Z}_\alpha \mathrm{d}\bar{Z}_\beta \cdots \mathrm{d}\bar{Z}_\gamma}{(Z^\gamma \bar{Z}_\gamma)^n}\,. \tag{30}$$

Clearly, we have $\mu(\Omega) = 1$. The associated transformation density is then given by

$$t^{\alpha\beta'}_{\beta\alpha'}(x) = n\frac{Z^{\alpha}(x)Z^{\beta'}(x)\bar{Z}_{\beta}(x)\bar{Z}_{\alpha'}(x)}{(Z^{\gamma}(x)\bar{Z}_{\gamma}(x))^2}. \tag{31}$$

The outcome of chance in such a measurement is a pure state. If the initial state is w^{α}_{β} and if the outcome of chance is the point x in $\mathbb{CP}^{n-1}$, the resulting normalized state transformation is

$$w^{\alpha}_{\beta} \to \frac{Z^{\alpha}(x)\bar{Z}_{\beta}(x)}{Z^{\gamma}(x)\bar{Z}_{\gamma}(x)}. \tag{32}$$

The probability that the outcome lies in a given Borel set A in the space of pure states is

$$\mathbb{P}(x \in A) = E^{\alpha}_{\beta}(A)\, w^{\beta}_{\alpha}, \tag{33}$$

where the "effect" $E^{\alpha}_{\beta}(A)$ associated with the set A is given by

$$E^{\alpha}_{\beta}(A) = n\int_{\Omega} \mathbb{1}\{x \in A\}\frac{Z^{\alpha}(x)\bar{Z}_{\beta}(x)}{Z^{\gamma}(x)\bar{Z}_{\gamma}(x)}\,\mu(\mathrm{d}x). \tag{34}$$

Note that $E^{\alpha}_{\beta}(\Omega) = \delta^{\alpha}_{\beta}$. It follows in particular that the function $\rho : \mathbb{CP}^{n-1} \to \mathbb{R}^{+}$ defined by

$$\rho(x) = n\frac{Z^{\alpha}(x)\, w^{\beta}_{\alpha}\, \bar{Z}_{\beta}(x)}{Z^{c}(x)\bar{Z}_{c}(x)} \tag{35}$$

is the probability density for the outcome $x \in \mathrm{d}x$. Thus we have

$$\mathbb{P}(x \in \mathrm{d}x) = \rho(x)\,\mu(\mathrm{d}x). \tag{36}$$

The significance of the factor of n in the expressions above is that it ensures that if the initial state is of the "totally unbiased" form

$$w^{\alpha}_{\beta} = \frac{1}{n}\delta^{\alpha}_{\beta}, \tag{37}$$

then the probability density is uniform, and we have $\rho(x) = 1$.

Now suppose we consider the situation where we have a large number of independent identical copies of the system, and we make a measurement of this type on each copy. If we analyze the statistics of the measurements, then we can to a good degree of accuracy determine $\rho(x)$, and hence determine the original state w^{α}_{β}. More precisely, the "ensemble" of measurement outcomes has the density $\rho(x)$, and therefore the state r^{α}_{β} of the ensemble representing the outcomes of the measurements is

$$r^{\alpha}_{\beta} = \int_{\Omega}\rho(x)\frac{Z^{\alpha}(x)\bar{Z}_{\beta}(x)}{Z^{\gamma}(x)\bar{Z}_{\gamma}(x)}\,\mu(\mathrm{d}x). \tag{38}$$

The integral can be worked out explicitly by use of the following identity:

$$\int_{\Omega}\frac{Z^{\alpha}(x)\,Z^{\beta'}(x)\,\bar{Z}_{\beta}(x)\,\bar{Z}_{\alpha'}(x)}{(Z^{\gamma}(x)\,\bar{Z}_{\gamma}(x))^2}\,\mu(\mathrm{d}x) = \frac{1}{n(n+1)}\left(\delta^{\alpha}_{\beta}\,\delta^{\beta'}_{\alpha'} + \delta^{\alpha}_{\alpha'}\,\delta^{\beta'}_{\beta}\right). \tag{39}$$

A calculation making use of (35) and (39) then shows that

$$r^{\alpha}_{\beta} = \frac{1}{n+1}\left(\delta^{\alpha}_{\beta} + w^{\alpha}_{\beta}\right). \tag{40}$$

We see that the original state w^{α}_{β} is in general "diluted" as a consequence of the measurement operation. But if the initial state is unbiased, then so is the final state. In all cases, nevertheless, we can recover the original state from the statistics of the measurement observations since

$$w^{\alpha}_{\beta} = (n+1)r^{\alpha}_{\beta} - \delta^{\alpha}_{\beta}. \tag{41}$$

We can call such an experiment a "universal quantum measurement" (UQM) since it can be applied to any finite-dimensional quantum system. No additional structure is required apart from what is already implicit in the original specification of the system $\mathcal{H}^{\alpha}$. Evidently, this is possible. One can envisage the construction of a machine with the property that given a sample consisting of a large number of identical "molecules" of some type, all in the same state, the structure of a typical molecule can be determined.

5. Disentangling operations

Universal quantum measurements of the type just discussed can form elements of composite operations. In that case, we introduce more structure on the Hilbert space, but typically not involving the choice of specific observables. An example is as follows. Consider the Hilbert space of a pair of qubits. The Hilbert space has dimension four, and the associated pure state space is $\mathbb{CP}^3$. The space of disentangled pure states is a quadric surface in $\mathbb{CP}^3$. The quadric is a doubly ruled surface, given by the product of two $\mathbb{CP}^1$s. Each of the $\mathbb{CP}^1$s is endowed with the Fubini-Study measure, so the quadric also has a natural uniform measure on it, given by the product measure. This gives rise to a class of UQMs that we can call "disentangling operations". Starting with a general (i.e. mixed) state of the two-qubit system, the outcome of a disentangling operation is a point on the quadric. The transformation density is the product of the transformation densities associated with the UQMs attached to the individual qubits.

In more detail we have the following. Write $\mathcal{H}^{AA'}$ $(A, A' = 1, 2)$ for the tensor product of the two qubit spaces $\mathcal{H}^{A}$ and $\mathcal{H}^{A'}$. A state vector $\xi^{AA'} \in \mathcal{H}^{AA'}$ is disentangled if it is of the form $\xi^{AA'} = \alpha^{A}\beta^{A'}$. The associated quadric $\mathcal{Q}$ in $\mathbb{CP}^3$ is the locus

$$\epsilon_{AB}\,\epsilon_{A'B'}\,\xi^{AA'}\xi^{BB'} = 0, \tag{42}$$

where $\epsilon_{AB} = -\epsilon_{BA}$. The resulting uniform measure on the quadric is given for $x \in \mathcal{Q}$ by

$$\mu_{\mathcal{Q}}(\mathrm{d}x) = \mu_{\alpha}(\mathrm{d}x)\mu_{\beta}(\mathrm{d}x), \tag{43}$$

where

$$\mu_{\alpha}(\mathrm{d}x) = \frac{\epsilon_{AB}\alpha^{A}(x)\mathrm{d}\alpha^{B}(x)\,\epsilon^{CD}\bar{\alpha}_{C}(x)\mathrm{d}\bar{\alpha}_{D}(x)}{(\alpha^{E}(x)\bar{\alpha}_{E}(x))^2} \tag{44}$$

and

$$\mu_{\beta}(\mathrm{d}x) = \frac{\epsilon_{A'B'}\beta^{A'}(x)\mathrm{d}\beta^{B'}(x)\,\epsilon^{C'D'}\bar{\beta}_{C'}(x)\mathrm{d}\bar{\beta}_{D'}(x)}{(\beta^{E'}(x)\bar{\beta}_{E'}(x))^2}. \tag{45}$$

The relevant transformation density has its mass concentrated entirely on the quadric, and is given for $x \in \mathcal{Q}$ by

$$t^{AA'CC'}_{BB'DD'}(x) = 4\,\frac{\alpha^{A}(x)\beta^{A'}(x)\alpha^{C}(x)\beta^{C'}(x)\ \bar{\alpha}_{B}(x)\bar{\beta}_{B'}(x)\bar{\alpha}_{D}(x)\bar{\beta}_{D'}(x)}{(\alpha^{E}(x)\bar{\alpha}_{E}(x))^2\ (\beta^{E'}(x)\bar{\beta}_{E'}(x))^2} \tag{46}$$

If the initial state is a prescribed density matrix of the form $w^{BB'}_{AA'}$, then the probability that the outcome lies in a measurable region $R \subset \mathcal{Q}$ is

$$\mathbb{P}(\omega \in R) = 4\int_{x\in R} w^{BB'}_{AA'}\,\frac{\alpha^{A}(x)\beta^{A'}(x)\bar{\alpha}_{B}(x)\bar{\beta}_{B'}(x)}{\alpha^{E}(x)\bar{\alpha}_{E}(x)\ \beta^{E'}(x)\bar{\beta}_{E'}(x)}\mu_{\mathcal{Q}}(\mathrm{d}x). \tag{47}$$

Journal of Physics: Conference Series **624** (2015) 012002 doi:10.1088/1742-6596/624/1/012002

One checks that if the initial state takes the unbiased form

$$w_{AA'}^{BB'} = \frac{1}{4}\delta_A^B \delta_{A'}^{B'}, \tag{48}$$

then the distribution of the outcome is uniform over the quadric. On the other hand, if the initial state is a pure singlet state of the form

$$w_{AA'}^{BB'} = \frac{1}{2}\epsilon_{AA'}\,\epsilon^{BB'}, \tag{49}$$

then the outcome probability is highest in regions of the quadric corresponding to EPR-Bohm pairs, and vanishes when the two particles have the same state. This type of operation may thus be useful as a model for the decay of a particle composed of spin one-half constituents.

A similar construction applies for entangled states of many-particle systems in higher dimensions, in which case the relevant outcome space is given by the Segre embedding of the product space of the pure state spaces associated with the constituent systems. Whether one views this experiment as a "measurement" or a "procedure" is to some extent a matter of taste. In any case, the effect of the operation is to disintegrate the system into its constituents.

6. Coherent measurements

Consider a three-dimensional Hilbert space, for which the space of pure states has the structure of the complex projective space $\mathbb{CP}^2$ endowed with the Fubini-Study metric. Let $\mathcal{C}$ be a real conic curve in $\mathbb{CP}^2$. By "real" we mean the following: we require that for any point x in $\mathcal{C}$ the associated complex conjugate line (representing pure states orthogonal to x) is tangent to the conic. Such a setup is equivalent to representing $\mathcal{H}^\alpha$ as a space of symmetric spinors $\mathcal{H}^{AB}$, with a typical element z^{AB} (where $A = 1, 2$) so $z^{AB} = z^{BA}$. The conic is given by

$$\epsilon_{AB}\,\epsilon_{CD}\,z^{AC}z^{BD} = 0\,, \tag{50}$$

where $\epsilon_{AB} = -\epsilon_{BA}$. The solution to this quadratic equation takes the form

$$z^{AB} = \phi^A\phi^B \tag{51}$$

for some ϕ^A. The associated complex conjugate line consists of all states x^{AB} such that

$$\bar{\phi}_A\bar{\phi}_B x^{AB} = 0. \tag{52}$$

Thus the pure states orthogonal to the point $z^{AB} = \phi^A\phi^B$ on the conic are of the form

$$x^{AB} = \bar{\phi}^{(A}\alpha^{B)} \tag{53}$$

for some α^A. But any such state lies on a line tangent to $\mathcal{C}$, the tangent point being $\bar{\phi}^A\bar{\phi}^B$. We can use $\mathcal{C}$ as the outcome space of a special class of measurements. For any initial spin-one mixed state w_{CD}^{AB} the outcome of the measurement is a point of the conic $\mathcal{C}$, that is to say, a pure spin state with a definite direction for the axis of spin. Thus, the state transformation is

$$w_{CD}^{AB} \to \phi^A\phi^B\bar{\phi}_C\bar{\phi}_D \;/\; (\phi^E\bar{\phi}_E)^2\,. \tag{54}$$

The probability that the outcome lies in a given Borel set $A \subset \mathcal{C}$ is given by

$$\mathbb{P}(\omega \in A) = 3\int_A \frac{w_{CD}^{AB}\;\phi^C(x)\phi^D(x)\bar{\phi}_A(x)\bar{\phi}_B(x)}{(\phi^E(x)\bar{\phi}_E(x))^2}\;\mu(\mathrm{d}x), \tag{55}$$

where $\mu(\mathrm{d}x)$ is the uniform probability measure on C induced by the Veronese embedding of $\mathbb{CP}^1$ in $\mathbb{CP}^2$ as a rational curve. Such an experiment on a spin-one system can be interpreted physically as a "measurement of the direction of the axis of the spin" of the particle. The result of the experiment is an answer to the question "what is the direction of the spin axis of the particle?". The state then transforms from the original state to a pure state, which is the unique state lying on the conic that has that axis of spin. Similar formulae apply for higher spin systems, in which case the defining structure involves a rational curve of degree $2s$ in $\mathbb{CP}^{2s}$ (the twisted cubic curve, the rational quartic curve, and so on). See Brody & Hughston (2001) for a discussion of the geometry of higher spin systems and the role played by rational curves.

More generally, one obtains a broader class of "coherent" measurements based on the Veronese embedding of $\mathbb{CP}^{n-1}$ as a rational variety of degree d in $\mathbb{CP}^{N-1}$, where N is given by (3). For example, if $n = 3$, we obtain the embedding of $\mathbb{CP}^2$ as a rational surface of degree d in $\mathbb{CP}^{N-1}$ for $N = \frac{1}{2}(d+1)(d+2)$. These varieties are the manifolds of so-called generalized coherent states or $\mathrm{SU}(n)$ coherent states (Brody & Graefe 2010). Hence starting from an arbitrary pure or mixed state of the system, the outcome of the measurement results in a coherent state.

Another class of UQM that one can consider, which we hope to discuss elsewhere, allows for a direct measurement of the "mean energy" of a quantum system. In this case, the additional structure required is the specification of a Hamiltonian operator. Unlike a standard projective measurement of the energy, the outcome of a mean-energy measurement lies in a continuum of possible values between the highest and lowest eigenvalues of the Hamiltonian. The mean-energy measurement operation is closely linked to the existence of the so-called mean-energy ensemble (Brody & Hughston 1998, 1999), and may be of some relevance in connection with the quantum thermodynamics of finite systems.

It is interesting to observe that nearly all of the examples we have considered rely rather heavily, or at least so it seems, on what Mielnik (2001) refers to as the "Ptolomean structure" of quantum mechanics—namely, the endless hierarchy of tensor products of Hilbert spaces, along with the various notions of entanglement thus entailed, upon which so many of the modern finite-dimensional applications of the theory appear to rest.

But it is an open question whether the Ptolomean structure really is an essential part of physics. Can it be softened somewhat, perhaps in the way in which the rigid Minkowskian geometry of special relativity survives in a weaker sense in the tangent space of a general relativistic space-time? This idea is one of the motivations for the geometric approach to quantum mechanics (see, e.g., Ashtekar & Schilling 1998, Bengtsson & Życzkowski 2006, Brody & Hughston 2001, Gibbons 1992, Hughston 1995, 1996, Kibble 1979, Mielnik 1968, 1974, and references cited therein). But whereas most of the attempts at generalizing quantum theory have focussed either on the consideration of generalized state spaces, or generalizations of the notion of observables as phase space functions, relatively little has been pursued so far in the direction of generalized measurements in the context of a nonlinear theory. It makes sense therefore to consider first those classes of operations that depend only on a minimal amount of structure.

References

[1] Adler, S. A., Brody, D. C., Brun, T. A. & Hughston, L. P. (2001) Martingale models for quantum state reduction. *J. Phys.* A **34**, 8795–8820.

[2] Ashtekar, A. & Schilling, T. A. (1998) Geometrical formulation of quantum mechanics. In: *On Einstein's Path: Essays in Honor of Engelbert Schucking*, A. Harvey, ed. (Berlin: Springer-Verlag).

[3] Bengtsson, I. & Życzkowski, K. (2006) *Geometry of Quantum States: An Introduction to Quantum Entanglement* (Cambridge: Cambridge University Press).

[4] Brody, D. C. & Graefe, E. M. (2010) Coherent states and rational surfaces. *J. Phys.* A **43**, 255205.

[5] Brody, D. C. & Hughston, L. P. (1998) The quantum canonical ensemble. *J. Math. Phys.* **39**, 6502–6508.

[6] Brody, D. C. & Hughston, L. P. (1999) Geometrisation of statistical mechanics. *Proc. R. Soc. Lond.* A **455**, 1683–1715.

[7] Brody, D. C. & Hughston, L. P. (2001) Geometric quantum mechanics. *J. Geom. Phys.* **38**, 19-53.

[8] Choi, M.-D.(1975) Completely positive linear maps on complex matrices. *Lin. Alg. Appl.* **10**, 285–290.

[9] Davies, E. B. (1976) *Quantum Theory of Open Systems* (New York: Academic Press).

[10] Davies, E. B. & Lewis, J. T. (1970) An operational approach to quantum probability. *Commun. Math. Phys.* **17**, 239–260.

[11] Geroch, R. (1970) *Lecture notes on particle physics*. Department of Physics, University of Texas at Austin (unpublished).

[12] Gibbons, G. W. (1992) Typical states and density matrices. *J. Geom. Phys.* **8**, 147-162.

[13] Haag, R. & Kastler, D. (1964) An algebraic approach to quantum field theory. *J. Math. Phys.* **5**, 848–861.

[14] Holevo, A. S. (1982) *Probabilistic and Statistical Aspects of Quantum Theory* (Amsterdam: North-Holland).

[15] Hughston, L. P., Jozsa, R. & Wootters, W. K. (1993) A complete classification of quantum ensembles having a given density matrix, *Phys. Lett. A***183**, 14–18.

[16] Hughston, L. P. (1995) Geometric aspects of quantum mechanics. In: *Twistor Theory*, S. Huggett, ed. (New York: Marcel Dekker).

[17] Hughston, L. P. (1996) Geometry of stochastic state reduction. *Proc. Roy. Soc. Lond.* A **452**, 953–979.

[18] Kibble, T. W. B. (1979) Geometrisation of quantum mechanics. *Commun. Math. Phys.* **65**, 189–201.

[19] Krauss, K. (1971) General state changes in quantum theory. *Ann. Phys.* **64**, 311–335.

[20] Krauss, K. (1983) *States, Effects, and Operations: Fundamental Notions of Quantum Theory*, Lecture Notes in Mathematics **190** (Berlin: Springer-Verlag).

[21] Lüders, G. (1951) Über die Zustandsänderung durch den Meßprozeß. *Ann Phys.* (Leipzig) **8**, 322–328.

[22] Mielnik, B. (1968) Geometry of quantum states. *Comm. Math. Phys.* **9**, 55–80.

[23] Mielnik, B. (1969) Theory of filters. *Comm. Math. Phys.* **15**, 1–46.

[24] Mielnik, B. (1974) Generalized quantum mechanics. *Comm. Math. Phys.* **37**, 221–256.

[25] Mielnik, B. (2001) Nonlinear quantum mechanics: a conflict with Ptolomean structure? *Phys. Lett.* A **289**, 1–8.

[26] Penrose, R. (1968) Structure of space-time. In: *Battelle Rencontres, 1967 Lectures in Mathematics and Physics*, C. M. DeWitt & J. A. Wheeler, eds. (New York: Benjamin).

[27] Penrose, R., & Rindler, W. (1984) *Spinors and Space-Time*, Volumes 1 and 2. (Cambridge: Cambridge University Press).

[28] Segal, I. E. (1947) Postulates for general quantum mechanics. *Ann. Math.* **48**, 930–948.

[29] von Neumann, J. (1932) *Mathematische Grundlagen der Quantenmechanik* (Berlin: Springer-Verlag). English translation by R. T. Beyer (1955) *Mathematical Foundations of Quantum Mechanics* (Princeton, New Jersey: Princeton University Press).

Journal of Physics: Conference Series **624** (2015) 012003 doi:10.1088/1742-6596/624/1/012003

Quantum communication protocols based on entanglement swapping

Guillermo Morales-Luna
Computer Science Department, CINVESTAV-IPN, Mexico City, Mexico
E-mail: gmorales@cs.cinvestav.mx

Abstract. We recall several cryptographic protocols based on entanglement alone and also on entanglement swapping. We make an exposition in terms of the geometrical aspects of the involved Hilbert spaces, and we concentrate on the formal nature of the used transformations.

1. Introduction

Entanglement has been widely exploited in the design of protocols and procedures for communication, cryptography and computation within quantum contexts. Quantum codes guarantee that information has been transmitted without any alteration. Cryptographic protocols aid for key agreement for secure communication, namely unconditionally secure information exchange. Entanglement has been used to implement and to speed-up paradigmatic quantum algorithms [1, 2].

Entanglement swapping may entangle two quantum systems without direct interaction among them, and this fact is exploited within several quantum cryptography schemes.

Here, we recall several well known cryptographic protocols using entanglement, alone, and entanglement swapping: the *Quantum Secure Direct Communication Protocol* (see Table 3 below) communicates securely bit strings with an even length, the *Quantum Bidirectional Communication Protocol* (see Table 4 below) is a generalization of the above protocol in which the communicating parts exchange simultaneously messages of even bit length, the *Quantum Multidirectional Communication Protocol* (see Table 8 below) allows the message exchange among three parts using the maximally entangled GHZ states, at Table 9 we recall a three parties protocol in which two correspondents communicate securely just after the authorisation of a third party (who does not catch the message exchange), and finally, the *Key Agreement Protocol Using Entanglement Swapping* is sketched at Table 10 in order to illustrate the use of entanglement swapping in cryptographic protocols.

We emphasize the algebraic aspect of the Hilbert space nature of all the involved protocols We establish a correspondence among unitary transforms, obtained as tensor products of Pauli maps, and permutations of basic vectors in the Hilbert spaces. These correspondence are summarized at Tables 2 and 6. Also, explicit expressions of the Bell basis, in terms of the Hadamard basis are given.

For any two integer numbers $i, j \in \mathbb{Z}^+$, with $i \leq j$, let us write

$$[\![i, j]\!] = \{i, i+1, \ldots, j-1, j\}.$$

Journal of Physics: Conference Series **624** (2015) 012003 doi:10.1088/1742-6596/624/1/012003

2. Qubits and Pauli transforms

Let $\mathbb{H}_1 = \mathbb{C}^2$ be the two-dimensional complex Hilbert space. The unit sphere of $\mathbb{H}_1$ is the set of *qubits*. The *canonical basis* consists of the vectors $\mathbf{e}_0 = [1\ 0]^T$ and $\mathbf{e}_1 = [0\ 1]^T$. Usually, it is written $|0\rangle = \mathbf{e}_0$ and $|1\rangle = \mathbf{e}_1$. Let $\mathbf{h}_0 = \frac{1}{\sqrt{2}}(|0\rangle + |1\rangle)$ and $\mathbf{h}_1 = \frac{1}{\sqrt{2}}(|0\rangle - |1\rangle)$ be the vectors forming the *Hadamard basis* at $\mathbb{H}_1$.

Let us consider the Pauli operators

$$\begin{aligned} \sigma_0 &= \begin{pmatrix} 1 & 0 \\ 0 & 1 \end{pmatrix}, \quad \sigma_x = \begin{pmatrix} 0 & 1 \\ 1 & 0 \end{pmatrix} \\ \sigma_y &= \begin{pmatrix} 0 & -i \\ i & 0 \end{pmatrix}, \quad \sigma_z = \begin{pmatrix} 1 & 0 \\ 0 & -1 \end{pmatrix} \end{aligned} \tag{1}$$

and let us number them as $[\sigma_0\ \ \sigma_1\ \ \sigma_2\ \ \sigma_3] = [\sigma_0\ \ \sigma_x\ \ \sigma_y\ \ \sigma_z]$. The action of these operators over the vectors at the canonical and the Hadamard basis is sketched at Table 1. At each entry is located the value of σ_i at the vector labeling the corresponding column. We see that, up to a factor which is a unitary complex number, the canonical basis remains fixed by σ_0 and σ_3 and is switched by σ_1 and σ_2 while the Hadamard basis remains fixed by σ_0 and σ_1 and is switched by σ_2 and σ_3. Thus, the operator σ_2 is switching both basis.

Table 1. *Action of the Pauli operators on basis vectors*

	$\|0\rangle$	$\|1\rangle$	$\mathbf{h}_0$	$\mathbf{h}_1$
σ_0	$\|0\rangle$	$\|1\rangle$	$\mathbf{h}_0$	$\mathbf{h}_1$
σ_1	$\|1\rangle$	$\|0\rangle$	$\mathbf{h}_0$	$-\mathbf{h}_1$
σ_2	$-i\,\|1\rangle$	$i\,\|0\rangle$	$i\,\mathbf{h}_1$	$-i\,\mathbf{h}_0$
σ_3	$\|0\rangle$	$-\,\|1\rangle$	$\mathbf{h}_1$	$\mathbf{h}_0$

Let $\mathbb{H}_2 = \mathbb{H}_1 \otimes \mathbb{H}_1$ be the Hilbert space containing the 2-quregisters. Any 2-quregister $\mathbf{x}$ has two components $\mathbf{x}_0$ and $\mathbf{x}_1$, each at the factor space $\mathbb{H}_1$, they are qubits. Let for $i, j \in [\![0, 1]\!]$,

$$\mathbf{b}_{ij} = \frac{1}{\sqrt{2}} \left(|0i\rangle + (-1)^j \,|1\bar{i}\rangle \right),$$

here the overline denotes complement modulus 2. Then $B = (\mathbf{b}_{ij})_{i,j \in [\![0,1]\!]}$ is the Bell basis of $\mathbb{H}_2$ and consists of four maximally entangled states. In terms of the Hadamard basis, the Bell vectors are expressed as follows:

$$\begin{aligned} \mathbf{b}_{00} &= \frac{1}{\sqrt{2}}(\mathbf{h}_0 \otimes \mathbf{h}_0 + \mathbf{h}_1 \otimes \mathbf{h}_1) \\ \mathbf{b}_{01} &= \frac{1}{\sqrt{2}}(\mathbf{h}_1 \otimes \mathbf{h}_0 + \mathbf{h}_0 \otimes \mathbf{h}_1) \\ \mathbf{b}_{10} &= \frac{1}{\sqrt{2}}(\mathbf{h}_0 \otimes \mathbf{h}_0 - \mathbf{h}_1 \otimes \mathbf{h}_1) \\ \mathbf{b}_{11} &= \frac{1}{\sqrt{2}}(\mathbf{h}_1 \otimes \mathbf{h}_0 - \mathbf{h}_0 \otimes \mathbf{h}_1) \end{aligned}$$

Any sequence $C = (\mathbf{c}_k)_{k \geq 0}$ whose terms are elements of B determines two sequences of qubits $C_0 = (\mathbf{c}_{k0})_{k \geq 0}$ and $C_1 = (\mathbf{c}_{k1})_{k \geq 0}$.

Through the radix expression of an index in base 2, we may number the Bell basis as $B = (\mathbf{b}_k)_{k\in[\![0,2^2-1]\!]}$. The tensor products $\sigma_{ij} = \sigma_i \otimes \sigma_j$ determine bijections $\alpha_{ij} : B \to B$, such that

$$\forall (i,j) \in [\![0,3]\!]^2\ \forall k \in [\![0,2^2-1]\!] : \quad \sigma_{ij}(\mathbf{b}_k) \in \mathcal{L}(\mathbf{b}_{\alpha_{ij}(k)}). \tag{2}$$

Let $A_2 : \sigma_{ij} \mapsto \alpha_{ij}$ be the map that associates to each tensor product σ_{ij} the corresponding permutation that it defines at the Bell basis. The image of A_2 consists of just $4 = 2^2$ permutations $(\beta_\ell)_{\ell\in[\![0,2^3-1]\!]}$, and each permutation is defined by 4 tensor products σ_{ij} as summarized at the Table 2: the first column displays the index ℓ, the second column the permutation β_ℓ and the third column the list of tensor product maps σ_{ij} producing β_ℓ under the map A_2.

Table 2. *Correspondence on permutations of the Bell basis and tensor products of Pauli operators*

ℓ	β_ℓ	$A_2^{-1}(\beta_\ell)$
0	[0 1 2 3]	$\{\sigma_{00}, \sigma_{01}, \sigma_{02}, \sigma_{03}\}$
1	[2 3 0 1]	$\{\sigma_{01}, \sigma_{00}, \sigma_{03}, \sigma_{02}\}$
2	[3 2 1 0]	$\{\sigma_{02}, \sigma_{03}, \sigma_{00}, \sigma_{01}\}$
3	[1 0 3 2]	$\{\sigma_{03}, \sigma_{02}, \sigma_{01}, \sigma_{00}\}$

Table 2 can in turn be summarized as

$$\begin{aligned} \forall i \in [\![0,3]\!] : \ A_2(\sigma_{i\beta_0(i)}) &= \beta_0, \quad & A_2(\sigma_{i\beta_3(i)}) &= \beta_1, \\ A_2(\sigma_{i\beta_1(i)}) &= \beta_2, \quad & A_2(\sigma_{i\beta_2(i)}) &= \beta_3. \end{aligned} \tag{3}$$

By looking at relation (2), we see that if i and k remain fixed, then the index j can be encoded by the value $\alpha_{ij}(k)$. This property can be exploited for secure communication purposes.

In Table 3 a *Quantum Secure Direct Communication Protocol* [3] is sketched. The purpose of this protocol is to communicate securely a word in $[\![0,3]\!]^*$. Alice should communicate a message $[\mu_\kappa]_{\kappa=0}^{m-1} \in [\![0,3]\!]^m$. Alice and Bob fix a Pauli transform σ_i, a Bell quregister $\mathbf{b}_k \in B$, an integer $n > m$ and an index set $J \in [\![0,n-1]\!]^{(m)}$. They share initially a constant sequence $C = (\mathbf{c}_\kappa)_{\kappa=0}^{n-1}$ whose terms coincide with $\mathbf{b}_k \in B$.

Using sequences of entangled quregisters it is also possible to build bidirectional communication protocols. In Table 4 a *Quantum Bidirectional Communication Protocol* is sketched [4]. The purpose of this protocol is to communicate securely two words in $[\![0,3]\!]^*$, one going from Alice to Bob and the other in the opposite direction. Alice and Bob should interchange messages in $[\![0,1]\!]^{2n}$, and they share initially a constant sequence $C = (\mathbf{c}_\kappa)_{\kappa=0}^{n-1}$.

3. Entanglement swapping

Entanglement swapping is a phenomenon which allows to put two particles into entangled states although these particles have not been close at any time. Departing from two pairs of entangled particles, a particle is chosen from each pair, then the joint pair of selected particles is measured with respect to the Bell basis, resulting in an entangled state. As a consequence, the pair consisting of the two partner particles is also entangled. This last pair is the result of the entanglement swapping beginning from the original two pairs.

In two 2-quregisters there are involved 4 qubits, let us identify them with the four indexes in the integer interval $[\![0,3]\!]$. Let us write $|\varepsilon\rangle_\mu$, for $\varepsilon \in \{0,1\}$, $\mu \in [\![0,3]\!]$, to denote the state of

Table 3. *Quantum Secure Direct Communication Protocol*

Alice		Bob
pads the message $[\mu_\kappa]_{\kappa=0}^{m-1}$ into a sequence $[j_\kappa]_{\kappa=0}^{n-1}$ by inserting the message into the positions at J		
codifies the message $[j_\kappa]_{\kappa=0}^{n-1}$ by calculating $D_1 = (\sigma_{j_\kappa}(\mathbf{c}_{\kappa 1}))_{\kappa=0}^{n-1}$		
sends D_1 through a quantum channel	$\xrightarrow{D_1}$	receives D_1
		calculates $D_0 = (\sigma_i(\mathbf{c}_{\kappa 0}))_{\kappa=0}^{n-1}$, which actually completes the calculation of $D = ((\sigma_i \otimes \sigma_{j_\kappa})(\mathbf{c}_\kappa))_{\kappa=0}^{n-1}$
		calculates E by measuring each term at D with respect to the Bell basis
		recovers the sequence $(\alpha_{ij_\kappa}(k))_{\kappa=0}^{n-1}$, consequently the padded sequence $[j_\kappa]_{\kappa=0}^{n-1}$, and the original message $[\mu_\kappa]_{\kappa=0}^{m-1}$

Table 4. *Quantum Bidirectional Communication Protocol*

Alice		Bob
by taking pairs of contiguous bits, she writes her message as a word $(i_\kappa)_{\kappa=0}^{n-1} \in [\![0,3]\!]^n$		by taking pairs of contiguous bits, he writes his message as a word $(j_\kappa)_{\kappa=0}^{n-1} \in [\![0,3]\!]^n$
codifies her message by calculating $D_0 = (\sigma_{i_\kappa}(\mathbf{c}_{\kappa 0}))_{\kappa=0}^{n-1}$		codifies his message by calculating $D_1 = (\sigma_{j_\kappa}(\mathbf{c}_{\kappa 1}))_{\kappa=0}^{n-1}$
sends D_0 through a quantum channel	$\xrightarrow{D_0}$	receives D_0
receives D_1	$\xleftarrow{D_1}$	sends D_1 through a quantum channel
for each $\kappa \in [\![0, n-1]\!]$ she measures the entangled quregister $\mathbf{d}_\kappa$ with respect to the Bell basis		for each $\kappa \in [\![0, n-1]\!]$ he measures the entangled quregister $\mathbf{d}_\kappa$ with respect to the Bell basis
since she knows i_κ, using (2), she recovers j_κ		since he knows j_κ, using (2), he recovers i_κ

the μ-th qubit. For any two different indexes $\mu, \nu \in [\![0,3]\!]$, let the respective Bell basis of the Hilbert space $\mathbb{H}_2$ be

$$\left(\mathbf{b}_{ij}^{(\mu\nu)} = \frac{1}{\sqrt{2}}\left(|0i\rangle_{\mu\nu} + (-1)^j\,|1\bar{i}\rangle_{\mu\nu}\right)\right)_{i,j\in[\![0,1]\!]}.$$

Let us assume that the 2-quregister consisting of the qubits 0 and 1 is entangled as well as the pair of qubits 2 and 3. Then a basis of the space $\mathbb{H}_4$ is

$$B_{(01)(23)} = \left(\mathbf{z}_{i_0 j_0 i_1 j_1} = \mathbf{b}_{i_0 j_0}^{(01)} \otimes \mathbf{b}_{i_1 j_1}^{(23)}\right)_{i_0,j_0,i_1,j_1\in[\![0,1]\!]}, \tag{4}$$

where $\forall i_0, j_0, i_1, j_1 \in [\![0,1]\!]$:

$$\mathbf{z}_{i_0 j_0 i_1 j_1} = \frac{1}{2}\Big(\left|0 i_0 0 i_1\right\rangle_{0123} + (-1)^{j_1}\left|0 i_0 1 \overline{i_1}\right\rangle_{0123} + (-1)^{j_0}\left|1 \overline{i_0} 0 i_1\right\rangle_{0123} + (-1)^{j_0+j_1}\left|1 \overline{i_0} 1 \overline{i_1}\right\rangle_{0123}\Big). \tag{5}$$

By rearranging the pairs and considering the pairs $(0,2)$ and $(1,3)$, we have that a second basis of $\mathbb{H}_4$ is

$$B_{(02)(13)} = \left(\mathbf{b}^{(02)}_{i_0 j_0} \otimes \mathbf{b}^{(13)}_{i_1 j_1}\right)_{i_0, j_0, i_1, j_1 \in [\![0,1]\!]}, \tag{6}$$

where $\forall i_0, j_0, i_1, j_1 \in [\![0,1]\!]$:

$$\mathbf{b}^{(02)}_{i_0 j_0} \otimes \mathbf{b}^{(13)}_{i_1 j_1} = \frac{1}{2}\Big(\left|0 i_0 0 i_1\right\rangle_{0213} + (-1)^{j_1}\left|0 i_0 1 \overline{i_1}\right\rangle_{0213} + (-1)^{j_0}\left|1 \overline{i_0} 0 i_1\right\rangle_{0213} + (-1)^{j_0+j_1}\left|1 \overline{i_0} 1 \overline{i_1}\right\rangle_{0213}\Big).$$

By swapping the middle qubits, the following 4-quregisters result: $\forall i_0, j_0, i_1, j_1 \in [\![0,1]\!]$,

$$\mathbf{y}_{i_0 j_0 i_1 j_1} = \frac{1}{2}\Big(\left|0 0 i_0 i_1\right\rangle_{0123} + (-1)^{j_1}\left|0 1 i_0 \overline{i_1}\right\rangle_{0123} + (-1)^{j_0}\left|1 0 \overline{i_0} i_1\right\rangle_{0123} + (-1)^{j_0+j_1}\left|1 1 \overline{i_0} \overline{i_1}\right\rangle_{0123}\Big). \tag{7}$$

Each 2-quregister $\mathbf{z}_{i_0 j_0 i_1 j_1}$ given by relation (5) can be expressed in terms of the 2-quregisters $\mathbf{y}_{i_0 j_0 i_1 j_1}$ given by relation (7), namely:

$$\mathbf{z}_{i_0 j_0 i_1 j_1} = \frac{1}{2}\Big(\mathbf{y}_{i_0 j_0 i_1 j_1} + (-1)^{j_1}\mathbf{y}_{i_0 \overline{j_0} \overline{i_1} \overline{j_1}} + (-1)^{j_0}\mathbf{y}_{\overline{i_0} j_0 \overline{i_1} j_1} + (-1)^{j_0+j_1}\mathbf{y}_{\overline{i_0 j_0 i_1 j_1}}\Big), \tag{8}$$

and this relation is symmetric:

$$\mathbf{y}_{i_0 j_0 i_1 j_1} = \frac{1}{2}\Big(\mathbf{z}_{i_0 j_0 i_1 j_1} + (-1)^{j_1}\mathbf{z}_{i_0 \overline{j_0} \overline{i_1} \overline{j_1}} + (-1)^{j_0}\mathbf{z}_{\overline{i_0} j_0 \overline{i_1} j_1} + (-1)^{j_0+j_1}\mathbf{z}_{\overline{i_0 j_0 i_1 j_1}}\Big), \tag{9}$$

In this way, the entanglement of the 4-registers $\mathbf{z}$ is reflected by the entanglement of the 4-registers $\mathbf{y}$, in other words, the entanglement of the pairs $(0,1)$ and $(2,3)$ is swapped into the entanglement of the pairs $(0,2)$ and $(1,3)$, and conversely.

4. Three-entanglement

Let us consider multi-party bidirectional protocols. In particular, we will illustrate these procedures with three communicating parties. A proper protocol considers maximally entangled 3-quregisters, members of $\mathbb{H}_3 = \mathbb{H}_1 \otimes \mathbb{H}_2$. Any 3-quregister $\mathbf{x}$ has three components $\mathbf{x}_0$, $\mathbf{x}_1$ and $\mathbf{x}_2$, each at the factor space $\mathbb{H}_1$, they are qubits. For $\varepsilon_1, \varepsilon_2, \varepsilon_3 \in [\![0,1]\!]$ let

$$\mathbf{b}_{\varepsilon_1 \varepsilon_2 \varepsilon_3} = \frac{1}{\sqrt{2}}\left(\left|0 \varepsilon_1 \varepsilon_2\right\rangle + (-1)^{\varepsilon_3}\left|1 \overline{\varepsilon_1} \overline{\varepsilon_2}\right\rangle\right).$$

These vectors form a basis, B_3, analogous to the Bell basis in $\mathbb{H}_3$, but they are called *Greensberger-Horne-Zeilinger* (GHZ) *states*. In terms of the Hadamard vectors, the GHZ states are expressed as shown at the Table 5.

Table 5. The GHZ states in terms of the Hadamard basis of qubits.

$$
\begin{aligned}
\mathbf{b}_{000} &= \frac{1}{2}(\mathbf{h}_0 \otimes (\mathbf{h}_0 \otimes \mathbf{h}_0 + \mathbf{h}_1 \otimes \mathbf{h}_1) + \mathbf{h}_1 \otimes (\mathbf{h}_0 \otimes \mathbf{h}_1 + \mathbf{h}_1 \otimes \mathbf{h}_0)) \\
\mathbf{b}_{001} &= \frac{1}{2}(\mathbf{h}_1 \otimes (\mathbf{h}_0 \otimes \mathbf{h}_0 + \mathbf{h}_1 \otimes \mathbf{h}_1) + \mathbf{h}_0 \otimes (\mathbf{h}_0 \otimes \mathbf{h}_1 + \mathbf{h}_1 \otimes \mathbf{h}_0)) \\
\mathbf{b}_{010} &= \frac{1}{2}(\mathbf{h}_0 \otimes (\mathbf{h}_0 \otimes \mathbf{h}_0 - \mathbf{h}_1 \otimes \mathbf{h}_1) + \mathbf{h}_1 \otimes (-\mathbf{h}_0 \otimes \mathbf{h}_1 + \mathbf{h}_1 \otimes \mathbf{h}_0)) \\
\mathbf{b}_{011} &= \frac{1}{2}(\mathbf{h}_1 \otimes (\mathbf{h}_0 \otimes \mathbf{h}_0 - \mathbf{h}_1 \otimes \mathbf{h}_1) + \mathbf{h}_0 \otimes (-\mathbf{h}_0 \otimes \mathbf{h}_1 + \mathbf{h}_1 \otimes \mathbf{h}_0)) \\
\mathbf{b}_{100} &= \frac{1}{2}(\mathbf{h}_0 \otimes (\mathbf{h}_0 \otimes \mathbf{h}_0 - \mathbf{h}_1 \otimes \mathbf{h}_1) + \mathbf{h}_1 \otimes (\mathbf{h}_0 \otimes \mathbf{h}_1 - \mathbf{h}_1 \otimes \mathbf{h}_0)) \\
\mathbf{b}_{101} &= \frac{1}{2}(\mathbf{h}_1 \otimes (\mathbf{h}_0 \otimes \mathbf{h}_0 - \mathbf{h}_1 \otimes \mathbf{h}_1) + \mathbf{h}_0 \otimes (\mathbf{h}_0 \otimes \mathbf{h}_1 - \mathbf{h}_1 \otimes \mathbf{h}_0)) \\
\mathbf{b}_{110} &= \frac{1}{2}(\mathbf{h}_0 \otimes (\mathbf{h}_0 \otimes \mathbf{h}_0 + \mathbf{h}_1 \otimes \mathbf{h}_1) + \mathbf{h}_1 \otimes (-\mathbf{h}_0 \otimes \mathbf{h}_1 - \mathbf{h}_1 \otimes \mathbf{h}_0)) \\
\mathbf{b}_{111} &= \frac{1}{2}(\mathbf{h}_1 \otimes (\mathbf{h}_0 \otimes \mathbf{h}_0 + \mathbf{h}_1 \otimes \mathbf{h}_1) + \mathbf{h}_0 \otimes (-\mathbf{h}_0 \otimes \mathbf{h}_1 - \mathbf{h}_1 \otimes \mathbf{h}_0))
\end{aligned}
$$

Through the radix expression of an index in base 2, we may number the Bell basis as $B_3 = (\mathbf{b}_k)_{k\in[\![0,2^3-1]\!]}$. The tensor products $\sigma_{ijk} = \sigma_i \otimes \sigma_j \otimes \sigma_k$ determine bijections $\alpha_{ijk} : B_3 \to B_3$ in an analogous way as in (2):

$$\forall (i,j,k) \in [\![0,3]\!]^3 \ \forall \ell \in [\![0,2^3-1]\!] : \ \ \sigma_{ijk}(\mathbf{b}_\ell) \in \mathcal{L}(\mathbf{b}_{\alpha_{ijk}(\ell)}). \qquad (10)$$

Let $A_3 : \sigma_{ijk} \mapsto \alpha_{ijk}$ be the map that associates to each tensor product σ_{ijk} the corresponding permutation at the Bell basis. The image of A_3 consists of just $8 = 2^3$ permutations $(\beta_\mu)_{\mu\in[\![0,2^3-1]\!]}$, and each permutation is defined by 8 tensor products σ_{ijk} as summarized in Table 6. As seen at the beginning of section 2, the operator σ_2 switches the canonical and the

Table 6. *Correspondence on permutations of the Bell basis and tensor products of Pauli operators*

μ	β_μ	$A_3^{-1}(\beta_\mu)$
0	[0 1 2 3 4 5 6 7]	$\{\sigma_{000}, \sigma_{033}, \sigma_{111}, \sigma_{122}, \sigma_{212}, \sigma_{221}, \sigma_{303}, \sigma_{330}\}$
1	[2 3 0 1 6 7 4 5]	$\{\sigma_{003}, \sigma_{030}, \sigma_{112}, \sigma_{121}, \sigma_{211}, \sigma_{222}, \sigma_{300}, \sigma_{333}\}$
2	[3 2 1 0 7 6 5 4]	$\{\sigma_{001}, \sigma_{032}, \sigma_{110}, \sigma_{123}, \sigma_{213}, \sigma_{220}, \sigma_{302}, \sigma_{331}\}$
3	[1 0 3 2 5 4 7 6]	$\{\sigma_{002}, \sigma_{031}, \sigma_{113}, \sigma_{120}, \sigma_{210}, \sigma_{223}, \sigma_{301}, \sigma_{332}\}$
4	[4 5 6 7 0 1 2 3]	$\{\sigma_{010}, \sigma_{023}, \sigma_{101}, \sigma_{132}, \sigma_{202}, \sigma_{231}, \sigma_{313}, \sigma_{320}\}$
5	[6 7 4 5 2 3 0 1]	$\{\sigma_{013}, \sigma_{020}, \sigma_{102}, \sigma_{131}, \sigma_{201}, \sigma_{232}, \sigma_{310}, \sigma_{323}\}$
6	[7 6 5 4 3 2 1 0]	$\{\sigma_{011}, \sigma_{022}, \sigma_{100}, \sigma_{133}, \sigma_{203}, \sigma_{230}, \sigma_{312}, \sigma_{321}\}$
7	[5 4 7 6 1 0 3 2]	$\{\sigma_{012}, \sigma_{021}, \sigma_{103}, \sigma_{130}, \sigma_{200}, \sigma_{233}, \sigma_{311}, \sigma_{322}\}$

Hadamard basis. Let us consider just operators of the form σ_{ijk} where $i \in [\![0,3]\!]$, $j,k \in \{0,2\}$. Then, the restriction of Table 6 to operators at the set $S = \{\sigma_{ijk} |\ i \in [\![0,3]\!], j,k \in \{0,2\}\}$ is shown at Table 7.

In Table 8 a *Quantum Multidirectional Communication Protocol* is sketched [4, 5]. The purpose of this protocol is to communicate securely four classical bits, two emitted by Alice,

Table 7. *Correspondence on permutations of the Bell basis and tensor products of Pauli operators at the set S, in alphabetical order according to the first element at $S \cap A_3^{-1}(\beta_\mu)$. Observe that if one index i, j, k is known, the other two can be deduced form this index and the 3-quregister $\sigma_{ijk}(\mathbf{b}_\ell)$.*

μ	$S \cap A_3^{-1}(\beta_\mu)$
0	$\{\sigma_{000}, \sigma_{122}\}$
3	$\{\sigma_{002}, \sigma_{120}\}$
5	$\{\sigma_{020}, \sigma_{102}\}$
6	$\{\sigma_{022}, \sigma_{100}\}$
7	$\{\sigma_{200}, \sigma_{322}\}$
4	$\{\sigma_{202}, \sigma_{320}\}$
2	$\{\sigma_{220}, \sigma_{302}\}$
1	$\{\sigma_{222}, \sigma_{300}\}$

one by Bob and another by Claire. By repeating the protocol the parties may exchange longer bit-strings. Alice, Bob and Claire should interchange four classical bits, two emitted by Alice, one by Bob and another by Claire. The parties share two GHZ states, $\mathbf{c}_0 = \mathbf{c}_1 = \mathbf{b}_\ell \in \mathbb{H}_3$ with respective components $\mathbf{c}_{00}, \mathbf{c}_{10}, \mathbf{c}_{20}$ and $\mathbf{c}_{01}, \mathbf{c}_{11}, \mathbf{c}_{21}$. The components $\mathbf{c}_{0k}$, $\mathbf{c}_{1k}$, $\mathbf{c}_{2k}$ are in possession of Alice, Bob and Claire respectively, $k \in [\![0, 1]\!]$. The quregister $\mathbf{c}_0$ is a record of the initial state $\mathbf{b}_\ell$, while the quregister $\mathbf{c}_1$ is to be transformed during the protocol.

Table 8. *Quantum Multidirectional Communication Protocol*

(i)	The two bits of Alice determine an index $i_A \in [\![0, 3]\!]$. She applies σ_{i_A} to her qubit $\mathbf{c}_{01}$.
(ii)	Bob applies either σ_0 or σ_2 to his qubit $\mathbf{c}_{11}$ according to the value of his bit.
(iii)	Claire applies either σ_0 or σ_2 to her qubit $\mathbf{c}_{21}$ according to the value of her bit.
(iv)	They take a measure of the transformed quregister with respect to the Bell basis.
(v)	Using table 7, since each participant knows his/her own message, they recover the transmitted bits.

Another bidirectional protocol [6] consists of three participants: Alice and Bob are the correspondents and Claire is the controller. The correspondents are able to communicate only after the authorization of the controller, but their correspondence should be kept in secret against the controller. The protocol is sketched at Table 9. Alice and Bob should interchange messages at $[\![0, 3]\!]^m$ after the authorization granted by Claire. The parties share a constant sequence $(\mathbf{c}_\nu)_{\nu=0}^{n-1}$ whose entries coincide with a GHZ initial state $\mathbf{b}_\ell$. The component sequence $(\mathbf{c}_{0\nu})_{\nu=0}^{n-1}$, let us say for ease of explanation, is owned by Claire, the component sequence $(\mathbf{c}_{1\nu})_{\nu=0}^{n-1}$ by Alice and the component sequence $(\mathbf{c}_{2\nu})_{\nu=0}^{n-1}$ by Bob.

A *Key Agreement Protocol Using Entanglement Swapping* is obtained [7] as follows. Let $B_2^{(\mu\nu)}$ be the Bell basis considering two qubits $\mu, \nu \in [\![0, 3]\!]$, $\mu \neq \nu$.

Let $\mathbf{c}^{(01)}, \mathbf{c}^{(23)}$ be two Bell 2-quregisters with respective components $\mathbf{c}^{(0)}, \mathbf{c}^{(1)}$ and $\mathbf{c}^{(2)}, \mathbf{c}^{(3)}$.

Alice may act on the pair $(\mathbf{c}^{(01)}, \mathbf{c}^{(23)}) = [\mathbf{c}^{(0)}, \mathbf{c}^{(1)}, \mathbf{c}^{(2)}, \mathbf{c}^{(3)}]$ either by (A_0:) doing nothing or by (A_1:) swapping the middle qubits, obtaining thus $[\mathbf{c}^{(0)}, \mathbf{c}^{(2)}, \mathbf{c}^{(1)}, \mathbf{c}^{(3)}]$.

Bob may act on the pair $(\mathbf{c}^{(01)}, \mathbf{c}^{(23)}) = [\mathbf{c}^{(0)}, \mathbf{c}^{(1)}, \mathbf{c}^{(2)}, \mathbf{c}^{(3)}]$ either by (B_0:) measuring $[\mathbf{c}^{(0)}, \mathbf{c}^{(1)}]$ with respect to the Bell basis $B_2^{(01)}$ and measuring $[\mathbf{c}^{(1)}, \mathbf{c}^{(2)}]$ with respect to the Bell

Table 9. *Quantum Controlled Bidirectional Communication Protocol*

(i)	Alice and Bob agree a set $J \subset [\![0, n-1]\!]$ of m positions among the index set $[\![0, m-1]\!]$.
(ii)	Alice codifies her message $(a_\mu)_{\mu=0}^{m-1} \in [\![0,3]\!]^{m-1}$ by applying σ_{a_μ} to her correspondent qubit $\mathbf{c}_{1\nu_\mu}$, with $\nu_\mu \in J$, and she applies arbitrary Pauli operators at her qubits with indexes not in J. Alice sends her codified sequence to Claire.
(iii)	Bob codifies his message $(b_\mu)_{\mu=0}^{m-1} \in [\![0,3]\!]^{m-1}$ by applying σ_{b_μ} to his correspondent qubit $\mathbf{c}_{2\nu_\mu}$, with $\nu_\mu \in J$, and he applies arbitrary Pauli operators at his qubits with indexes not in J. Bob sends his codified sequence to Claire.
(iv)	Claire receives the component sequences $\left(\mathbf{c}_{1\nu_\mu}\right)_{\nu=0}^{n-1}$ and $\left(\mathbf{c}_{2\nu_\mu}\right)_{\nu=0}^{n-1}$, and she measures the whole sequence $\left(\mathbf{c}_{\nu_\mu}\right)_{\nu=0}^{n-1}$ with respect to the basis B_3. She sends her results to Alice and Bob as an authorization to proceed the transaction.
(v)	Using the table 6, her knowledge of her own message and the index set J, Alice recovers Bob's message $(b_\mu)_{\mu=0}^{m-1} \in [\![0,3]\!]^{m-1}$.
(vi)	Using the table 6, his knowledge of his own message and the index set J, Bob recovers Alice's message $(a_\mu)_{\mu=0}^{m-1} \in [\![0,3]\!]^{m-1}$.

basis $B_2^{(23)}$ or $(B_1:)$ by measuring $[\mathbf{c}^{(0)}, \mathbf{c}^{(2)}]$ with respect to the Bell basis $B_2^{(02)}$ and measuring $[\mathbf{c}^{(1)}, \mathbf{c}^{(3)}]$ with respect to the Bell basis $B_2^{(13)}$.

If the chosen actions are (A_0, B_0) or (A_1, B_1), the actions are said to be *correlated*, otherwise, they are *anticorrelated*.

Table 10. *Key Agreement Protocol Using Entanglement Swapping*

(i)	Alice selects a sequence $\mathbf{B} = \left(\mathbf{b}_{\alpha(k)}\right)_{k=0}^{2m-1}$ of entangled states at the Bell basis. Each pair of two such states $\left(\mathbf{b}_{\alpha(2k)}, \mathbf{b}_{\alpha(2k+1)}\right)$, $k \in [\![0, m-1]\!]$, involves 4 qubits, say $[\mathbf{c}^{(0k)}, \mathbf{c}^{(1k)}, \mathbf{c}^{(2k)}, \mathbf{c}^{(3k)}]$
(ii)	For each $k \in [\![0, m-1]\!]$, Alice applies an operation A_0 or A_1 to the current pair $\left(\mathbf{b}_{\alpha(2k)}, \mathbf{b}_{\alpha(2k+1)}\right)$. She obtains the sequence $\mathbf{C}$ and she sends this sequence to Bob through a public quantum channel.
(iii)	Bob receives the sequence $\mathbf{C}$ and for any block of 4 consecutive qubits, say $[\mathbf{d}^{(0k)}, \mathbf{d}^{(1k)}, \mathbf{d}^{(2k)}, \mathbf{d}^{(3k)}]$, he selects randomly an operation B_0 or B_1 and he applies it to $[\mathbf{d}^{(0k)}, \mathbf{d}^{(1k)}, \mathbf{d}^{(2k)}, \mathbf{d}^{(3k)}]$.
(iv)	Alice and Bob exchange through a classic channel the lists of their corresponding selected operations.
(v)	The common key is established by selecting the 4 blocks measurements corresponding to the correlated pairs of operations.
(vi)	It is worth to remark that at the anticorrelated positions, both Alice and Bob may recover two common bits, corresponding to the initial state of Alice for the current 4-block.

In the protocol, the agreed common key is the juxtaposition of the measures obtained at the positions in which correlated operators do occur. When looking for a greater efficiency it is possible to recover also not 4, for 2 bits at any block corresponding to an anti correlated operator.

References

[1] Gisin N, Ribordy G, Tittel W and Zbinden H 2002 *Rev. Mod. Phys.* **74**(1) 145–195 URL http://link.aps.org/doi/10.1103/RevModPhys.74.145

[2] Lanyon B P, Weinhold T J, Langford N K, Barbieri M, James D F V, Gilchrist A and White A G 2007 *Phys. Rev. Lett.* **99**(25) 250505 URL http://link.aps.org/doi/10.1103/PhysRevLett.99.250505

[3] Deng F G, Long G L and Liu X S 2003 *Phys. Rev. A* **68**(4) 042317 URL http://link.aps.org/doi/10.1103/PhysRevA.68.042317

[4] Gao Fei, Guo FenZhuo, Wen QiaoYan and Zhu FuChen 2008 *Science China Physics, Mechanics & Astronomy* **51** 559 (pages 0) URL http://phys.scichina.com:8083/sciGe/EN/abstract/article_408422.shtml

[5] Song-Kong Chong and Tzonelih Hwang 2011 *Optics Communications* **284** 515 – 518 ISSN 0030-4018 URL http://www.sciencedirect.com/science/article/pii/S0030401810008928

[6] Man Zhong-Xiao and Xia Yun-Jie 2006 *Chinese Physics Letters* **23** 1680 URL http://stacks.iop.org/0256-307X/23/i=7/a=007

[7] Fei Gao, Fen-Zhuo Guo, Qiao-Yan Wen and Fu-Chen Zhu 2010 *International Journal of Modern Physics B* **24** 4611–4618 (*Preprint* http://www.worldscientific.com/doi/pdf/10.1142/S021797921005661X) URL http://www.worldscientific.com/doi/abs/10.1142/S021797921005661X

Mielnik50 IOP Publishing
Journal of Physics: Conference Series **624** (2015) 012004 doi:10.1088/1742-6596/624/1/012004

Note on the quantum correlations of two qubits coupled to photon baths

Claudia Quintana and Oscar Rosas-Ortiz
Physics Department, Cinvestav, A.P. 14-740, México D.F. 07000, Mexico
E-mail: lquintana@fis.cinvestav.mx, orosas@fis.cinvestav.mx

Abstract. The time-evolution of the quantum correlations between two qubits that are coupled to a pair of photon baths is studied. We show that conditioned transitions occurring in the entire system have influence on the time-evolution of the subsystems. Then, we show that the study of the population inversion of each of the qubits is a measure of the correlations between them that is in agreement with the notion of concurrence.

1. Introduction

Coherent superpositions in quantum mechanics give rise to correlations between the parts of a given system [1,2]. Such correlations can be classic or quantum, and both of them may coexist for a given system [3]. Entanglement was proposed as a manifestation of quantum correlations [2], though not all quantum correlations are associated to entanglement since separable states can be quantum correlated [4, 5] (see also [6, 7]). Diverse measures of entanglement have been introduced over the time, examples are the concurrence [8, 9] and the negativity [10]. More general measures, as the quantum discord [11–13], quantify quantum correlations without the requirement of entanglement.

In this contribution we analyze the time-evolution of the correlation between two qubits that are coupled to two independent photon baths. It is assumed that the systems qubit+bath are one isolated from the other; that is, they are in cavities for which no communication is allowed. In this form, each qubit interacts with its environment (the photon bath) and decoherence results. The initial correlation between the qubits is then lost and recovered in time by time because the entire system is closed. We investigate the coherences of the entire system (two qubits, two photon baths) that are missed when information of one of its subsystems (the qubits) is required by summing up (partial tracing) over the degrees of freedom of the other parts (the photon baths). Such coherences include information of conditioned transitions between the states of the entire system that is lost, in a first sight, as a consequence of looking at the subsystems. However, this information can be recovered by analyzing the state of the parts in proper form. Indeed, we shall show that the study of the population inversion of the qubits represents a measure of quantum correlations that is in agreement with the concept of concurrence.

In Section 2 we introduce notation and general properties of a bipartite system formed of two qubits. We show that the partial trace of the density matrix produces leaky information about the transitions between the states of the subsystems. The notion of concurrence is recovered from basic properties of the reduced matrices. In Section 3 we analyze the time-evolution of the qubit bipartite system when it is coupled to photon baths and discuss about the behavior of the correlations with time. We show that the population inversions and the concurrence lead to the same information about the correlation that the qubits present at a given time. Some concluding remarks are given at the very end of the paper.

2. Qubit bipartite systems

Consider a system $\mathcal{S}$ formed of two qubits $\mathcal{S}_\ell$, $\ell = A, B$, in two separated (and isolated) cavities. Let us represent each qubit as a two level system with ground and excited states, $|-\rangle$ and $|+\rangle$ respectively, separated by an energy difference $\Delta E = \hbar\omega_q$. Using the simplest representation of the vector state basis

$$|+\rangle = \begin{pmatrix} 1 \\ 0 \end{pmatrix}, \quad |-\rangle = \begin{pmatrix} 0 \\ 1 \end{pmatrix}, \tag{1}$$

we can write the creation and annihilation operators for the qubit excitation as follows:

$$\sigma^+ = |+\rangle\langle -| \equiv X^{+-} = \begin{pmatrix} 0 & 1 \\ 0 & 0 \end{pmatrix}, \qquad \sigma^- = |-\rangle\langle +| \equiv X^{-+} = \begin{pmatrix} 0 & 0 \\ 1 & 0 \end{pmatrix}. \tag{2}$$

The dyads X^{+-} and X^{-+} are two of the four *Hubbard operators* associated to the qubit vector spaces [14], and (2) is called the *X-representation* of the operators σ^+ and σ^-. These last, together with the Pauli operator

$$\sigma_3 = [\sigma^+, \sigma^-] \equiv X^{++} - X^{--} = \begin{pmatrix} 1 & 0 \\ 0 & -1 \end{pmatrix}, \tag{3}$$

are the generators of the $su(2)$ Lie algebra. Notice that the basis elements (1) are eigenvectors of this last operator $\sigma_3|\pm\rangle = \pm|\pm\rangle$. As the state space for each qubit is two-dimensional

$$\mathcal{H}_\ell = \text{Span}\{|+\rangle, |-\rangle\}, \quad \ell = A, B, \tag{4}$$

the state space of the qubit bipartite system is four-dimensional

$$\mathcal{H} = \text{Span}\{|+\rangle \otimes |+\rangle, |+\rangle \otimes |-\rangle, |-\rangle \otimes |+\rangle, |-\rangle \otimes |-\rangle\}. \tag{5}$$

In (5) the left (right) element in each tensor product refers to the qubit A (B). Besides this last representation we shall use the following simplified notation

$$\begin{aligned} &|+\rangle \otimes |+\rangle = |++\rangle = |\varphi_1\rangle, \qquad |+\rangle \otimes |-\rangle = |+-\rangle = |\varphi_2\rangle, \\ &|-\rangle \otimes |+\rangle = |-+\rangle = |\varphi_3\rangle, \qquad |-\rangle \otimes |-\rangle = |--\rangle = |\varphi_4\rangle. \end{aligned} \tag{6}$$

Any state of the bipartite system $\mathcal{S}$ can be represented by a density operator of the form

$$\rho = \sum_{i,j=1}^{4} r_{ij} X^{ij}; \quad r_{ij} \in \mathbb{C}, \quad X^{ij} = |\varphi_i\rangle\langle\varphi_j|, \tag{7}$$

where the (sixteen) dyads X^{ij} are the Hubbard operators associated to the four-dimensional space (5). The matrix elements r_{ij} are defined according to the specific state ρ of the system $\mathcal{S}$ and satisfy the following equations

$$\begin{aligned} &\sum_{k=1}^{4} r_{kk} = 1, \quad r_{kj} = \bar{r}_{kj} \in \mathbb{C}, \quad r_{kk} \in \mathbb{R}, \\ &0 \leq \sum_{k=1}^{4} r_{kk}^2 + 2(|r_{12}|^2 + |r_{13}|^2 + |r_{14}|^2 + |r_{23}|^2 + |r_{24}|^2 + |r_{34}|^2) \leq 1, \end{aligned} \tag{8}$$

with $\bar{z}$ the complex conjugate of $z \in \mathbb{C}$. The diagonal elements r_{kk} are the (populations) probabilities of finding the system $\mathcal{S}$ in the state $|\varphi_k\rangle$. In turn, the off-diagonal elements $r_{k,j\neq k}$ are associated to the coherence between different states. That is, $r_{k,j\neq k}$ and its complex conjugate $\bar{r}_{k,j\neq k}$ are the probability amplitudes of the transitions

$$|\varphi_k\rangle \leftrightarrow |\varphi_j\rangle, \quad j \neq k. \tag{9}$$

No transitions between the states $|\varphi_k\rangle$ are allowed if the density matrix (7) is diagonal (this condition means the absence of interference terms in the time-depending case). Thus, the system $\mathcal{S}$ is in an *incoherent* (*coherent*) *superposition* of the basis states $|\varphi_k\rangle$ if the density matrix (7) is (not) diagonal with more than one non vanishing element. Notice, however, that the concept of *coherent superposition* depends on the representation for the density matrix (see, e.g. [15]).

2.1. *Missed correlations in reduced density operators*

To get information of the subsystem $\mathcal{S}_A$ we calculate a partial trace of ρ by summing up over all the degrees of freedom of $\mathcal{S}_B$. Let us rewrite (7) in the form

$$\rho = \begin{pmatrix} \overset{r_{11}}{X^{11} = X^{++} \otimes X^{++}} & \overset{r_{12}}{X^{12} = X^{++} \otimes X^{+-}} & \overset{r_{13}}{X^{13} = X^{+-} \otimes X^{++}} & \overset{r_{14}}{X^{14} = X^{+-} \otimes X^{+-}} \\ \overset{r_{21}}{X^{21} = X^{++} \otimes X^{-+}} & \overset{r_{22}}{X^{22} = X^{++} \otimes X^{--}} & \overset{r_{23}}{X^{23} = X^{+-} \otimes X^{-+}} & \overset{r_{24}}{X^{24} = X^{+-} \otimes X^{--}} \\ \overset{r_{31}}{X^{31} = X^{-+} \otimes X^{++}} & \overset{r_{32}}{X^{32} = X^{-+} \otimes X^{+-}} & \overset{r_{33}}{X^{33} = X^{--} \otimes X^{++}} & \overset{r_{34}}{X^{34} = X^{--} \otimes X^{+-}} \\ \overset{r_{41}}{X^{41} = X^{-+} \otimes X^{-+}} & \overset{r_{42}}{X^{42} = X^{-+} \otimes X^{--}} & \overset{r_{43}}{X^{43} = X^{--} \otimes X^{-+}} & \overset{r_{44}}{X^{44} = X^{--} \otimes X^{--}} \end{pmatrix}, \tag{10}$$

where we have made explicit the relationship between the Hubbard operator X^{kj} corresponding to r_{kj} and the tensor product of the qubit Hubbard operators $X^{\alpha\beta}$, $\alpha, \beta = \pm$, associated to $\mathcal{S}_A$ and $\mathcal{S}_B$. Notice that the elements in red are of the form $X^{kj} = X^{\alpha\beta} \otimes X^{\gamma\gamma}$, so that $\text{Tr}_{s_B} X^{kj} = X^{\alpha\beta}(\text{Tr} X^{\gamma\gamma}) = X^{\alpha\beta}$ will preserve the element r_{kj} in the position (α, β) of the reduced matrix ρ_{s_A}. The elements in black, in counterposition, make no contribution. Therefore, we get

$$\rho_{s_A} = \text{Tr}_{s_B}\rho = \begin{pmatrix} r_{11} + r_{22} & r_{13} + r_{24} \\ \bar{r}_{13} + \bar{r}_{24} & r_{33} + r_{44} \end{pmatrix}. \tag{11}$$

In a similar form, the terms in blue of the matrix array

$$\rho = \begin{pmatrix} \overset{r_{11}}{X^{11} = X^{++} \otimes X^{++}} & \overset{r_{12}}{X^{12} = X^{++} \otimes X^{+-}} & \overset{r_{13}}{X^{13} = X^{+-} \otimes X^{++}} & \overset{r_{14}}{X^{14} = X^{+-} \otimes X^{+-}} \\ \overset{r_{21}}{X^{21} = X^{++} \otimes X^{-+}} & \overset{r_{22}}{X^{22} = X^{++} \otimes X^{--}} & \overset{r_{23}}{X^{23} = X^{+-} \otimes X^{-+}} & \overset{r_{24}}{X^{24} = X^{+-} \otimes X^{--}} \\ \overset{r_{31}}{X^{31} = X^{-+} \otimes X^{++}} & \overset{r_{32}}{X^{32} = X^{-+} \otimes X^{+-}} & \overset{r_{33}}{X^{33} = X^{--} \otimes X^{++}} & \overset{r_{34}}{X^{34} = X^{--} \otimes X^{+-}} \\ \overset{r_{41}}{X^{41} = X^{-+} \otimes X^{-+}} & \overset{r_{42}}{X^{42} = X^{-+} \otimes X^{--}} & \overset{r_{43}}{X^{43} = X^{--} \otimes X^{-+}} & \overset{r_{44}}{X^{44} = X^{--} \otimes X^{--}} \end{pmatrix}, \tag{12}$$

lead to the reduced density operator

$$\rho_{s_B} = \text{Tr}_{s_A}\rho = \begin{pmatrix} r_{11} + r_{33} & r_{12} + r_{34} \\ \bar{r}_{12} + \bar{r}_{34} & r_{22} + r_{44} \end{pmatrix}. \tag{13}$$

Remark that (i) the anti-diagonal elements of ρ are missing in the reduced densities ρ_{s_A}, ρ_{s_B}, and (ii) all the diagonal elements of ρ are included in the diagonal of both reduced matrices. The latter property means that the populations r_{kk} represent information of the entire system $\mathcal{S}$ that can be obtained from any of the subsystems $\mathcal{S}_\ell$. In contrast, property (i) means that the probability amplitudes associated to the anti-diagonal coherences of ρ are not directly recoverable from neither $\mathcal{S}_A$ nor $\mathcal{S}_B$ since the elements r_{41}, r_{32}, and their complex conjugates, are missing in (11) and (13). According to (9), such amplitudes describe the transitions

$$|\varphi_4\rangle \leftrightarrow |\varphi_1\rangle, \qquad |\varphi_3\rangle \leftrightarrow |\varphi_2\rangle, \tag{14}$$

which in the qubit basis read as

$$|--\rangle \leftrightarrow |++\rangle, \qquad |-+\rangle \leftrightarrow |+-\rangle. \tag{15}$$

The first expression in (14)-(15) represents a transition $|-\rangle_B \leftrightarrow |+\rangle_B$ in $\mathcal{S}_B$ that is conditioned to the transition $|-\rangle_A \leftrightarrow |+\rangle_A$ in $\mathcal{S}_A$ and vice versa. Graphically,

$$|\varphi_4\rangle \leftrightarrow |\varphi_1\rangle \text{ means either } \begin{cases} |-\rangle_A \leftrightarrow |+\rangle_A & \Rightarrow \quad |-\rangle_B \leftrightarrow |+\rangle_B \\ & \text{or} \\ |-\rangle_B \leftrightarrow |+\rangle_B & \Rightarrow \quad |-\rangle_A \leftrightarrow |+\rangle_A \end{cases} \tag{16}$$

The second expression in (14)-(15) includes the following information

$$|\varphi_3\rangle \leftrightarrow |\varphi_2\rangle \text{ means either } \begin{cases} |-\rangle_A \leftrightarrow |+\rangle_A & \Rightarrow \quad |+\rangle_B \leftrightarrow |-\rangle_B \\ & \text{or} \\ |+\rangle_B \leftrightarrow |-\rangle_B & \Rightarrow \quad |-\rangle_A \leftrightarrow |+\rangle_A \end{cases} \tag{17}$$

Besides the anti-diagonal elements of ρ, other coherences $r_{k,j\neq k}$ are missed in ρ_{s_A} and ρ_{s_B}. Namely, the trace over the degrees of freedom of $\mathcal{S}_A$ eliminates the coherences included in the anti-diagonal 2×2 sub-matrices of ρ, see Eq. (12). That is, the probability amplitudes of the transitions $|\varphi_4\rangle \leftrightarrow |\varphi_2\rangle$ and $|\varphi_3\rangle \leftrightarrow |\varphi_1\rangle$ are also absent in ρ_{s_B}. In turn, according to (10), the tracing over $\mathcal{S}_B$ overrides the information of the transitions $|\varphi_2\rangle \leftrightarrow |\varphi_1\rangle$ and $|\varphi_4\rangle \leftrightarrow |\varphi_3\rangle$ in ρ_{s_A}.

For an arbitrary state ρ the off-diagonal elements of ρ_{s_A} have not connections a priori with the off-diagonal elements of ρ_{s_B}. However, if $\mathcal{S}$ is in a pure state, the elements of ρ_{s_A} and ρ_{s_B} are interrelated in such a manner that these last matrices have the same determinant. As we shall see, this fact is useful in recovering the information of the transitions (14)-(17) by using the information that is accessible from either ρ_{s_A} or ρ_{s_B}.

2.2. Pure and mixed states

Some relationships between ρ_{s_A} and ρ_{s_B} are achievable by considering a general pure state of the system $\mathcal{S}$, we take

$$|\psi\rangle = a|\varphi_1\rangle + b|\varphi_2\rangle + c|\varphi_3\rangle + d|\varphi_4\rangle, \tag{18}$$

with $|a|^2+|b|^2+|c|^2+|d|^2=1$. The pure state $|\psi\rangle$ would be the initial condition of the qubit bipartite system $\mathcal{S}$, assuming that the coupling of the qubits to the radiation fields in the cavities is activated at $t=0$. Using this state to construct the density operator $\rho_\psi = |\psi\rangle\langle\psi|$, one can verify that the reduced densities

$$\rho_{s_A} = \begin{pmatrix} |a|^2+|b|^2 & a\bar{c}+b\bar{d} \\ \bar{a}c+\bar{b}d & |c|^2+|d|^2 \end{pmatrix}, \quad \rho_{s_B} = \begin{pmatrix} |a|^2+|c|^2 & a\bar{b}+c\bar{d} \\ \bar{a}b+\bar{c}d & |b|^2+|d|^2 \end{pmatrix}, \tag{19}$$

have the determinant [16]:

$$\begin{aligned}\mathrm{Det}\,\rho_{s_A} = \mathrm{Det}\,\rho_{s_B} = |ad-bc|^2 &= |ad|^2+|cb|^2-[ad(\overline{bc})+(\overline{ad})bc]\\ &= |r_{14}|^2+|r_{32}|^2-2\mathrm{Re}\left(r_{14}r_{32}\,\mathrm{e}^{i2(\theta_d-\theta_c)}\right),\end{aligned} \tag{20}$$

where $d=|d|\,\mathrm{e}^{i\theta_d}$, $c=|c|\,\mathrm{e}^{i\theta_c}$, and θ_d, $\theta_c\in\mathbb{R}$. The first two additive terms of (20) are the probability densities associated to the transitions (14) while the third one corresponds to the interference between such transitions. Remark that the determinant (20) is not null if at least two coefficients in the superposition (18) are different from zero, provided they are such that either $ad\neq 0$ or $bc\neq 0$.

Equation (20) makes clear that, besides the populations $|a|^2$, $|b|^2$, $|c|^2$, $|d|^2$, one can obtain some other information of the entire state ρ_ψ by studying the state of any of the subsystems $\mathcal{S}_A$ and $\mathcal{S}_B$. In particular, the determinant (20) is a measure of the transitions (14)-(17) that occur in the entire system. For instance, let us assume that the transition $|-\rangle_A\leftrightarrow|+\rangle_A$ occurs at a given time t. According to (16) and (17), this last implies a transition in the state of $\mathcal{S}_B$ that is ruled by either $|-\rangle_B\leftrightarrow|+\rangle_B$ or $|+\rangle_B\leftrightarrow|-\rangle_B$. In the former case the transition in $\mathcal{S}_A$ cancels the coherence r_{32} so that $\mathrm{Det}\,\rho_{s_{A,B}}=|r_{14}|^4\equiv|da|^2$. In the second case the coherence r_{14} is nulled and $\mathrm{Det}\,\rho_{s_{A,B}}=|r_{32}|^4\equiv|bc|^2$. A similar description is obtained when either $|-\rangle_B\leftrightarrow|+\rangle_B$ or $|+\rangle_B\leftrightarrow|-\rangle_B$ activate the transition process in the entire system. Remark that these conditioned transitions in the subsystems $\mathcal{S}_A$ and $\mathcal{S}_B$ lead to maxima values of the determinant (20). Transitions in $\mathcal{S}_A$ that make no constraints in the transitions of its counterpart $\mathcal{S}_B$, and viceversa, give rise to interference phenomena in the process. Indeed, from the previous example, the (not binding) activation $|-\rangle_A\leftrightarrow|+\rangle_A$ produces a superposition state in $\mathcal{S}_B$ that includes both transitions, $|-\rangle_B\leftrightarrow|+\rangle_B$ and $|+\rangle_B\leftrightarrow|-\rangle_B$, with a probability different from zero each one. Then, for not conditioned transitions, the interference term in (20) is relevant since this can even cancel the determinant $\mathrm{Det}\,\rho_{s_{A,B}}$. In summary, the transitions in $\mathcal{S}_{A,B}$ will be occasionally conditioned by the transitions in $\mathcal{S}_{B,A}$, so that $\mathrm{Det}\,\rho_{s_{A,B}}$ will oscillate between minima and maxima values over the time.

As we can see, the determinant (20) represents a measure of the correlations that are lost after the partial tracing of the entire state ρ_ψ. Such correlations are represented by the conditioned transitions (16)-(17) and produce maxima values in $\mathrm{Det}\,\rho_{s_{A,B}}$. Since this determinant takes nonnegative values, one can introduce the normalized quantity [16]:

$$C(\rho_\psi)=\kappa_0\sqrt{\mathrm{Det}\,\rho_{s_{A,B}}}=\kappa_0|ad-bc|, \tag{21}$$

where the normalization constant κ_0 (in our case $\kappa_0=2$) makes $C=1$ for the density operators associated to the Bell basis vectors:

$$\begin{aligned}|\beta_1\rangle&=\tfrac{1}{\sqrt{2}}(|++\rangle+|--\rangle), &\quad |\beta_2\rangle&=\tfrac{1}{\sqrt{2}}(|++\rangle-|--\rangle),\\ |\beta_3\rangle&=\tfrac{1}{\sqrt{2}}(|+-\rangle+|-+\rangle), &\quad |\beta_4\rangle&=\tfrac{1}{\sqrt{2}}(|+-\rangle-|-+\rangle).\end{aligned} \tag{22}$$

In the literature, the quantity $C(\rho_\psi)$ is known as *concurrence* and represents a measure of the *entanglement* between the states of $\mathcal{S}_A$ and those of $\mathcal{S}_B$ whenever the entire system $\mathcal{S}$ is in a pure state (see [8, 9] and references quoted therein). This quantity takes the values $0\leq C\leq 1$, with $C=1$ for fully entangled states and $C=0$ for *factorizable* (not entangled) states. In our case, C gives information of the correlation between the states $|\varphi_4\rangle$ and $|\varphi_1\rangle$ ($|\varphi_3\rangle$ and $|\varphi_2\rangle$) of the entire system. If these are strongly correlated then, according to the cases discussed above, one has

$$C(\rho_\psi)=2\sqrt{\mathrm{Det}\,\rho_{s_{A,B}}}=2|ad|. \tag{23}$$

The correlation is maximum if $|ad| = \frac{1}{2}$. Then, as the vector (18) is normalized, we can take $a = d = 1/\sqrt{2}$ to get $|\psi^{max}_{b=c=0}\rangle = |\beta_1\rangle$. That is, the Bell vector $|\beta_1\rangle$ defined in (22) corresponds to a pure state of $\mathcal{S}$ for which the correlation between the basis vectors $|\varphi_4\rangle$ and $|\varphi_1\rangle$ is maximum. Other values of this correlation are feasible by using $a = \cos\theta$ and $b = \sin\theta$, with $\theta \in \mathbb{R}$, for which

$$|\psi_{1,4}(\theta)\rangle = \cos\theta|\varphi_1\rangle + \sin\theta|\varphi_4\rangle \quad \Rightarrow \quad C(\rho_{\psi(\theta)}) = |\sin 2\theta|. \tag{24}$$

Notice that $|\psi_{1,4}(\frac{\pi}{4})\rangle = |\beta_1\rangle$ and $|\psi_{1,4}(\frac{3\pi}{4})\rangle = |\beta_2\rangle$ give $C = 1$. In similar form, for $|\varphi_3\rangle$ and $|\varphi_2\rangle$ we have

$$|\psi_{2,3}(\alpha)\rangle = \cos\alpha|\varphi_2\rangle + \sin\alpha|\varphi_3\rangle \quad \Rightarrow \quad C(\rho_{\psi(\alpha)}) = |\sin 2\alpha|, \quad \alpha \in \mathbb{R}. \tag{25}$$

In this case $|\psi_{2,3}(\frac{\pi}{4})\rangle = |\beta_3\rangle$ and $|\psi_{2,3}(\frac{3\pi}{4})\rangle = |\beta_4\rangle$ give $C = 1$.

The notion of concurrence can be generalized to include mixed states, this is accomplished by transforming the density operator ρ_ψ as follows [17]:

$$\rho^*_\psi = \sigma_2 \otimes \sigma_2 \, \rho_\psi \, \sigma_2 \otimes \sigma_2, \tag{26}$$

and calculating the trace of the product $\rho^*_\psi \, \rho_\psi$ to get

$$C^2(\rho_\psi) = \mathrm{Tr}(\rho^*_\psi \, \rho_\psi). \tag{27}$$

The product $\sigma_2 \otimes \sigma_2$ in (26) stands for the spin-flip operator [18]. Assuming that (27) holds for an arbitrary state ρ, pure or mixed, the concurrence can be written as [9]:

$$C(\rho) = \max\{0, \lambda_1 - \lambda_2 - \lambda_3 - \lambda_4\}, \quad \lambda_j > \lambda_{j+1}\,, \; j = 1, 2, 3, \tag{28}$$

where the nonnegative numbers λ_j are the square-roots of the eigenvalues of the operator $\rho^*\rho$.

In principle, the expression (28) applies not only to pure states but also for mixed states. This is remarkable because, once the coupling between the qubits and the radiation fields is initiated, each cavity includes a system (the qubit) in an active electromagnetic environment (the photon bath). Therefore, the study of the qubit bipartite state requires a summation over the degrees of freedom of the fields and, as we have discussed, this last produces leaky in the information of the system. Hence, the description of the qubits system is necessarily done in terms of mixed states.

3. Two entangled qubits coupled to two independent quantum oscillators

Now let us assume that each of the qubits of the previous section is coupled to an independent single mode radiation field. The latter will be represented as a quantum harmonic oscillator of frequency ω_b and ladder operators $a^\dagger$, $a = (a^\dagger)^\dagger$. Using the dipole and rotating wave approximations, and considering $\omega_q \approx \omega_b$, the (dimensionless) Hamiltonian of the entire system $H = H_A + H_B$ has the components

$$H_\ell = H_{0;\ell} + H_{I;\ell}, \quad H_{0;\ell} = N_\ell + \frac{1}{2} + \frac{1}{2}\sigma_3, \quad H_{I;\ell} = \gamma_\ell(\sigma_+ a_\ell + \sigma_- a^\dagger_\ell), \quad \ell = A, B, \tag{29}$$

with $\{a_\ell, a^\dagger_\ell, N_\ell\}$ the generators of the photon algebra associated to the ℓth radiation field, γ_ℓ the coupling strength in the ℓth cavity, and $\{\sigma_3, \sigma_\pm\}$ the generators of the $su(2)$ Lie algebra associated to the excitation of the qubits in the cavities. Notice that we have dropped the sub-label ℓ from these last operators for simplicity in notation. Hereafter we shall assume that there is a bath of r (s) photons in cavity A (B), and $\gamma_A = \gamma_B = \gamma$.

The Hamiltonian H of the entire system $\mathcal{S}'$ is then formed of a free part $H_{0;A} + H_{0;B}$ and a coupling term $H_{I;A} + H_{I;B}$ (no communication between cavities A and B is allowed, so that a coupling between them is not considered). It is straightforward to verify that $[H_{0;\ell}, H_{I;\ell'}] = 0$, so there is a common vector basis for the free and coupling parts of the Hamiltonian.

The system $\mathcal{S}'$ we consider is tetra-partite: it is formed of two independent qubits $\mathcal{S}_\ell$ coupled to two independent photon baths $\mathcal{S}_{b_\ell}$, $\ell = A, B$. Then, the state space of the entire system is the direct sum [19, 20]:

$$\mathcal{H} = \bigoplus_{n,m} \Xi_n^{(A)} \otimes \Xi_m^{(B)}, \tag{30}$$

where, provided n and m, the subspaces $\mathcal{H}_{nm} = \Xi_n^{(A)} \otimes \Xi_m^{(B)}$ are given by

$$\mathcal{H}_{nm} = \text{Span}\{|+,n\rangle \otimes |+,m\rangle, |+,n\rangle \otimes |-,m+1\rangle, |-,n+1\rangle \otimes |+,m\rangle, |-,n+1\rangle \otimes |-,m+1\rangle\}.$$

The time-evolution operator of the entire system is $U(t) = \mathrm{e}^{-iHt}$, since H is time-independent. Considering that the cavities are one independent of the other, and the commutation properties of $H_{0;\ell}$ and $H_{I;\ell}$, one can write

$$U(t) = U_A(t) \otimes U_B(t) = \mathrm{e}^{-iH_{I;A}t}\,\mathrm{e}^{-iH_{0;A}t} \otimes \mathrm{e}^{-iH_{I;B}t}\,\mathrm{e}^{-iH_{0;B}t}, \qquad U(0) = \mathbb{I}_A \otimes \mathbb{I}_B = \mathbb{I}. \tag{31}$$

In order to analyze the effect that the photon baths $\mathcal{S}_b = \mathcal{S}_{b_A} + \mathcal{S}_{b_B}$ produce in the correlations between the states of the qubits subsystem $\mathcal{S} = \mathcal{S}_A + \mathcal{S}_B$, let us assume that the initial state of $\mathcal{S}$ is the vector $|\psi_{2,3}(\alpha)\rangle$ defined in (25). That is, the bipartite qubit system $\mathcal{S}$ is initially correlated in the states $|\varphi_2\rangle$ and $|\varphi_3\rangle$ by an amount $C_0 = |\sin 2\alpha|$. As the cavities have r and s photons respectively, for $\mathcal{S}'$ we have the initial state

$$|\Phi(0)\rangle = |\psi_{2,3}(\alpha)\rangle \otimes |r\rangle \otimes |s\rangle = \cos\alpha|+,-,r,s\rangle + \sin\alpha|-,+,r,s\rangle. \tag{32}$$

To rewrite $|\Phi(0)\rangle$ in the representation of $\mathcal{H}$ given in (30) we use the fact that the direct product $A \otimes B$ is permutation equivalent to $B \otimes A$, see Theorem M4 of [14]. Therefore, up to a permutation, we have

$$|\Phi(0)\rangle = \cos\alpha|+,r\rangle \otimes |-,s\rangle + \sin\alpha|-,r\rangle \otimes |+,s\rangle. \tag{33}$$

Notice that the initial state $|\Phi(0)\rangle$ is in the subspace $\mathcal{H}_{r,s-1} \oplus \mathcal{H}_{r-1,s} \subset \mathcal{H}$. Applying (31) in (33) we get, up to a global phase $\mathrm{e}^{-i(r+s+1)t}$, the time-evolved state of the tetra-partite system

$$\begin{aligned}
|\Phi(t)\rangle = \cos\alpha\, &\{i\cos(\gamma t\sqrt{r+1})\sin(\gamma t\sqrt{s})|+,r\rangle \otimes |+,s-1\rangle \\
&+ \cos(\gamma t\sqrt{r+1})\cos(\gamma t\sqrt{s})|+,r\rangle \otimes |-,s\rangle - \sin(\gamma t\sqrt{r+1})\sin(\gamma t\sqrt{s})|-,r+1\rangle \otimes |+,s-1\rangle \\
&+ i\sin(\gamma t\sqrt{r+1})\cos(\gamma t\sqrt{s})|-,r+1\rangle \otimes |-,s\rangle\} \\
+\sin\alpha\, &\{i\cos(\gamma t\sqrt{s+1})\sin(\gamma t\sqrt{r})|+,r-1\rangle \otimes |+,s\rangle - \sin(\gamma t\sqrt{r})\sin(\gamma t\sqrt{s+1})|+,r-1\rangle \otimes |-,s+1\rangle \\
&+ \cos(\gamma t\sqrt{r})\cos(\gamma t\sqrt{s+1})|-,r\rangle \otimes |+,s\rangle + i\cos(\gamma t\sqrt{r})\sin(\gamma t\sqrt{s+1})|-,r\rangle \otimes |-,s+1\rangle\}.
\end{aligned} \tag{34}$$

Remarkably, the time-evolved state $|\Phi(t)\rangle$ is in the subspace $\mathcal{H}_{r,s-1} \oplus \mathcal{H}_{r-1,s} \oplus \mathcal{H}_{r,s+1} \subset \mathcal{H}$; its eight components give rise to a pure state $\rho(t) = |\Phi(t)\rangle\langle\Phi(t)|$ that is expressed in terms of *sixty four* Hubbard operators [16]. Tracing over $\mathcal{S}_b$, for the qubit bipartite system $\mathcal{S}$ one gets the mixed state

$$\rho_q(\alpha, r, s; t) = \begin{pmatrix} \Omega_{11} & 0 & 0 & 0 \\ 0 & \Omega_{22} & \Omega_{23} & 0 \\ 0 & \Omega_{32} & \Omega_{33} & 0 \\ 0 & 0 & 0 & \Omega_{44} \end{pmatrix}, \tag{35}$$

where the time-dependent matrix elements $\Omega_{kj}(\alpha, r, s; t)$ are given by

$$\begin{aligned}
\Omega_{11} &= \cos^2\alpha \cos^2(\gamma t\sqrt{r+1})\sin^2(\gamma t\sqrt{s}) + \sin^2\alpha\cos^2(\gamma t\sqrt{s+1})\sin^2(\gamma t\sqrt{r}),\\
\Omega_{22} &= \cos^2\alpha \cos^2(\gamma t\sqrt{r+1})\cos^2(\gamma t\sqrt{s}) + \sin^2\alpha\sin^2(\gamma t\sqrt{s+1})\sin^2(\gamma t\sqrt{r}),\\
\Omega_{33} &= \cos^2\alpha \sin^2(\gamma t\sqrt{r+1})\sin^2(\gamma t\sqrt{s}) + \sin^2\alpha\cos^2(\gamma t\sqrt{s+1})\cos^2(\gamma t\sqrt{r}),\\
\Omega_{44} &= \cos^2\alpha \sin^2(\gamma t\sqrt{r+1})\cos^2(\gamma t\sqrt{s}) + \sin^2\alpha\sin^2(\gamma t\sqrt{s+1})\cos^2(\gamma t\sqrt{r}),\\
\Omega_{23} &= \Omega_{32} = \cos\alpha\sin\alpha\cos(\gamma t\sqrt{r+1})\cos(\gamma t\sqrt{r})\cos(\gamma t\sqrt{s+1})\cos(\gamma t\sqrt{s}).
\end{aligned}$$

The mixed state (35) is a coherent superposition of the basis kets $|\varphi_k\rangle$ for $\alpha \in (0, 2\pi)$, with $\alpha \neq \frac{\pi}{2}, \pi, \frac{3\pi}{2}$, and $r \neq 0$, $s \neq 0$. This has only two nonzero off-diagonal elements $\Omega_{23} = \Omega_{32} \in \mathbb{R}$ that arise from the operation

$$\text{Tr}_{s_b}(|+,r\rangle \otimes |-,s\rangle)(\langle -,r| \otimes \langle +,s|) = X^{+-} \otimes X^{-+} = X^{23}.$$

The latter is the trace over $\mathcal{S}_b$ of the outer product between the second and seventh kets of (34). Such outer product corresponds to a coherence in the entire tetra-partite state $\rho(t)$ that involves the probability amplitude of the conditioned transition

$$|+,r\rangle_{\text{cav A}} \to |-,r\rangle_{\text{cav A}} \quad \Leftrightarrow \quad |-,s\rangle_{\text{cav B}} \to |+,s\rangle_{\text{cav B}}. \tag{36}$$

As Ω_{23} is the qubit coherence involving the bipartite states $|\varphi_2\rangle$ and $|\varphi_3\rangle$, Eq. (36) means that the qubit transitions (17) are performed without a change in the number of photons in the cavities! Thus, there is a missing photon in cavity A and an additional photon in cavity B. The former is, in principle, released by $\mathcal{S}_A$ in its decaying from $|+\rangle_A$ to $|-\rangle_A$. The second is required by $\mathcal{S}_B$ in its excitation from $|-\rangle_B$ to $|+\rangle_B$. However, there is no communication between cavities A and B, so that the photon released by $\mathcal{S}_A$ is, with certainty, not the photon that increases the energy of $\mathcal{S}_B$. The puzzle is best analyzed by considering the initial state (33) of the entire system $\mathcal{S}'$. This is a linear combination of two orthogonal vectors $|+,r,-,s\rangle$ and $|-,r,+,s\rangle$ that share the same eigenvalue of the (free of interaction) energy $E_{0;A+B} = r+s+1$. Then, at $t=0$, the system $\mathcal{S}'$ is in the state $|+,r-,s\rangle$ with probability $\cos^2\alpha$, and in the state $|-,r,+,s\rangle$ with probability $\sin^2\alpha$. Therefore, from the very beginning, we do not know whether the system of cavity A is in the state $|+,r\rangle$ or in the state $|-,r\rangle$. A similar argument holds for the state of the system of cavity B. This fact is evidence of the non-locality of the quantum states involved since, if we avoid the coherence of $|\Psi(0)\rangle$ by making $\alpha = 0$ ($\alpha = \frac{\pi}{2}$), then we know with certainty that the system A is initially in the state $|+,r\rangle$ ($|-,r\rangle$) while the system B is in the state $|-,s\rangle$ ($|+,s\rangle$), and then the state ρ_q in (35) is an incoherent superposition of the four basis vectors (6). The transition (36) is, in this form, a consequence of the coherence Ω_{23} between the states $|+,r,-,s\rangle$ and $|-,r,+,s\rangle$, established at $t=0$ by the degree of correlation $C_0 = |\sin 2\alpha|$ of the qubits (see Figure 1). The other non-zero coherence $\Omega_{32} = \Omega_{23}$ in (35) refers to the reversed process $|-,r,+,s\rangle \to |+,r,-,s\rangle$.

On the other hand, the straightforward calculation leads to the diagonal form of the qubit density matrix, $\rho_q = \text{diag}(\lambda_1, \lambda_2, \lambda_3, \lambda_4)$, where λ_k are its eigenvalues

$$\begin{gathered}
\lambda_1 = \Omega_{11}, \quad \lambda_2 = \frac{(\Omega_{22}+\Omega_{33}) + \sqrt{(\Omega_{22}+\Omega_{33})^2 + 4(\Omega_{23}^2 - \Omega_{22}\Omega_{33})}}{2},\\
\lambda_3 = \Omega_{22} + \Omega_{33} - \lambda_2, \quad \lambda_4 = \Omega_{44}.
\end{gathered} \tag{37}$$

The related eigenvectors read as follows

$$|\phi_{\lambda_1}\rangle = |\varphi_1\rangle, \quad |\phi_{\lambda_2}\rangle = |\psi_{2,3}(\alpha)\rangle, \quad |\phi_{\lambda_3}\rangle = |\psi_{2,3}(-\alpha)\rangle, \quad |\phi_{\lambda_4}\rangle = |\varphi_4\rangle. \tag{38}$$

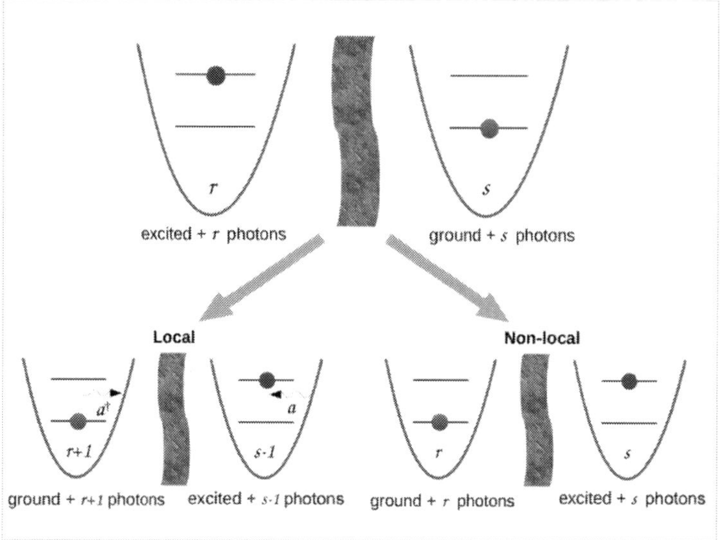

Figure 1. Schematic representation of the states transition (36), in all cases the parabolic curve represents the photon bath, blue and red dots mean respectively occupation of the excited and ground energy levels of the qubit (horizontal lines), and cavity A (left) is separated from cavity B (right) by an *impenetrable barrier*. The coherence Ω_{23} of the density operator (35) implies the change from the configuration depicted in the upper line to the one in the lower line. There are two possible forms to get such a transition. A local phenomenon (lower line, left) requires at least two photons that are not included in the initial configuration: one released by the qubit $\mathcal{S}_A$ (represented by $a^\dagger$) and one *absorbed* by the qubit $\mathcal{S}_B$ (represented by a). As the impenetrable barrier forbids communication between the cavities, the photon created in cavity A is, with certainty, not the one annihilated in cavity B. On the other hand, a non-local phenomenon (lower line, right) requires no photons of interchange. This is a consequence of the linear superposition in the initial state (33). That is, the transition (36) corresponds to transforming the configuration $|+,r,-,s\rangle$, that initially occurs with probability $\cos\alpha^2$, into the configuration $|-,r,+,s\rangle$, that occurs with probability $\sin\alpha^2$ at $t=0$. The other not null coherence $\Omega_{32}=\Omega_{23}$ in (35) is associated to the reversed process $|-,r,+,s\rangle \rightarrow |+,r,-,s\rangle$.

The four vectors (38) are orthonormal so that ρ_q can be expressed as follows

$$\rho_q = \text{diag}(\lambda_1,\lambda_2,\lambda_3,\lambda_4) = \lambda_1 X^{11} + \lambda_2 X^{22}(\alpha) + \lambda_3 X^{33}(\alpha) + \lambda_4 X^{44}$$

with $X^{22}(\alpha) = |\psi_{2,3}(\alpha)\rangle\langle\psi_{2,3}(\alpha)|$ and $X^{33}(\alpha) = |\psi_{2,3}(-\alpha)\rangle\langle\psi_{2,3}(-\alpha)|$. The diagonal form of ρ_q verifies that it is the probabilistic mixture of two disentangled pure states $|\varphi_1\rangle$, $|\varphi_4\rangle$, and two partially entangled pure states $|\psi_{2,3}(\alpha)\rangle$, $|\psi_{2,3}(-\alpha)\rangle$. It is now easy to verify that $\text{Tr}(\rho_q)=1$, and $\text{Tr}(\rho_q^2)<1$, as expected.

In order to analyze the qubits' transitions we calculate the population inversion by using $\mathcal{P}_{+-}^{(A,B)} = \text{Tr}(\rho_{q_{A,B}}\sigma_3)$, with $\rho_{q_{A,B}} = \text{Tr}_{s_{B,A}}\rho_q$ the trace of ρ_q over the degrees of freedom of $\mathcal{S}_B$ ($\mathcal{S}_A$). We obtain

$$\begin{aligned}\mathcal{P}_{+-}^{(A)} &= \cos^2\alpha\cos(2\gamma t\sqrt{r+1}) - \sin^2\alpha\cos(2\gamma t\sqrt{r}),\\ \mathcal{P}_{+-}^{(B)} &= \sin^2\alpha\cos(2\gamma t\sqrt{s+1}) - \cos^2\alpha\cos(2\gamma t\sqrt{s}).\end{aligned} \tag{39}$$

At $t=0$ one gets the inversions defined by the initial state $|\Phi(0)\rangle$ in (33), namely $\mathcal{P}_{+-}^{(A)} = \cos^2\alpha - \sin^2\alpha = -\mathcal{P}_{+-}^{(B)}$. For fully entangled qubits ($\alpha=\pi/4$ or $\alpha=3\pi/4$) the populations in each energy level are the same and the inversion is equal to zero in both cavities (see Figure 2, left). In turn, partially entangled qubits (i.e., for other allowed values of α) produce inversions

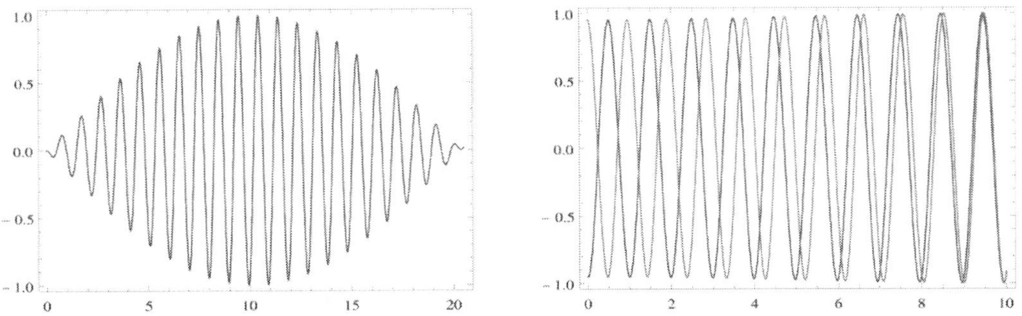

Figure 2. The time-dependence of the population inversions (39), qubit A in blue and qubit B in red. In both cavities the number of photons is the same $r = s = 10$. For fully entangled initial qubits with $\alpha = \pi/4$ (left), the time-evolution is the same in both cavities and the correlation between the qubits is smaller as larger is the absolute value of the inversion. For initial partially entangled qubits with $\alpha = 9\pi/20$ (right), the inversions have opposite sign and the correlation is smaller as larger is the absolute value of $\mathcal{P}_{+-}^{(A)} + \mathcal{P}_{+-}^{(B)}$. In both cases the initial correlation is lost in the vicinity of $t = 10$.

that differ by a sign in the cavities (Figure 2, right). The sinusoidal time-dependence of the population inversion in both cavities leads to a periodic repetition of the initial configuration. That is, the initial configuration will be lost and recovered periodically with time; the times at which it is lost are the times at which there is no correlation between the qubit in cavity A and the one in cavity B. In the case of fully entangled initial states, the correlation between the qubits is smaller as larger are the absolute values of the population inversions. For partially entangled states, the correlation is smaller as larger is the absolute value of the quantity $\mathcal{P}_{+-}^{(A)} + \mathcal{P}_{+-}^{(B)}$. In Figure 2 we have depicted the time-evolution of the population inversions (39) for a fully entangled state ($\alpha = \pi/4$) and a partially entangled state ($\alpha = 9\pi/20$) as initial condition, in both cases the number of photons is the same $r = s = 10$. Notice that the fully entangled case (Figure 2, left) departs from zero at $t = 0$ and reaches the same value at $t = 20$ (approx.). Besides, it oscillates between 1 and -1 in the vicinity of $t = 10$; in this last interval of time $t \in 10 \pm \delta$ the correlations between the qubit A and the qubit B are lost. In turn, the partially entangled case (Figure 2, right) departs from $\mathcal{P}_{+-}^{(A)} + \mathcal{P}_{+-}^{(B)} = 0$ at $t = 0$ and is such that $\mathcal{P}_{+-}^{(A)} + \mathcal{P}_{+-}^{(B)} = \pm 2$ in the vicinity of $t = 10$. As in the previous case, the correlations between the qubits are lost for $t \in 10 \pm \delta$.

The above description of the time-evolution of the correlation between the qubits is in agreement with the notion of concurrence discussed in the previous section. In Figure 3 we compare the time dependence of the population inversion and the concurrence for the fully entangled state $|\beta_3\rangle$ as initial condition. As it can be appreciated, the concurrence is minimum when the population inversion is in either its minimum (-1) or its maximum (1) value. In that case, we know that the two qubits are respectively in either the state $|+\rangle$ or $|-\rangle$, so that the initial configuration (the same probability to be in $|+\rangle$ as in $|+\rangle$) is lost. In turn, the concurrence is maximum when the population inversion cancels (the initial configuration!). Therefore, the collapses and revivals of the population inversion are a direct measure of the entanglement between the qubits, these last are maximally entangled at the time a revival occurs and they are disentangled when the inversion collapses. This oscillatory behavior of the concurrence is associated to the concepts of *sudden dead* and recovery of entanglement in the literature [21].

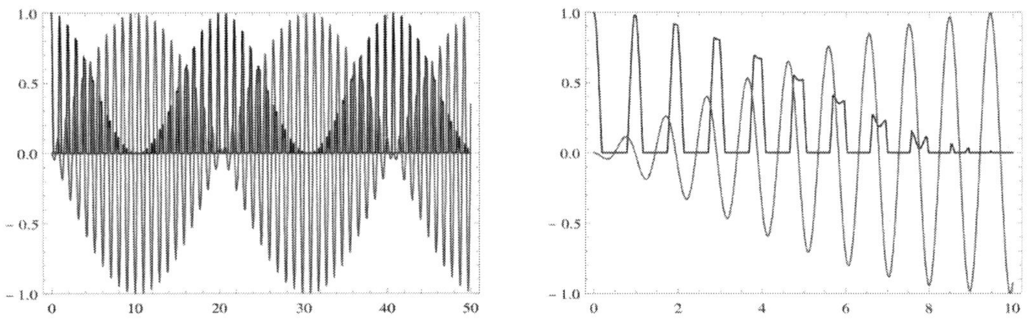

Figure 3. Time-evolution of the concurrence (28) and population inversions (39) for a pair of fully entangled qubits ($\alpha = \pi/4$) coupled to two photon baths (10 photons each one) in isolated cavities. The maxima of the absolute values of the population inversion (blue) represent the lost of correlations between the qubits and coincide with the minima of the concurrence (black). At the right we have depicted a detail of the curves in a shorter interval of time for comparison.

4. Concluding remarks

We have studied the time-evolution of the correlations between a pair of qubits that are coupled to independent photon baths. We have shown that the population inversion of each of the qubits gives the same information as the concurrence. This last because the partial trace of the density matrix associated to the entire system produces leaky information that can be recovered by analyzing the transitions between the energy states of the qubits.

Acknowledgments

CQH gratefully acknowledges the funding received through the CONACyT Scholarship.

References

[1] A. Einstein, B. Podolsky and N. Rose, Phys. Rev. **47** (1935) 777
[2] E. Schrödinger, Die Naturwissenschaften **23** (1935) 807 (English translation in Wheeler J. and Zurek W., Quantum theory and measurement, Princeton University Press, 1983); Math. Proc. of the Cambridge Philosophical Society **31** (1935) 555; Math. Proc. of the Cambridge Philosophical Society **32** (1935) 446
[3] A. Auyuanet and L. Davidovich, Phys. Rev. A **82** (2010) 032112
[4] E. Knill and R. Laflamme, Phys. Rev. Lett. **81** (1998) 5672
[5] B.P. Lanyon, M. Barbieri, M.P. Almeida and A.G.White, Phys. Rev. Lett. **101** (2008) 200501
[6] L.K. Grover, Phys. Rev. Lett. **79** (1997) 325
[7] J. Ahn, T.C. Weinacht and P/H/ Bucksbaum, Science **287** (200) 463
[8] S. Hill and W.K. Wooters, Phys. Rev. Lett. **78** (1997) 5022
[9] W.K. Wooters, Phys. Rev. Lett. **80** (1998) 2245
[10] K. Zyczkowski, P. Horodecki, A. Sampera and M. Lewenstein, Phys. Rev. A **58** (1998) 883
[11] H. Olliver and W.H. Zurek, Phys. Rev. Lett. **88** (2001) 017901
[12] L. Handerson and V. Vedral, J. Phys. A: Math. Gen. **34** (2001) 6899
[13] W.H. Zurek, Phys. Rev. A **67** (2003) 012320
[14] M. Enríquez and O. Rosas-Ortiz, Ann. Phys. **339** (2013) 218
[15] K. Blum, *Density Matrix, Theory and Applications*, Third Edition, Springer, New York, 2012
[16] C. Quintana, *Quantum entanglement: time-evolution of bipartite systems*, MSc Thesis (in Spanish), Physics Department, Cinvestav, México, 2014.
[17] J. Audretsch, *Entangled Systems*, Wiley, Berlin, 2005
[18] M.N. Nielsen and I.S. Chuang, *Quantum Computation and Quantum Information*, Cambridge University Press, Cambridge, 2000
[19] M. Enríquez, C. Quintana and O. Rosas-Ortiz, J. Phys.: Conf. Ser. **512** (2014) 012022
[20] M. Enríquez and O. Rosas-Ortiz, in preparation
[21] M. Yonac, T. Yu and J.H. Eberly, J. Phys. B: At. Mol. Opt. Phys. **39** (2006) S621

Conjugate states to the energy eigenstates: The continuous energy spectrum case

G Torres-Vega
Physics Department, Cinvestav, Apdo. Postal 14-740, 07300 México D.F., México
E-mail: gabino@fis.cinvestav.mx

Abstract. The usual method for obtaining the eigenstates of an operator is to solve the corresponding eigenvalue equation. This procedure cannot be applied when the operator of interest is not known at all. We develop a method which generates the eigenstates of an operator, an the operator itself, which will be conjugate to a given known operator. This is particularly useful for the case of the time operator in Quantum Mechanics. We illustrate the method by obtaining time eigenstates for the free particle.

1. Introduction

The purpose of this talk is to propose a way to handle the time variable in Classical and Quantum Mechanics. Our point of view would be that of considering time as the coordinate which is conjugate to the energy variable. The usual set of conjugate coordinates is position-momentum and time-energy could be an alternative symplectic set of coordinates with which we can describe the dynamics of a classical or quantum system.

Quantum time functions usually are found by quantizing the classical expression for the time of a particle, leading to many inherent difficulties like arriving to quantities which are not well defined, and facing the problem of symmetrization of products of functions. Consequently, only in few cases this can be done explicitly. For instance, for a free particle, the classical expression for time is

$$t = t_0 + m\frac{q}{p},$$

a quantity which is not defined at $p = 0$, and with many possible quantum versions of it [1]. Time eigenstates for other systems are hard, if not impossible, to find with this method.

There is also a method of time propagation of the coordinate eigenstate $\delta(q)$ multiplied by $\sqrt{p/m}$, generating that way the required time states [2]. This method is appropriate for the free particle and it was suggested to use it also for any potential function [3–6].

One of the proposed methods for finding of time operators is solving of the differential equation that results from the conjugate condition $[\hat{T}, \hat{H}] = i\hbar\hat{I}$. This method requires the solution of complicated differential equations [7].

Another proposal for time operators and Lyapunov functions makes use of the sum of terms of the form $|E'\rangle\langle E|/(E - E')$. One limitation of this proposal is that the domain of the resulting operators is a subset of the Hilbert space [7–11].

2. The method

In this section, we define and find the properties of the conjugate states to the energy eigenstates. We expect that the resulting states, and the operators that can be obtained from them, be related to the time operator and to its eigenstates.

The conjugate states that will be generated can be used as another coordinate, as an alternative representation for quantum states and operators, and, in particular, they can be used to construct a quantum time operator

We will illustrate the use of the generated states with the free particle model, for quantum systems and with a nonlinear oscillator model for classical systems.

2.1. Classical systems. The normal direction to the energy shells

Let us consider first the case of Classical systems. This will make more understandable the quantum case.

From the euclidean geometry point of view, the vanishing of the Poisson bracket for the same function, in this case the Hamiltonian of the system,

$$\{H(p,q), H(p,q)\} = \nabla H \cdot \mathbf{X}_H = 0$$

defines a vector field $\mathbf{X}_{\perp H} := \nabla H$ which is normal to the energy shells. We can take advantage of this vector field to generate a conjugate variable to the energy variable. The main characteristic of the new variable is that it crosses all of the energy shells. Thus one of the variables is well defined but the corresponding conjugate variable is not, just what is needed for a coordinate system.

Thus, the method is to take any of the normal shells to the energy shells as a zero-time shell and, by means of time propagation of them, generate a conjugate, or symplectic orthogonal, time variable. This process will generate a coordinate set in which the description of the system is the most simple possible because the set of variables needed to follow the evolution of a given system is reduced, from (q, p, t) to only (t, E), i.e. we are replacing a parametric description of the system with a non parametric one.

2.2. Motion along the normal direction to the constant energy shells

To generate the curves which are normal to the constant energy shells, in the euclidean sense, we define a dynamical system for the phase space point $z^{\perp H}$ with the vector field $\mathbf{X}_{\perp H}$ and parameter ϵ,

$$\frac{dz^{\perp H}}{d\epsilon} = \mathbf{X}_{\perp H}, \quad \text{i.e.} \quad \frac{dq^{\perp H}}{d\epsilon} = \frac{\partial H}{\partial q}, \quad \frac{dp^{\perp H}}{d\epsilon} = \frac{\partial H}{\partial p}$$

To illustrate how these curves look like, let us consider a particular system: a nonlinear oscillator with Hamiltonian

$$H(z) = \frac{p^2}{2} + \frac{k}{2}\left(\sqrt{a^2+q^2} - \ell\right)^2, \tag{1}$$

where $k, a, \ell \in (0, \infty)$ are the parameters of the model. The normal curves to the constant energy shells for this system are shown in Fig. 1.

The curves generated with the vector field $\mathbf{X}_\perp$ are normal to the constant energy curves in the Euclidean geometry sense. However, the time curves should be normal to the constant energy shells in the symplectic geometry sense. These normal curves have the nice property that they cross all of the constant energy shells, thus they are very useful for generating the time curves.

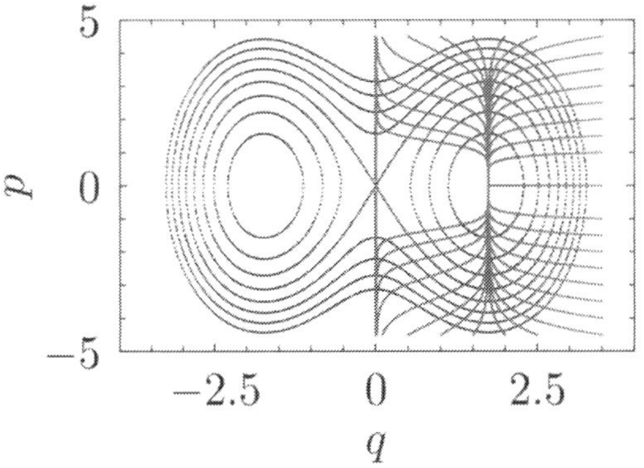

Figure 1. Curves (in red) which are normal to the constant energy shell curves (in back) for the nonlinear oscillator with $\ell = 2$, $k = 9.8$ and $a = 1$. Dimensionless units.

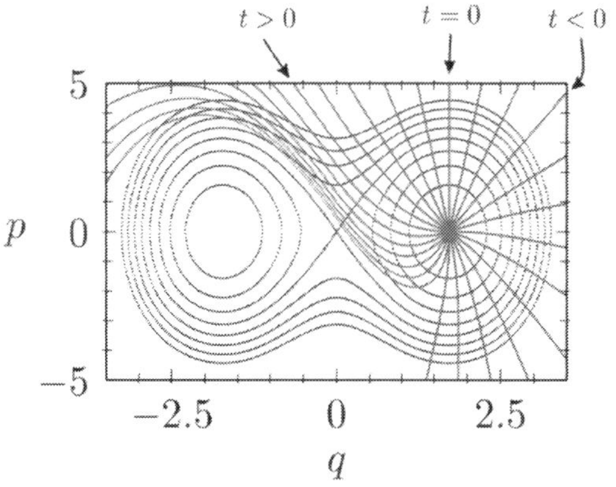

Figure 2. Time-energy coordinates for the nonlinear oscillator generated from the time evolution of the curve $q = q_0$ where q_0 is the location of the minimum of the potential function with positive coordinate. For this calculation, we have set $\ell = 2$, $k = 9.8$ and $a = 1$. Dimensionless units.

2.3. Generating a time coordinate

Taking advantage of the fact that the origin of time is arbitrary, we proceed to generate a time coordinate by time evolving any of normal curves to the constant energy shells. The simplest curves are the ones that cross the location of the extremal points of the potential function, or the curve of zero momentum. Those curves are just straight lines passing through the fixed points.

For instance, for the nonlinear oscillator, we can take as the zero-time curve the constant coordinate curve that crosses the extremal point of the potential function with positive coordinate. After time evolution of that curve, we obtain the time coordinate system shown in Fig. 2.

Since now we have a time coordinate, we can assign a time-energy set of values to each point in the phase space. The time values so obtained are shown in Fig. 3, the surface of time values. The domain of the time function is not the whole of the phase space with the separatrix removing the regions which do not belong to the domain of $t(p,q)$.

2.4. Unfolding the degeneracy of the energy values and classical distributions in energy-time

One of the advantages of having a time-energy coordinate system is that we can unfold the degeneracy of the energy values; to each value of energy correspond two values of the momentum of a particle. What we can do to unfold the energy values degeneracy is to extend the values of the energy to the negative side of the real axes, with the sign of the energy values corresponding to the sign of the momentum values.

We illustrate this procedure by considering a Gaussian probability density in phase space centered at $(q,p) = (0,-1)$,

$$\rho(q,p) = \frac{1}{2\pi}\,\mathrm{e}^{-(q-q_0)^2/2\sigma-\sigma(p-p_0)^2/2}\,. \tag{2}$$

This density has a part with negative momentum. So, in the time-energy coordinates, with $\epsilon = \mathrm{sign}(p)E$, the phase-space Gaussian looks like the one shown in Fig. 4.

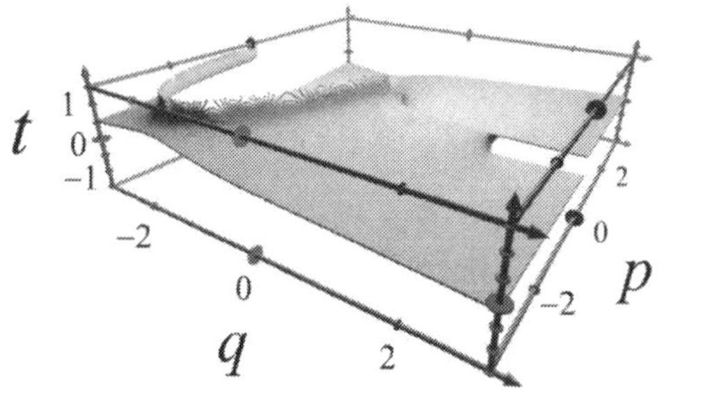

Figure 3. Time values for the nonlinear oscillator when the zero-time curve is the $q = q_0$ line, where q_0 is the location of the minimum of the potential function located at the positive part of the real axes. Here, $\ell = 2$, $k = 9.8$ and $a = 1$. Dimensionless units.

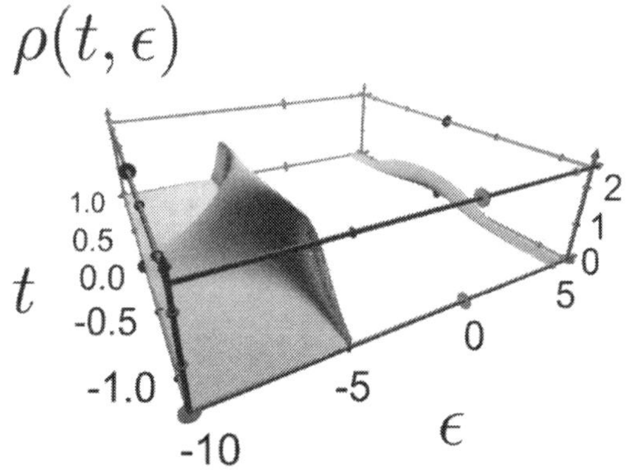

Figure 4. Classical time-energy representation of a phase-space Gaussian density centered at $(q, p) = (0, -1)$, for the nonlinear oscillator with $a = 1$, $\ell = 2$, $k = 9.8$, and $\sigma = .8$. Dimensionless units.

3. Quantum systems

We will now consider the quantum analogue of the procedure just outlined for classical systems. We will assume that energy spectrum $\{E\}_0^\infty$ of the given quantum system is continuous and that the minimum value that the energy can take is zero.

3.1. Unfolding of the energy degeneracy

Let us start the quantum treatment by rewriting a given time dependent wave function in terms of energy and momentum eigenstates as

$$\begin{aligned}|\psi(t)\rangle &:= \mathrm{e}^{-it\hat{H}/\hbar}|\psi\rangle \\ &= \int_{-\infty}^{0} dp \int_{0}^{\infty} dE\ \mathrm{e}^{-itE/\hbar}|E\rangle\langle E|p\rangle\langle p|\psi\rangle + \int_{0}^{\infty} dp \int_{0}^{\infty} dE\ \mathrm{e}^{-itE/\hbar}|E\rangle\langle E|p\rangle\langle p|\psi\rangle.\end{aligned}$$

After the replacements $\epsilon = \mathrm{sign}(p)E$, we get

$$\begin{aligned}|\psi(t)\rangle := -\int_{-\infty}^{0} dp \int_{0}^{-\infty} d\epsilon\ \mathrm{e}^{it\epsilon/\hbar}|-\epsilon\rangle\langle -\epsilon|p\rangle\langle p|\psi\rangle + \int_{0}^{\infty} dp \int_{0}^{\infty} d\epsilon\ \mathrm{e}^{-it\epsilon/\hbar}|\epsilon\rangle\langle\epsilon|p\rangle\langle p|\psi\rangle \\ = \int_{-\infty}^{0} d\epsilon\ \mathrm{e}^{-i(-t)\epsilon/\hbar}\langle\epsilon|\psi\rangle_{-}|\epsilon\rangle + \int_{0}^{\infty} d\epsilon\ \mathrm{e}^{-it\epsilon/\hbar}\langle\epsilon|\psi\rangle_{+}|\epsilon\rangle,\end{aligned}$$

where $|-E\rangle := |E\rangle$, and

$$\langle\epsilon|\psi\rangle_{-} = \int_{-\infty}^{0} dp\ \langle\epsilon|p\rangle\langle p|\psi\rangle \ \text{ and } \ \langle\epsilon|\psi\rangle_{+} := \int_{0}^{\infty} dp\ \langle\epsilon|p\rangle\langle p|\psi\rangle,$$

3.2. A spectral decomposition of the identity operator
Let us introduce an spectral decomposition of the identity operator in terms of the energy eigenstates, and let us rewrite it as

$$\begin{aligned}\hat{I} = \int_0^\infty dE\, |E\rangle\langle E| = \int_0^\infty dE' dE\; \delta(E-E')|E'\rangle\langle E| \\ = \int_0^\infty dE' dE\, \frac{1}{2\pi\hbar}\int_{-\infty}^{\infty} dt\;\mathrm{e}^{it(E-E')/\hbar}\,|E'\rangle\langle E| \\ = \int_{-\infty}^{\infty} dt \int_0^\infty dE' dE\, \frac{\mathrm{e}^{-itE'/\hbar}}{\sqrt{2\pi\hbar}}|E'\rangle\langle E|\frac{\mathrm{e}^{itE/\hbar}}{\sqrt{2\pi\hbar}} := \int_{-\infty}^{\infty} dt\,|t\rangle\langle t| \quad (3)\end{aligned}$$

where we have defined the kets $|t\rangle$ as

$$|t\rangle := \int_0^\infty dE\,\frac{\mathrm{e}^{-itE/\hbar}}{\sqrt{2\pi\hbar}}|E\rangle. \quad (4)$$

These are the kets that will be taken as time eigenkets in what follows. It contains all of the energy eigenkets, it is conjugate to them, and they are the time shifting of the zero-time ket

$$|t=0\rangle = \frac{1}{\sqrt{2\pi\hbar}}\int_0^\infty dE\,|E\rangle. \quad (5)$$

They are a good candidate for a conjugate coordinate to the energy eigenstates in Quantum Mechanics.

Recall that the commutator between $\hat{I}$ and $\hat{H}$ vanishes.

3.3. A time operator
Now, let us consider the operator

$$\begin{aligned}\hat{T} := \int_{-\infty}^{\infty} dt\; t\,|t\rangle\langle t| = \int_0^\infty dE' dE\,|E'\rangle\langle E|\frac{1}{2\pi\hbar}\int_{-\infty}^{\infty} dt\; t\;\mathrm{e}^{it(E-E')/\hbar} \\ = \int_0^\infty dE' dE\,|E'\rangle\langle E|\delta'(E-E'). \quad (6)\end{aligned}$$

The commutator between $\hat{T}$ and the Hamiltonian is

$$\begin{aligned}\left[\hat{T},\hat{H}\right] = \int_0^\infty dE' dE\frac{1}{2\pi\hbar}\int_{-\infty}^{\infty} dt\; t\;\mathrm{e}^{it(E-E')/\hbar}\left[|E'\rangle\langle E|,\hat{H}\right] \\ = \int_0^\infty dE' dE(E-E')|E'\rangle\langle E|\left(-i\hbar\frac{\partial}{\partial E}\right)\delta(E-E') \\ = \int_0^\infty dE' dE\;\delta(E-E')\left(i\hbar\frac{\partial}{\partial E}\right)(E-E')|E'\rangle\langle E| \\ - i\hbar\int_0^\infty dE'(E-E')|E'\rangle\langle E|\delta(E-E')\Big|_{E=0}^{\infty} \\ = i\hbar\int_0^\infty dE' dE\;\delta(E-E')|E'\rangle\langle E| = i\hbar\hat{I}. \quad (7)\end{aligned}$$

where we have made use of the integration by parts. This is the result expected for the commutator between a time operator and the Hamiltonian

Note that the time operator $\hat{T}$ is symmetric and its domain is the Hilbert space; the envelope of the energy eigenstates $\{|E\rangle\}_0^\infty$.

4. States which are conjugate to the energy eigenstates

Based on the above results, we define a time eigenstate as

$$|t\rangle \; := \int_{-\infty}^{\infty} d\epsilon \, \frac{e^{-it\epsilon/\hbar}}{\sqrt{2\pi\hbar}} ||\epsilon|\rangle = \int_{0}^{\infty} dE \, \frac{e^{itE/\hbar}}{\sqrt{2\pi\hbar}} \hat{P}_{-}|E\rangle + \int_{0}^{\infty} dE \, \frac{e^{-itE/\hbar}}{\sqrt{2\pi\hbar}} \hat{P}_{+}|E\rangle.$$

where $\epsilon \in \mathbb{R}$, $E = |\epsilon|$, $||\epsilon|\rangle = |E\rangle$, and

$$\hat{P}_{-} = \int_{-\infty}^{0} dp|p\rangle\langle p|, \quad \text{and} \quad \hat{P}_{+} = \int_{0}^{\infty} dp|p\rangle\langle p|. \tag{8}$$

This a state with a part evolving forward and another part evolving backwards in time. This definition unfolds the energy degeneracy due to that for a given value of the energy there correspond two values of the momentum.

4.1. Evolution equations for the transformation functions

The transformation function between the time and energy representations is,

$$\langle t|E\rangle := \frac{e^{itE/\hbar}}{\sqrt{2\pi\hbar}}. \tag{9}$$

This function is an eigenfunction of the time operator in the energy representation

$$-i\hbar\frac{\partial}{\partial E}\langle E|t\rangle = -i\hbar\frac{\partial}{\partial E}\frac{e^{-itE/\hbar}}{\sqrt{2\pi\hbar}} = -t\,\langle E|t\rangle, \tag{10}$$

and it is also an eigenfunction of the energy operator in the time representation

$$i\hbar\frac{\partial}{\partial t}\langle t|E\rangle = i\hbar\frac{\partial}{\partial t}\frac{e^{itE/\hbar}}{\sqrt{2\pi\hbar}} = -E\langle t|E\rangle. \tag{11}$$

The negative signs on the right hand sides of these equations are due to that we are dealing with the basis states and not with wave functions.

5. The free particle

We will consider the free particle model, a simple system with a continuous energy spectrum. We will illustrate how the time eigenstates defined in previous section look like for this system.

5.1. Energy eigenstates

A set of eigenfunctions of the one-dimensional free-particle Hamiltonian operator

$$-\frac{\hbar^2}{2m}\frac{d^2}{dq^2}, \tag{12}$$

is

$$\langle q|E\rangle_{\pm} = \frac{e^{\pm i\sqrt{2mE}\,q/\hbar}}{\sqrt{2\pi\hbar}}. \tag{13}$$

An alternative set of eigenfunctions is given by

$$\langle q|E\rangle_{o} = \frac{1}{\sqrt{2\pi\hbar}}\sin\left(\sqrt{2mE}\,\frac{q}{\hbar}\right) \quad \text{and} \quad \langle q|E\rangle_{e} = \frac{1}{\sqrt{2\pi\hbar}}\cos\left(\sqrt{2mE}\,\frac{q}{\hbar}\right). \tag{14}$$

We combine pairs of solutions into one written in terms of the variable $\epsilon \in \mathbb{R}$, $E = |\epsilon|$,

$$\langle q|\epsilon\rangle_e = \cos\left(\frac{\sqrt{2m|\epsilon|}}{\hbar}q\right), \quad \langle q|\epsilon\rangle_o = \mathrm{sgn}(\epsilon)\sin\left(\frac{\sqrt{2m|\epsilon|}}{\hbar}q\right), \tag{15}$$

$$\langle q|\epsilon\rangle_\pm = \frac{\mathrm{e}^{\pm\,\mathrm{sgn}(\epsilon)i\sqrt{2m|\epsilon|}\;q/\hbar}}{\sqrt{2\pi\hbar}}, \tag{16}$$

where

$$\hat{H}|\epsilon\rangle = E\,|\epsilon\rangle. \tag{17}$$

5.2. Time eigenstates

Different time eigenstates for the free particle are then defined as

$$\langle q|t\rangle_e = \int_{-\infty}^{\infty} d\epsilon \frac{\mathrm{e}^{it\epsilon/\hbar}}{\sqrt{\pi\hbar}} \cos\left(\mathrm{sgn}(\epsilon)\frac{\sqrt{2m|\epsilon|}}{\hbar}q\right), \tag{18}$$

$$\langle q|t\rangle_o = \int_{-\infty}^{\infty} d\epsilon \frac{\mathrm{e}^{it\epsilon/\hbar}}{\sqrt{\pi\hbar}} \sin\left(\mathrm{sgn}(\epsilon)\frac{\sqrt{2m|\epsilon|}}{\hbar}q\right), \tag{19}$$

and

$$\langle q|t\rangle_\pm = \int_{-\infty}^{\infty} d\epsilon \frac{\mathrm{e}^{it\epsilon/\hbar}}{\sqrt{2\pi\hbar}}\,\mathrm{e}^{\pm\,\mathrm{sgn}(\epsilon)i\sqrt{2m|\epsilon|}\;q/\hbar}. \tag{20}$$

5.3. Orthogonality of conjugate states

It is easy to see that the conjugate states are orthogonal indeed

$$\begin{aligned}
{}_o\langle t'|t\rangle_o &= \int_{-\infty}^{\infty} dq\; {}_o\langle t'|q\rangle\langle q|t\rangle_o \\
&= \int_{-\infty}^{\infty} dq \int_{-\infty}^{\infty} d\epsilon' \frac{\mathrm{e}^{-it'\epsilon'/\hbar}}{\sqrt{\pi\hbar}} \sin\left(\mathrm{sgn}(\epsilon')\frac{\sqrt{2m|\epsilon'|}}{\hbar}q\right) \int_{-\infty}^{\infty} d\epsilon \frac{\mathrm{e}^{it\epsilon/\hbar}}{\sqrt{\pi\hbar}} \sin\left(\mathrm{sgn}(\epsilon)\frac{\sqrt{2m|\epsilon|}}{\hbar}q\right) \\
&= \frac{1}{\pi\hbar}\int_{-\infty}^{\infty} d\epsilon' d\epsilon\; \mathrm{e}^{i(t\epsilon - t'\epsilon')/\hbar} \int_{-\infty}^{\infty} dq\; \sin\left(\mathrm{sgn}(\epsilon')\frac{\sqrt{2m|\epsilon'|}}{\hbar}q\right) \sin\left(\mathrm{sgn}(\epsilon)\frac{\sqrt{2m|\epsilon|}}{\hbar}q\right) \\
&= -\frac{1}{\pi\hbar}\int_{-\infty}^{\infty} d\epsilon' d\epsilon\; \mathrm{e}^{i(t\epsilon - t'\epsilon')/\hbar} \frac{1}{4}\int_{-\infty}^{\infty} dq\; \left(\mathrm{e}^{-i(p+p')q/\hbar} - \mathrm{e}^{i(p-p')q/\hbar} - \mathrm{e}^{i(p'-p)q/\hbar} + \mathrm{e}^{i(p+p')q/\hbar}\right) \\
&= -\frac{2\pi\hbar}{4\pi\hbar}\int_{-\infty}^{\infty} d\epsilon' d\epsilon\; \mathrm{e}^{i(t\epsilon - t'\epsilon')/\hbar} \left(\delta(p+p') - \delta(p-p') - \delta(p'-p) + \delta(p+p')\right) \\
&= \int_{-\infty}^{\infty} d\epsilon' d\epsilon\; \mathrm{e}^{i(t\epsilon - t'\epsilon')/\hbar} \left(\delta(p-p') - \delta(p+p')\right) = \int_{-\infty}^{\infty} d\epsilon\; \mathrm{e}^{i(t-t')\epsilon/\hbar} - \int_{-\infty}^{\infty} d\epsilon\; \mathrm{e}^{i(t+t')\epsilon/\hbar} \\
&= \delta(t-t') - \delta(t+t')
\end{aligned} \tag{21}$$

$${}_e\langle t'|t\rangle_e = \int_{-\infty}^{\infty} dq\; {}_e\langle t'|q\rangle\langle q|t\rangle_e = \delta(t-t') + \delta(t+t') \tag{22}$$

$${}_-\langle t'|t\rangle_- = \int_{-\infty}^{\infty} dq\; {}_-\langle t'|q\rangle\langle q|t\rangle_- = \delta(t-t'). \tag{23}$$

$${}_+\langle t'|t\rangle_+ = \int_{-\infty}^{\infty} dq\; {}_+\langle t'|q\rangle\langle q|t\rangle_+ = \delta(t-t'). \tag{24}$$

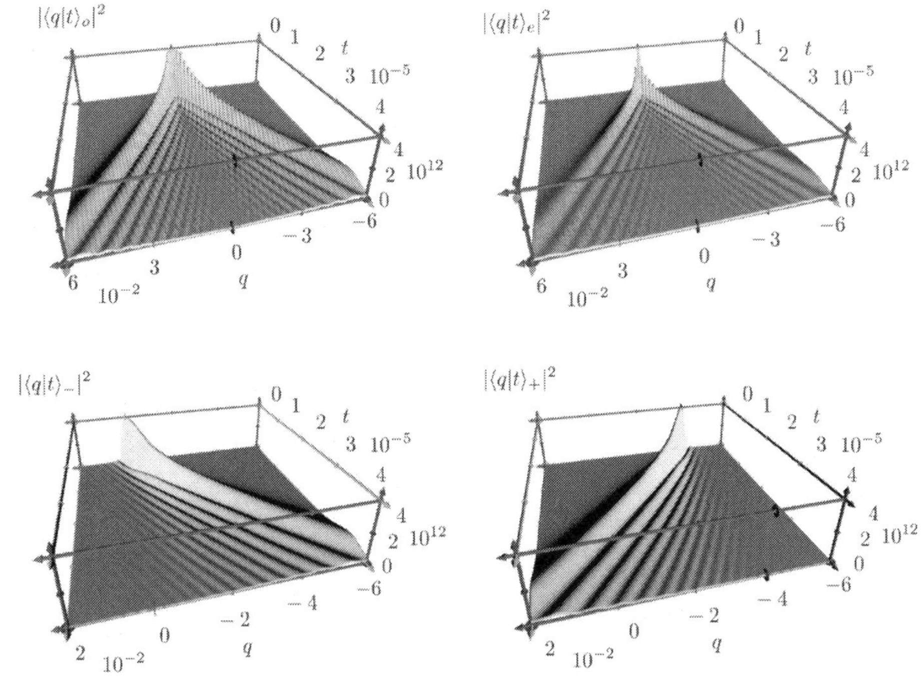

Figure 5. Time evolution of some time eigenstates for the free particle. Approximated expressions for $\mathcal{E} = 10$. Dimensionless units.

6. Evolution of time eigenstates

In order to have an idea of how these states look like, we consider the approximation obtained by calculating the integrals with finite limits

$$\langle q|t\rangle_e = \lim_{\mathcal{E}\to\infty} \frac{1}{\sqrt{2\pi\hbar}} \int_{-\mathcal{E}}^{\mathcal{E}} d\epsilon\ \mathrm{e}^{it\epsilon/\hbar} \cos\left(\mathrm{sign}(\epsilon)\sqrt{2m|\epsilon|}\ \frac{q}{\hbar}\right)$$
$$\langle q|t\rangle_o = \lim_{\mathcal{E}\to\infty} \frac{1}{\sqrt{2\pi\hbar}} \int_{-\mathcal{E}}^{\mathcal{E}} d\epsilon\ \mathrm{e}^{it\epsilon/\hbar} \sin\left(\mathrm{sign}(\epsilon)\sqrt{2m|\epsilon|}\ \frac{q}{\hbar}\right), \tag{25}$$

and

$$\langle q|t\rangle_\pm = \lim_{\mathcal{E}\to\infty} \frac{1}{\sqrt{2\pi\hbar}} \int_{-\mathcal{E}}^{\mathcal{E}} d\epsilon \frac{\mathrm{e}^{it\epsilon/\hbar}}{\sqrt{2\pi\hbar}}\, \mathrm{e}^{\pm\mathrm{sign}(\epsilon) i\sqrt{2m|\epsilon|}\ q/\hbar}. \tag{26}$$

Plot of some of these functions are shown in Fig. 5. These functions are initially highly concentrated around the origin and, as time goes by, they depart from the origin and spread.

The time operator can be approximated in the same way and it is shown in Fig. 6

7. Conclusion

We have found symmetric operators with domain the envelope of the energy eigenstates $\{|E\rangle\}_0^\infty$, i.e. the whole of the Hilbert space, and with orthogonal eigenstates which are conjugate to the

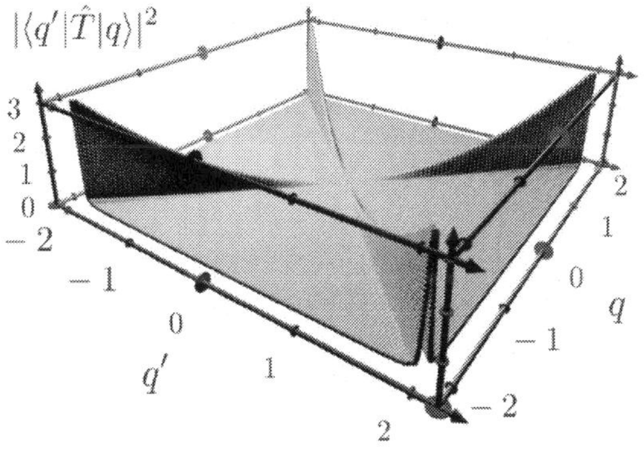

Figure 6. Matrix elements of the time operator for a free particle. Approximated expressions for $\mathcal{E} = 10$. Dimensionless units.

energy eigenstates. Thus, they are good candidates for being time operators because they comply with the requirements for such operators.

As we have seen, in fact there are several time operators which can be used in different situations. There are time operators which differentiate between the sign of the momentum, and there are other time operators that include both signs of the momentum into them.

We are optimistic about the operators introduced in this paper to be good candidates for time operators.

References

[1] Aharonov Y and Bohm D 1961 *Phys. Rev.* **122** 1649
[2] Kijowski J 1974 *Rep. Math. Phys.* **6** 361
[3] Baute A D, Sala-Mayato R, Palao J P, Muga J G and Egusquiza I L 2000 *Phys. Rev. A* **61** 022118
[4] Baute A D, Egusquiza I L and Muga J G 2001 *Phys. Rev. A.* **64** 012501
[5] Torres-Vega G and García-Jiménez M N 2013 *Entropy* **15** 2415
[6] Torres-Vega G 2013 *Entropy* **15** 4105
[7] Sombillo D L B and Galapon E A 2012 *J. Math. Phys.* **53** 043702
[8] Hall M J W 2008 Comment on "An Arrow of Time Operator for Standard Quantum Mechanics" (a sign of the time!) (*Preprint* `quant-ph 0802.2682`)
[9] Hegerfeldt G C, Muga J G and Muñoz J 2010 *Phys. Rev. A* **82** 012113
[10] Arai A and Matsuzawa Y 2008 *Rev Math Phys* **20** 951
[11] Arai A 2009 *Lett. Math. Phys.* **87** 67

Journal of Physics: Conference Series **624** (2015) 012006 doi:10.1088/1742-6596/624/1/012006

Teleportation based on control of anisotropic Ising interaction in three dimensions

Francisco Delgado

Departamento de Física y Matemáticas, Escuela de Ingeniería y Ciencias, Tecnológico de Monterrey, Campus Estado de México, Atizapán, Estado de México, CP. 52926, México.

E-mail: `fdelgado@itesm.mx`

Abstract. Teleportation is possibly the most representative quantum algorithm in the public domain. Nevertheless, though this quantum procedure transmits only information and not objects, its coverage is still very limited and easily subject to errors. Based on a fine control of quantum resources, particularly those entangled, the research to extend its coverage and flexibility is open, in particular on matter based quantum systems. This work shows how anisotropic Ising interactions could be used as a natural basis for this procedure, based on a sequence of magnetic pulses driving Ising interaction, stating results in specialized quantum gates designed for magnetic systems.

1. Introduction

Quantum computation is possibly the upmost application goal of quantum mechanics nowadays. Feynman devised this relation based on use of quantum properties to speed up the simulation of computational problems by physical systems [1, 2]. Thus, based on quantum algorithms, useful developments using quantum information processing have been effectively developed theoretically and experimentally: quantum dense coding [3, 4], quantum key distribution [5, 6], quantum computation [7, 8] and quantum teleportation [9].

Quantum gate array computation is the most common, direct and clear approach in terms of proximity with classical computing. Their similitude is based on the use of computer gates replicated from classical programming. These gates are reproduced by several designs in terms of physical resources where they have been carried out: ion traps and electromagnetic cavities [10, 11], Josephson junctions [12], nuclear magnetic resonance [13] and spins [14–16]. Nevertheless, their translation to theoretical gates is not always immediate, requiring control or iterative procedures.

Teleportation is a physical process now in the public domain by its attractiveness. In it, a quantum state can be transferred to other using a previously shared entangled pair, with assistance of classical communications and local operations. In [9], a Bell state is used to this goal together with Hadamard, C^a-NOT$_b$, X and Z gates. Departing from this development, many proposals and goals have been made to transfer a multi-qubit state, inclusively in the experimental terrain [17–19]. Teleportation has been accomplished among optical systems [17,20,21], photons and a single atomic ensemble [22–24] and trapped atomic ions using Coulomb interaction [25–27]. Therefore, teleportation is commonly considered as one of the most striking progresses of quantum information theory.

Ising model [28–30] is used as a simple approach to magnetic interaction between quantm objects (electronic gases, quantum dots, ions, etc.). Nielsen [31] was the first reporting studies of entanglement between magnetic systems based on a two spin systems driven with an external magnetic field. One property of this model is that it generates entanglement, one of the more interesting properties of quantum mechanics [32–36]. This property is a central aspect in most of quantum applications, because its non local properties improve capacity and speed of information processing [37–39]. Control of entanglement is achievable in Ising model through the driven magnetic fields being introduced on the physical system. This is the case for teleportation. Different models of Ising interaction (XX, XY, XYZ depending on interest of each author and physical systems being considered) are used to reproduce effects related with bipartite or multipartite systems [4, 40, 41] and quantum dots [42, 43]).

Nowadays, quantum gate array computation is being experimentally explored to adapt it to stuff in which it can be settled, particularly in terms of noise control and reproduction of computational gates, i.e., the interactions capable to reproduce them [10–16]. Quantum dots and electronic gases are developments towards a scalable spin-based quantum computer, which can be controlled with electromagnetic interactions among neighboring spins to obtain universal quantum operations [44–46] in terms of DiVincenzo criteria [47].

The aim of this paper is to apply some non local properties recently reported by [48] in the anisotropic Ising model and some control procedures on it [49], which naturally reproduce non local gates useful for teleportation. These gates reduce teleportation algorithm to a driven magnetic interaction based on matter to transfer spin quantum states. Several variants of this process are presented, as well their extension to multiple qubits teleportation.

2. Analytic evolution for anisotropic Ising model in three dimensions

The use of magnetic systems as quantum resources is a basis on which quantum applications could be settled. From quantum memories to quantum processors, matter susceptible of magnetic control is considered as stuff for quantum computation or quantum information processing. Particularly, Ising model, is a simple model of interaction bringing an easy basis to generate and manipulate quantum states and entanglement particularly. In this model, as is shown in the upper part of Figure 1, two qubits interact via Ising interaction using additional local magnetic fields as driven elements.

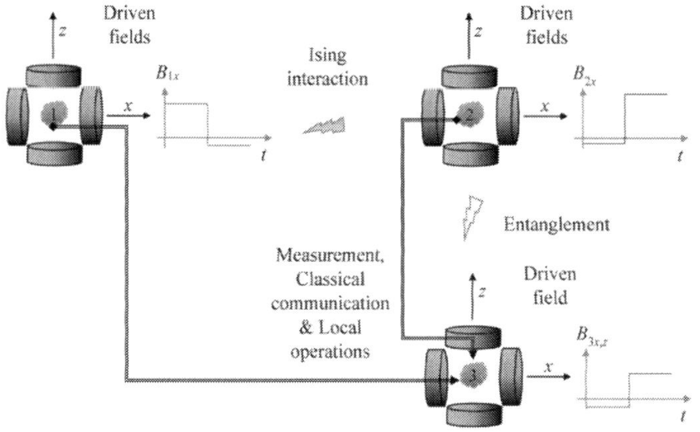

Figure 1. Ising interaction between two qubits with driven local magnetic fields to produce teleportation on a third qubit entangled with the system.

Journal of Physics: Conference Series **624** (2015) 012006 doi:10.1088/1742-6596/624/1/012006

2.1. Three dimensional anisotropic Ising model and notation

Some results about anisotropic Ising model generalize their treatment and suggest an algebraic structure when it is depicted in Bell basis [48]. Using this work, we focus on the following Hamiltonian for the bipartite anisotropic Ising model including an inhomogeneous magnetic field restricted to the h-direction ($h = 1, 2, 3$ corresponding to x, y, z respectively):

$$H_h = -\sigma_1 \cdot \mathbf{J} \cdot \sigma_2 + \mathbf{B_1} \cdot \sigma_1 + \mathbf{B_2} \cdot \sigma_2 = -\sum_{k=1}^{3} J_k \sigma_{1k} \sigma_{2k} + B_{1h} \sigma_{1h} + B_{2h} \sigma_{2h} \tag{1}$$

which includes several models considered in the previously cited works. Following the definitions and the notation adopted in [48], we introduce the scaled parameters:

$$b_{h\pm} = \frac{B_{h\pm}}{R_{h\pm}}, \quad j_{h\pm} = \frac{J_{\{h\}\mp}}{R_{h\pm}} \quad \in [-1, 1] \tag{2}$$

$$\begin{aligned} \text{with:} \quad & J_{\{h\}\pm} \equiv J_{i,j\pm} = J_i \pm J_j \\ & B_{h\pm} = B_{1_h} \pm B_{2_h} \\ & R_{h\pm} = \sqrt{B_{h\pm}^2 + J_{\{h\}\mp}^2} \end{aligned}$$

being h, i, j a cyclic permutation of $1, 2, 3$ and simplified by $\{h\} \equiv i, j$. Additionally, we introduce the same nomenclature introduced in [48] based on different subscripts than those inherited by the computational basis: greek scripts for $-1, +1$ or $-, +$, for scripts in states and operators (meaning $-1, +1$ in mathematical expressions respectively); capital scripts $A, B, ...$ for $0, 1$ as in the computational basis; and latin scripts $h, i, j, k, ...$ for spatial directions x, y, z or $1, 2, 3$. Bell states become in latter notation:

$$\begin{aligned} |\beta_{--}\rangle &\equiv |\beta_{00}\rangle, \qquad & |\beta_{-+}\rangle &\equiv |\beta_{01}\rangle \\ |\beta_{+-}\rangle &\equiv |\beta_{10}\rangle, \qquad & |\beta_{++}\rangle &\equiv |\beta_{11}\rangle \end{aligned} \tag{3}$$

2.2. Algebraic structure of evolution operator

In agreement with latter notation, energy levels in (1) are denoted by eigenvalues $E_{h\mu\nu}(E_{h--}, E_{h-+}, E_{h+-}, E_{h++})$ and they are:

$$E_{h\mu\nu} \equiv \mu J_h + \nu R_{h-\mu} = \mu J_h + \nu \sqrt{B_{h-\mu}^{\,2} + J_{\{h\}\mu}^2} \tag{4}$$

with their eigenvectors reported in [48]. As there, by introducing the following definitions:

$$\Delta_{h\mu}^{\nu} = \frac{t}{2}(E_{h\mu+} + \nu E_{h\mu-}) = \begin{cases} \mu J_h t & \text{if} \quad \nu = + \\ R_{h-\mu} t & \text{if} \quad \nu = - \end{cases} \tag{5}$$

$$\begin{aligned} e_{h\alpha}^{\beta} &= \cos \Delta_{h\alpha}^{-} + i\beta j_{h-\alpha} \sin \Delta_{h\alpha}^{-} \\ d_{h\alpha} &= b_{h-\alpha} \sin \Delta_{h\alpha}^{-} \end{aligned} \tag{6}$$

the evolution operators in Bell basis are in the matrix form:

$$U_1(t)=\left(\begin{array}{c|c|c|c} e^{i{\Delta_1}^{+}_{-}}\,{e_1}^{-*}_{-} & i\,e^{i{\Delta_1}^{+}_{-}}\,d_{1-} & 0 & 0 \\ \hline i\,e^{i{\Delta_1}^{+}_{-}}\,d_{1-} & e^{i{\Delta_1}^{+}_{-}}\,{e_1}^{-}_{-} & 0 & 0 \\ \hline 0 & 0 & e^{i{\Delta_1}^{+}_{+}}\,{e_1}^{+*}_{+} & -i\,e^{i{\Delta_1}^{+}_{+}}\,d_{1+} \\ \hline 0 & 0 & -i\,e^{i{\Delta_1}^{+}_{+}}\,d_{1+} & e^{i{\Delta_1}^{+}_{+}}\,{e_1}^{+}_{+} \end{array}\right)\in \mathbf{S}_1^*, \tag{7}$$

$$U_2(t)=\left(\begin{array}{c|c|c|c} e^{i{\Delta_2}^{+}_{+}}\,{e_2}^{+*}_{+} & 0 & 0 & -\,e^{i{\Delta_2}^{+}_{+}}\,d_{2+} \\ \hline 0 & e^{i{\Delta_2}^{+}_{-}}\,{e_2}^{+*}_{-} & -\,e^{i{\Delta_2}^{+}_{-}}\,d_{2-} & 0 \\ \hline 0 & e^{i{\Delta_2}^{+}_{-}}\,d_{2-} & e^{i{\Delta_2}^{+}_{-}}\,{e_2}^{+}_{-} & 0 \\ \hline e^{i{\Delta_2}^{+}_{+}}\,d_{2+} & 0 & 0 & e^{i{\Delta_2}^{+}_{+}}\,{e_2}^{+}_{+} \end{array}\right)\in \mathbf{S}_2^*, \tag{8}$$

$$U_3(t)=\left(\begin{array}{c|c|c|c} e^{i{\Delta_3}^{+}_{-}}\,{e_3}^{+*}_{-} & 0 & i\,e^{i{\Delta_3}^{+}_{-}}\,d_{3-} & 0 \\ \hline 0 & e^{i{\Delta_3}^{+}_{+}}\,{e_3}^{+*}_{+} & 0 & i\,e^{i{\Delta_3}^{+}_{+}}\,d_{3+} \\ \hline i\,e^{i{\Delta_3}^{+}_{-}}\,d_{3-} & 0 & e^{i{\Delta_3}^{+}_{-}}\,{e_3}^{+}_{-} & 0 \\ \hline 0 & i\,e^{i{\Delta_3}^{+}_{+}}\,d_{3+} & 0 & e^{i{\Delta_3}^{+}_{+}}\,{e_3}^{+}_{+} \end{array}\right)\in \mathbf{S}_3^*, \tag{9}$$

where clearly $U_h(t)$ have a 2×2 block structure, having each one the semidirect product structure: $U(2)=U(1)\times SU(2)$. Thus, $U_h(t)$ belongs to subgroups: $\mathbf{S}_1^*, \mathbf{S}_2^*, \mathbf{S}_3^* \subset SU(4)$ (defined in [48]). These subgroups are characterized by their block structure and properties inherited from (7-9) when $b_{h\pm\alpha}, j_{h\pm\alpha}$ are fixed. Then, these subgroups have a semidirect product structure: $S_h^* = U(1)\times SU(2)\times SU(2)$. Thus, identity and inverses are included in each subgroup, while there is closure in the product. At the same time, quantum states in $\mathcal{H}^{\otimes 2}$ become split in a direct sum of two subspaces, each one generated by pairs of Bell states. This structure is essential in the current work because it assures the existence of solutions in their control.

2.3. *Block structure*

Blocks are elements of $U(2)$ as is reported in [48, 49]. Their general structure is:

$$\begin{aligned} s_{h_j} =& e^{i\Delta_{h\alpha}^{+}}\begin{pmatrix} e_{h\alpha}^{\beta *} & -qi^{h}d_{h\alpha} \\ qi^{*h}d_{h\alpha} & e_{h\alpha}^{\beta} \end{pmatrix} \\ \text{with: } \alpha =&(-1)^{h+j+1} \\ \beta =&(-1)^{j(h+l_j-k_j+1)} \\ q =&\beta(-1)^{h+1} \end{aligned} \tag{10}$$

being h the spatial direction of associated magnetic field; $j=1,2$ an ordering label for each block as it appears in the rows of the evolution matrix; k_j, l_j are the labels for its rows in $U_h(t)$ (by example, in s_{21}, $k_2=2, l_2=3$ are labels of the rows of second block, $j=2$, in $U_{h=2}(t)$). Note that $\det(s_{h_j}) = e^{2i\Delta_{h\alpha}^{+}}$ is unitary. Last structure lets to introduce the generation of control operations in terms of factorization of special unitary matrices in $SU(4)$ [50].

3. Controlled gates generated by anisotropic Ising interaction driven by magnetic fields

As was reported in [49], controlled gates in each block could be constructed in two pulses:

$$U_h(T=t+t') = U_h(t')U_h(t) \tag{11}$$

For that purpose, one block should be diagonalized into I_2 and another block antidiagonalized into σ_1 or $i\sigma_2$ (these forms are straight for the block, any basis change from the computational basis to the Bell basis is required). Prescriptions to achieve last forms on each block were obtained by [49] and they are:

$$\begin{aligned} \frac{B_{h\alpha}}{J_{\{h\}_{-\alpha}}} = \xi, \frac{B'_{h\alpha}}{J'_{\{h\}_{-\alpha}}} &= -\xi^{-1}, \frac{B_{h-\alpha}}{J_{\{h\}_{\alpha}}} = \chi, \frac{B'_{h-\alpha}}{J'_{\{h\}_{\alpha}}} = \chi \\ \frac{|J_{\{h\}_{-\alpha}}|t}{(2n_{-\alpha}+1)} = \frac{|J'_{\{h\}_{-\alpha}}|t'}{(2n'_{-\alpha}+1)|\xi|} &= \frac{\pi}{2\sqrt{\xi^2+1}} \end{aligned} \tag{12}$$

$$\begin{aligned} \text{where: } |\xi| &= \frac{-AB \pm \sqrt{A^2+B^2-1}}{B^2-1} \\ \chi^2 &= \left(\frac{2n_\alpha \sqrt{\xi^2+1}}{S_\alpha(2n_{-\alpha}+1) + P_\alpha S'_\alpha(2n'_{-\alpha}+1)|\xi|} \right)^2 - 1 \\ A &= \frac{(2n_{-\alpha}+1)J_h}{2(m_\alpha+n_\alpha)|J_{\{h\}_{-\alpha}}|} \\ B &= \frac{(2n'_{-\alpha}+1)J'_h}{2(m_\alpha+n_\alpha)|J'_{\{h\}_{-\alpha}}|} \\ P_\alpha &= \text{sign}(J'_{\{h\}_\alpha} J_{\{h\}_\alpha}) \\ S_\alpha &= |\frac{J_{\{h\}_\alpha}}{J_{\{h\}_{-\alpha}}}|, S'_\alpha = |\frac{J'_{\{h\}_\alpha}}{J'_{\{h\}_{-\alpha}}}| \end{aligned} \tag{13}$$

and α corresponds to diagonal block and $-\alpha$ to antidiagonal one. Also, $n_{\pm\alpha}, n'_\alpha, m_\alpha \in \mathbf{Z}$. These prescriptions let obtain the design parameters $B_{\pm\alpha}, B'_{\pm\alpha}, t, t'$ in terms of strengths $J_h, J'_h, J_{\{h\}_{\pm\alpha}}, J'_{\{h\}_{\pm\alpha}}$. Note that parameters $|\xi|$, χ^2 should fulfill positivity restrictions.

If $s_{hj}^{\pm\alpha}, s'^{\pm\alpha}_{hj}$ are the respective blocks α (diagonal) and $-\alpha$ (antidiagonal) for $U_h(t), U_h(t')$, then, last prescriptions generate the combined blocks in $U_h(T)$:

$$s'^{\alpha}_{hj} s_{hj}^{\alpha} = (-1)^{m_\alpha} I_2 \tag{14}$$

$$s'^{-\alpha}_{hj} s_{hj}^{-\alpha} = \begin{pmatrix} 0 & (-1)^{s_\alpha} \\ (-1)^{h+s_\alpha} & 0 \end{pmatrix} = (-1)^{s_\alpha} i^{h \bmod 2} \sigma_{1+h \bmod 2}, \tag{15}$$

with s_α becoming an integer or a semi-integer in terms of the additional condition:

$$2(m_\alpha + n_\alpha) = -(h + \text{sign}(q\beta b'_{h_\alpha} j_{h_\alpha}) + \quad 2(n_{-\alpha} + n'_{-\alpha} - s_{-\alpha} + 1)) \tag{16}$$

which shows that s_α is a semi-integer for $h = 2$.

4. Quantum teleportation based on Ising interaction

4.1. One qubit state teleportation using the $\mathcal{A}_{1,2}^{0,\frac{\pi}{2}}$ *gate*

Driven Ising interaction generate generalized controlled gates presented in [49], $\mathcal{D}_h^\phi$, $\mathcal{A}_{h,j}^{\phi,\varphi_{h_\alpha}}$. First ones are closely related with the identity combined with phase gates. Next ones are more like to C^a-NOT$_b$ gates used in traditional gate array quantum computation. Anyway, both effects are direct expressions of spin version phenomenon noted by Mielnik [51] for partial or complete Evolution Loops in control processes.

When $\phi = 0, \varphi = \pi/2$, they are simply the matrices $\mathcal{A}_{h,j}^{0,\frac{\pi}{2}\alpha}$ obtained from combination (in the structure of $U_h(t)$) of exactly each one of blocks in (14-15) with m_α, s_α even:

$$\mathcal{A}_{1,1}^{0,\frac{\pi}{2}} = \begin{pmatrix} 0 & 1 & 0 & 0 \\ -1 & 0 & 0 & 0 \\ 0 & 0 & 1 & 0 \\ 0 & 0 & 0 & 1 \end{pmatrix}, \mathcal{A}_{2,1}^{0,\frac{\pi}{2}} = \begin{pmatrix} 0 & 0 & 0 & i \\ 0 & 1 & 0 & 0 \\ 0 & 0 & 1 & 0 \\ i & 0 & 0 & 0 \end{pmatrix}, \mathcal{A}_{3,1}^{0,\frac{\pi}{2}} = \begin{pmatrix} 1 & 0 & 0 & 0 \\ 0 & 0 & 0 & 1 \\ 0 & 0 & 1 & 0 \\ 0 & -1 & 0 & 0 \end{pmatrix}$$
$$\mathcal{A}_{1,2}^{0,\frac{\pi}{2}} = \begin{pmatrix} 1 & 0 & 0 & 0 \\ 0 & 1 & 0 & 0 \\ 0 & 0 & 0 & 1 \\ 0 & 0 & -1 & 0 \end{pmatrix}, \mathcal{A}_{2,2}^{0,\frac{\pi}{2}} = \begin{pmatrix} 1 & 0 & 0 & 0 \\ 0 & 0 & i & 0 \\ 0 & i & 0 & 0 \\ 0 & 0 & 0 & 1 \end{pmatrix}, \mathcal{A}_{3,2}^{0,\frac{\pi}{2}} = \begin{pmatrix} 0 & 0 & 1 & 0 \\ 0 & 1 & 0 & 0 \\ -1 & 0 & 0 & 0 \\ 0 & 0 & 0 & 1 \end{pmatrix}. \quad (17)$$

Eq. (17) suggests they can be used as natural operations replacing some standard gates if computing is done using non local quantum resources (remember that last expressions are based on non local Bell basis).

One immediate application is the teleportation algorithm. We begin, as commonly, with the state $(\alpha|0\rangle + \beta|1\rangle) \otimes |\beta_{--}\rangle$ with the first qubit to teleport and in possession of Alice. Bell state is first in possession of Bob, who sends his first part to Alice. Then, Alice drives an Ising interaction as in our model, applying for example, $\mathcal{A}_{1,2}^{0,\frac{\pi}{2}}$ operation between the qubits in their current possession. With just this, they almost obtain the standard teleportation algorithm for one qubit [9]:

$$\begin{aligned}(\alpha|0\rangle + \beta|1\rangle) \otimes |\beta_{00}\rangle = & \frac{\alpha}{2}\Big((|\beta_{00}\rangle + |\beta_{10}\rangle) \otimes |0\rangle + (|\beta_{01}\rangle + |\beta_{11}\rangle) \otimes |1\rangle \Big) \\ & + \frac{\beta}{2}\Big((|\beta_{01}\rangle - |\beta_{11}\rangle) \otimes |0\rangle + (|\beta_{00}\rangle - |\beta_{10}\rangle) \otimes |1\rangle \Big) \\ \xrightarrow{\mathcal{A}_{1,2}^{0,\frac{\pi}{2}}} & \frac{\alpha}{2}\Big((|\beta_{00}\rangle - |\beta_{11}\rangle) \otimes |0\rangle + (|\beta_{01}\rangle + |\beta_{10}\rangle) \otimes |1\rangle \Big) \\ & + \frac{\beta}{2}\Big((|\beta_{01}\rangle - |\beta_{10}\rangle) \otimes |0\rangle + (|\beta_{00}\rangle + |\beta_{11}\rangle) \otimes |1\rangle \Big), \quad (18)\end{aligned}$$

where for simplicity, to identify easily the results in computational basis, we come back at this point to the traditional notation for Bell states. Figures 2 and 3 show two alternatives of the algorithm using the gate $\mathcal{A}_{1,2}^{0,\frac{\pi}{2}}$, which is substituting the traditional $H_1 C^1\text{-NOT}_2$ operation nn computational basis.

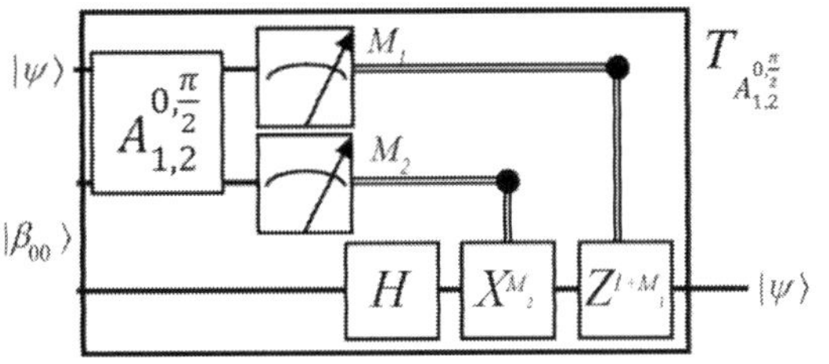

Figure 2. Ising teleportation based on computational measurements.

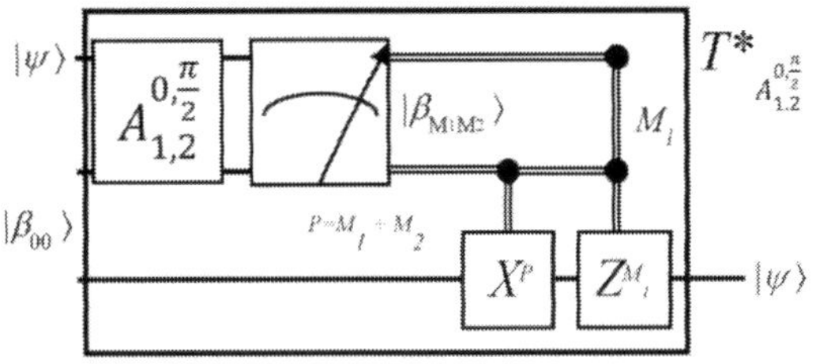

Figure 3. Ising teleportation based on non local measurements.

After to apply Ising interaction, Alice makes a measurement, for example in the computational basis, of the qubits. Results are listed in Table 1, where $|-\rangle, |+\rangle$ are the eigenstates of X (Teleportated state column in Table 1). Clearly a Hadamard gate is necessary at the end and in addition, if measurement result is $|M_1 M_2\rangle$, then it is required to apply $X^{M_2} Z^{1+M_1}$ to teleportate the original qubit 1 on qubit 3 ($M_1, M_2 = 0, 1$ are in the traditional computational basis, not in the $+, -$ notation). Thus, Alice uses classical communication to send her outcomes to Bob, who finally applies the adequate gates (complementary gates column in Table 1) to obtain exactly (until unitary factors) the original state $\alpha|0\rangle + \beta|1\rangle$, as is shown in Table 1 and Figure 2.

There, I, X, Y, Z, H are the traditional gates in computational basis, which should be applied on qubit 3 as is depicted in Table 1. We use these symbols to distinguish from $I_2, \sigma_1, \sigma_2, \sigma_3$ previously used (with exception of Hamiltonian (1)), which are only matrix forms written for the Bell states corresponding to each subspace in the direct sum decomposition of $\mathcal{H}^{\otimes 2}$.

Another alternative for Alice and Bob is to make a measurement in Bell basis $|\beta_{\alpha\beta}\rangle = |\beta_{M_1 M2}\rangle$ (for both notations). In such a case, Bob will need to apply $Z^{\frac{1+\alpha}{2}} X^{\frac{1-\alpha\beta}{2}}$ (with $\alpha, \beta = +, -$) or $Z^{M_1} X^{M_1+M_2}$ (in the traditional notation $M_1, M_2 = 0, 1$). Figure 3 shows this procedure.

Table 1. Measurement outcomes in two different basis, output state and complementary gates for teleportation algorithm based on $\mathcal{A}_{1,2}^{0,\frac{\pi}{2}}$.

Basis	Measurement	Teleportated state	Complementary gates
Computational	$\|00\rangle$	$\alpha\|+\rangle - \beta\|-\rangle$	$Z_3 H_3$
	$\|01\rangle$	$-\alpha\|-\rangle + \beta\|+\rangle$	$X_3 Z_3 H_3$
	$\|10\rangle$	$\alpha\|+\rangle + \beta\|-\rangle$	H_3
	$\|11\rangle$	$\alpha\|-\rangle + \beta\|+\rangle$	$X_3 H_3$
Non local	$\|\beta_{--}\rangle$	$\alpha\|0\rangle + \beta\|1\rangle$	I_3
	$\|\beta_{-+}\rangle$	$\alpha\|1\rangle + \beta\|0\rangle$	X_3
	$\|\beta_{+-}\rangle$	$\alpha\|1\rangle - \beta\|0\rangle$	$Z_3 X_3$
	$\|\beta_{++}\rangle$	$-\alpha\|0\rangle + \beta\|1\rangle$	Z_3

4.2. General teleportation with other Ising gates and multiqubit states

Last example has shown that gate $\mathcal{A}_{1,2}^{0,\frac{\pi}{2}}$ reproduce the traditional teleportation algorithm. Nevertheless, it can be achieved with any $\mathcal{A}_{h,j}^{0,\frac{\pi}{2}}$ gate. Following the steps in the last subsection, it is possible to show that with any of these gates (for arbitrary $h = 1, 2, 3$ and $j = 1, 2$) and inclusively if initial Bell state used is $|\beta_{AB}\rangle$ instead of $|\beta_{--}\rangle = |\beta_{00}\rangle$ as before, teleportation procedure can be achieved if we use the following complementary gates (assuming non local measurements based on Bell states, $|\beta_{M_1 M_2}\rangle$):

$$Z^{a_{h,j;A,B}^{M_1,M_2}} X^{b_{h,j;A,B}^{M_1,M_2}}, \tag{19}$$

$$\begin{aligned} \text{with: } a_{h,j}^{M_1,M_2} &= A + \frac{1}{8}\left(4(h-\frac{5}{2})^2 - 9\right)(j - 2 - M_2) + (h-2)^2 M_1, \\ b_{h,j}^{M_1,M_2} &= B + \frac{1}{8}\left(4(h-\frac{3}{2})^2 - 9\right)(j - 2 - M_1) + (h-2)^2 M_2. \end{aligned} \tag{20}$$

The generalization of last procedure for a multiqubit state is easy. In Figures 2 and 3, we define the procedure T as a map from $\mathcal{H}^{\otimes 3}$ on $\mathcal{H}$ (here, $\mathcal{H}$ is a two level Hilbert space):

$$\begin{aligned} &T: \mathcal{H}_1 \otimes (\mathcal{H}^{\otimes 2})_{2,3} \to \mathcal{H}_3, \\ &T((\alpha \left|\psi_\alpha\right>_1 + \beta \left|\psi_\beta\right>_1) \otimes \left|\phi\right>_{2,3}) = \alpha \left|\psi_\alpha\right>_3 + \beta \left|\psi_\beta\right>_3 . \end{aligned} \tag{21}$$

We can note that it is linear in the original state:

$$T((\alpha \left|\psi_\alpha\right>_1 + \beta \left|\psi_\beta\right>_1) \otimes \left|\phi\right>_{2,3}) = \alpha T(\left|\psi_\alpha\right>_1 \otimes \left|\phi\right>_{2,3}) + \beta T(\left|\psi_\beta\right>_1 \otimes \left|\phi\right>_{2,3}), \tag{22}$$

while, state $\alpha \left|\psi_\alpha\right>_1 + \beta \left|\psi_\beta\right>_1$ is separable from the remaining state $\left|\phi\right>_{2,3}$. In the last expressions, the subscripts are the labels for each one of the three qubits involved. Then, generalization is obvious. If we have the n-qubit state:

$$\left|\psi\right>_{1,\ldots,n} = \sum_{i_1,\ldots,i_n=0}^{1} \alpha_{\{i_1,\ldots,i_n\}} \bigotimes_{j=1}^{n} \left|i_j\right>_j \tag{23}$$

to teleportate, then, if we use the following state for the ancilla qubits:

$$\left|\phi\right>_{n+1,\ldots,2n} = \bigotimes_{j=1}^{n} \left|\beta_{A_j B_j}\right>_{n-1+2j,n-2+2j} \tag{24}$$

and we apply the teleportation protocol T_j on qubits $j, n-1+2j, n-2+2j$ (here, each T_j could be $T_{j\,A_{1,2}^{0,\frac{\pi}{2}}}$ or $T^*_{j\,A_{1,2}^{0,\frac{\pi}{2}}}$):

$$\begin{aligned} \bigotimes_{j}^{n} T_j(\left|\psi\right>_{1,\ldots,n} \otimes \left|\phi\right>)_{n+1,\ldots,2n} &= \sum_{i_1,\ldots,i_n=0}^{1} \alpha_{\{i_1,\ldots,i_n\}} \bigotimes_{j=1}^{n} T_j(\left|i_j\right>_j \otimes \left|\beta_{A_j B_j}\right>_{n-1+2j,n-2+2j}) \\ &= \sum_{i_1,\ldots,i_n=0}^{1} \alpha_{\{i_1,\ldots,i_n\}} \bigotimes_{j=1}^{n} \left|i_j\right>_{n-2+2j}) = \left|\psi\right>_{n+2,n+4,\ldots,2n-2,2n} \end{aligned} \tag{25}$$

Note in particular that not all single qubit teleportation algorithms can be the same type as well as the Bell state $\left|\beta_{A_j B_j}\right>$ being used for each one. Equal signs used in (25) are until unitary factors. These results show that controlled Ising model reproduces equivalent standard gates proposed in quantum teleportation.

5. Conclusions

Optics has been partially a dominant arena of developments in quantum information and quantum computation. Nevertheless, quantum storage and massive processing of information are tasks more probable to develop on matter based quantum systems. Systems based on trapped ions, e-Helium, nuclear magnetic resonance, superconductors, doped silicon and quantum dots are developments in this trend, showing possibilities to set up stable and efficient implementations of those technologies [52]. Thus, spin-based quantum computing has been developed through experimental implementations using mainly magnetic systems and exploiting Ising interactions with different approaches [53], together with control on quantum states, in particular entanglement control.

The procedure analyzed in this paper shows how Ising interaction could be controlled to generate gates capable to produce teleportation. Because of its linearity, this scheme shows the same generalization as traditional algorithm [9]. In addition, the different paths provided in the analysis, in terms of the Bell states used as resources and the field direction stated to generate the controlled gates, open other future research directions. One, is the analysis of control in the process and its fidelity in terms of control parameters (interaction and fields strengths, time intervals). The next one is about flexibility, as was stated in [54], in terms of using a broad spectrum of non-local resources which can be discriminated in parallel with the teleportation process during the measurement. It means, to achieve teleportation independently of initial knowledge of what kind of non local resource is being used. Finally, extensions to optimize the economy of qubits (in terms of resources used in the process) is open as it is in general in teleportation.

References

[1] Feynman R 1982 *Int. J. Theor. Phys.* **21** 467.
[2] Feynman R 1985 *Optics News* **11** (2) 11
[3] Bennett C H and Wiesner S J 1992 *Phys. Rev. Lett* **69** 2881
[4] Wang X 2001 *Phys. Lett. A* **281** 101
[5] Ekert A 1991 *Phys. Rev. Lett* **67** 661
[6] Bennet C H and DiVincenzo P 2000 *Nature* (London) **404** 247
[7] Nielsen M A and Chuang I L 2000 *Quantum Computation and Quantum Information* (Cambridge: Cambridge University Press).
[8] Ren B C and Deng F G 2014 *Sci. Rep.* **4** 4623
[9] Bennet C H, Brassard G and Mermin N D 1992 *Phys. Rev. Lett.* **68** 557
[10] Cirac J I and Zoller P 1995 *Phys. Rev. Lett.* **74** 4091
[11] Turchette Q A, Hood C J, Lange W, Mabuchi H, and Kimble H J 1995 *Phys. Rev. Lett.* **75** 4710
[12] Shnirman A, Schäon G and Hermon Z 1997 *Phys. Rev. Lett.* **79** 2371
[13] Chuang I L, Gershenfeld N and Kubinec M 1998 *Phys. Rev. Lett.* **80** 3408
[14] Kane B E 1998 *Nature* **393** 133
[15] Loss D and DiVincenzo D P 1998 *Phys. Rev. A* **57** 120
[16] Vrijen R, Yablonovitch E, Wang K L, Jiang H W, Balandin A, Roychowdhury V, Mor T and DiVincenzo D 2000 *Phys. Rev. A* **62**012306
[17] Bouwmeester D, Pan J W, Mattle K, Eibl M, Weinfurter H and Zeilinger A 1997 *Nature* **390** 575
[18] Karlsson A and Bourennane A 1998 *Phys. Rev. A* **58** 4394
[19] Nielsen M A, Knill E and Laflamme R 1998 *Nature* **396** 52
[20] Boschi D, Branca S, De Martini F, Hardy L and Popescu S 1998 *Phys. Rev. Lett.* **80** 1121
[21] Furusawa A, Sørensen J L, Braunstein S L, Fuchs C A, Kimble H J and Polzik E S 1998 *Science* **282** 706
[22] Sherson J F, Krauter H, Olsson R K, Julsgaard B, Hammerer K, Cirac I and Polzik E S 2006 *Nature* **443** 557
[23] Chen Y A, Chen S, Yuan Z S, Zhao B, Chuu C S, Schmiedmayer J and Pan J W 2008 *Nature* 4 103
[24] Olmschenk S, Matsukevich D N, Maunz P, Hayes D, Duan L M and Monroe C 2009 *Science* **323** 486
[25] Riebe M, Häffner H, Roos C F, Hänsel W, Benhelm J, Lancaster G P T, Körber T W, Becher C, Schmidt-Kaler F, James D F V 2004 *et al, Nature* **429** 734
[26] Barrett M D, Chiaverini J, Schaetz T, Britton J, Itano W M, Jost J D, Knill E, Langer E, Leibfried D, Ozeri R *et al* 2004 *Nature* **429** 737
[27] Riebe M, Chwalla M, Benhelm J, Häffner H, Hänsel W, Roos C F and Blatt R 2007 *New J. Phys.* **9** 211
[28] Ising E 1925 *Z. Phys.* **31** 253
[29] Brush S G 1967 *Rev. Mod. Phys.* **39** 883
[30] Baxter R J 1982 *Exactly solved models in statistical mechanics* (London: Acad. Press).
[31] Nielsen M A 1998 *Ph. D. Thesis University of New Mexico* (e-print: quant-ph/0011036)
[32] von Neumann J 1932 *Mathematische Grundlagen der Quantenmechanic* (Springer, Berlin)
[33] Schrödinger E and Born M 1935 *Proc. Cambridge Phil. Soc.* **31** (4) 555
[34] Schrödinger E 1935 *Naturwissenschften* **23** 807
[35] Einstein E, Podolsky B and Rosen N 1935 *Phys. Rev.* **47** 777
[36] Schrödinger E and Dirac P A M 1936 *Mathematical Proceedings of the Cambridge Philosophical Society* **32** (3) 446
[37] Jozsa R and Linden N 2022 *Proc. Royal Soc. A: Mathematical, Physical and Engineering Sciences* **459** (2036) 2011
[38] Jozsa R *Entanglement and quantum computation* (e-print quant-ph/9707034)

[39] Benett C H, DiVincenzo D P, Smolin J A and Wooters W K 1996 *Phys. Rev. A* **54** 3824
[40] Berman G P, Doolen G D, López G V and Tsifrinovich V I *Generalized Quantum Control-Not Gate in Two-Spin Ising System* (quant-ph/9802013v1)
[41] D'Alessandro D 2007 *Introduction to Quantum Control and Dynamics* (Boca Raton: Chapman Hall Applied Mathematics Nonlinear Science)
[42] Brunner R, Shin Y S, Obata T, Pioro-Ladriere M, Kubo T, Yoshida K, Taniyama T, Tokura Y and Tarucha S 2011 *Phys. Rev. Lett.* **107** 146801
[43] Gaudreau L, Granger G, Kam A, Aers G C, Studenikin S A, Zawadzki P, Pioro-Ladriere M, Wasilewski Z R and Sachrajda A S 2012 *Nature* **8** 54
[44] Recher P and Loss D 2002 *Proceedings of SPINTRONICS 2001: International Conference on Novel Aspects of Spin-Polarized Transport and Spin Dynamics* vol 15 (New York: Springer) p 49
[45] Saraga D S, Altshuler B L, Loss D and Westervelt R M 2004 *Phys. Rev. Lett.* **92** 246803
[46] Koppens F H K, Buizert C, Tielrooij K J, Vink I T, Nowack K C, Meunier T, Kouwenhoven L P and Vandersypen L M K 2006 *Nature* **442** (7104) 766
[47] DiVincenzo D P 1997 *Mesoscopic Electron Transport of NATO Advanced Study Institute, Series E: Applied Sciences* vol 345 (Curaçao, Kluwer Dordrecht,) p 657.
[48] Delgado F 2014 *Algebraic and group structure for bipartite three dimensional anisotropic Ising model on a non-local basis* (arXiv: quant-ph/1410.5148)
[49] Delgado F 2014 *Generation of non-local evolution loops and exchange operations for quantum control in three dimensional anisotropic Ising model* (arXiv: quant-ph/1410.5515)
[50] Delgado F 2014 *Factorization of unitary matrices induced by 3D anisotropic Ising interaction* (arXiv: quant-ph/1410.5891)
[51] Mielnik B 1986 *Journal of Mathematical Physics* **27** 2290
[52] Kloeffel C and Loss D 2013 *Annu. Rev. Condens. Matter Phys.* **4** 51
[53] Johnson M W *et al* 2011 *Nature* **473** 194
[54] Delgado F 2011 Teleportation algorithm using two species of entangled pairs *International Conference on Quantum Information* (Washington: Optical Society of America) QMI40

Quantum self-controlling free-falling cats

C. Chryssomalakos[1], **H. Hernández-Coronado**[2], **E. Serrano-Ensástiga**[1]

[1] Instituto de Ciencias Nucleares, Universidad Nacional Autónoma de México, Apartado Postal 70-543, México D.F. 04510, México.

[2] Departamento de Física, CINVESTAV, A.P. 14-740, México D.F. 07000, México.

E-mail: chryss@nucleares.unam.mx, hhernandez@fis.cinvestav.mx, eduardo.serrano@correo.nucleares.unam.mx

Abstract. In the separation of rotations from internal motions in the n-body problem, there appear some gauge fields which physically represent Coriolis effects. These fields are also present in the "falling cat" problem: at the kinematical level they map changes in the cat's shape to changes in its orientation whereas at the dynamical level they show up as gauge potentials in the Hamiltonian. Classically, the vanishing angular momentum condition allows for the orientation degrees of freedom to decouple from the internal ones and the cat's re-orientation can be accounted for at the kinematical level, partially. In the quantum case the cat's re-orientation description requires to be done on dynamical grounds. In this paper we explore the quantum version of the falling cat modelled as a three body problem.

1. Introduction

As it is familiar to almost everyone from everyday life experience, a free falling cat lands on its feet more often than not, regardless of its initial orientation and in spite of the fact that it might have zero angular momentum and nothing to push against. This phenomenon is counter-intuitive as it may seem to violate conservation of angular momentum, however, as it turns out cats can change their orientation by changing its shape precisely *because* of conservation of angular momentum [1–3]. In the description of the cat's reorientation, it is particularly useful the use of geometrical methods applied to the n-body problem [4].

Now, let us imagine that our cat is somehow "promoted" to a quantum cat (*i.e.*, a cat following the laws of quantum mechanics) and that it is freely released at a given height. At this point the cat may understandably feel quite uneasy and it may feel the need to consult an expert on quantum control, say professor Bogdan Mielnik [5–8]. Can the quantum cat control its muscles so that it lands on its feet as its classical analogue?

In this work we focus on the analysis of the above question in terms of geometrical methods. The structure of the paper is as follows: in Section 2 we introduce notation and discuss the general case of n-body problem in terms of geometric language, which follows closely Ref. [9], in Section 3 we present the main results of our work through the simplified case of a body composed only of three point-like masses. We give a brief conclusion in Section 4.

2. A cat as an n-body

Let us assume the cat is described by an isolated deformable body modelled by a system of n point-like particles interacting only between themselves through a potential V. Clearly, the body's configuration space is $\mathcal{C}_{\rm tot} = \mathbb{R}^{3n}$. If $\mathbf{r}_{s,\alpha} \in \mathbb{R}^3$ represents the components of α-th particle's position vector with respect to an inertial or space frame (indicated by the s sub-index) and m_α its mass ($\alpha = 1, 2, \ldots, n$), then the Lagrangian of the system is given by:

$$L_{\rm tot} = \frac{1}{2}\sum_{\alpha=1}^{n} m_\alpha |\dot{\mathbf{r}}_{s\alpha}|^2 - V(\mathbf{r}_{s,1}, \ldots, \mathbf{r}_{s,n}), \tag{1}$$

where "$\cdot$" $= d/dt$. Since there are not external forces acting on the system, it is convenient to eliminate the translational degrees of freedom by introducing relative coordinates, *e.g.*, the mass-weighted Jacobi coordinates $\boldsymbol{\rho}_{s\alpha}$, $\alpha = 1, \ldots, n-1$. In this way, the Lagrangian can be written as $L_{\rm tot} = L_{\rm CM} + L$, with $L_{\rm CM} = M|\dot{\mathbf{R}}_s|^2/2$ and L given by:

$$L = \frac{1}{2}\sum_{\alpha=1}^{n-1} |\dot{\boldsymbol{\rho}}_{s\alpha}|^2 - V(\boldsymbol{\rho}_{s,1}, \ldots, \boldsymbol{\rho}_{s,n-1}). \tag{2}$$

Likewise, in these coordinates it is clear that the total configuration space can be written as $\mathcal{C}_{\rm tot} = \mathbb{R}^3 \times \mathcal{C}$, where $\mathcal{C} = \mathbb{R}^{3n-3}$ is the translation-reduced configuration space upon which $(\boldsymbol{\rho}_{s,1}, \ldots, \boldsymbol{\rho}_{s,n-1})$ are coordinates. Since the dynamics of the center of mass is trivial, without loss of generality we can assume that $\mathbf{R}_s = 0$ for the rest of the paper and refer to $\mathcal{C}$ simply as *the* configuration space. A point in $\mathcal{C}$ is defined by an n-body's shape and an orientation. There are $3n-6$ coordinates needed to specify the body's shape q^μ, which correspond to independent functions on configuration space invariant under proper rotations, *i.e.*, $q^\mu(\boldsymbol{\rho}_{s,1}, \ldots, \boldsymbol{\rho}_{s,n-1}) = q^\mu(\mathsf{Q}\boldsymbol{\rho}_{s,1}, \ldots, \mathsf{Q}\boldsymbol{\rho}_{s,n-1})$ for all $\mathsf{Q} \in SO(3)$. The body's orientation, on the other hand, is defined by *a)* a convention of body frame given by the relations

$$\boldsymbol{\rho}_\alpha = \boldsymbol{\rho}_\alpha(q^\mu), \qquad \alpha = 1, \ldots, n-1, \tag{3}$$

and *b)* the orientation coordinates in $SO(3)$ defining a rotation $\mathsf{R}(\theta^i)$ that maps the body frame into the space frame. In the above relations we introduce the convention that a vector appearing without sub-index s is referred to the body frame. The convention of body frame is a choice of gauge, it consists of specifying the reference orientation in which both, the body and the space frames coincide, and it is defined by smoothly (and otherwise arbitrarily) attaching a frame to the body for each shape. For the non-collinear configurations considered in this work, this separation between rotations and internal motions allows us to regard $\mathcal{C}$ as an $SO(3)$ principal fibre bundle, where the base space is the "shape space" $SS = \mathbb{R}^{3n-3}/SO(3)$ and the fibre is isomorphic to $SO(3)$. In this geometric language the choice of gauge (3) defines a section $\mathcal{S}$, *i.e.*, a transversal embedding of shape space in the configuration space $\mathcal{C}$.

In terms of these shape and orientation coordinates, a point in $\mathcal{C}$ is defined according to:

$$\boldsymbol{\rho}_{s,\alpha} = \mathsf{R}(\theta^i)\boldsymbol{\rho}_\alpha(q^\mu), \qquad \alpha = 1, \ldots, n-1. \tag{4}$$

The above relation expresses the fact that given a shape of the body q^μ and a reference orientation $\boldsymbol{\rho}_\alpha(q^\mu)$, any configuration of the system can be reached through a rotation $\mathsf{R}(\theta^i)$. Similarly a velocity vector on the configuration's tangent space can be expressed in the angular velocity and shape anoholonomic basis as $v^a = (\boldsymbol{\omega}, \dot{q}^\mu)$, where $\omega^i = -\epsilon^{ijk}(\mathsf{R}^T \cdot \dot{\mathsf{R}})_{jk}/2$ is the i-th component of angular velocity of the body frame w.r.t. the space frame, referred to the body frame as indicated by the absence of subscript s. In the previous expression and in what follows we assume Einstein summation convention for repeated indices. Accordingly, the Lagrangian (2) can be recast as:

$$L = \frac{1}{2}G_{ab}v^a v^b - V(q), \tag{5}$$

where

$$G_{ab} = \begin{pmatrix} \mathsf{M} & \mathsf{M}\mathbf{A}_\nu \\ \mathbf{A}_\nu^T\mathsf{M} & g_{\mu\nu} + \mathbf{A}_\mu \cdot \mathsf{M} \cdot \mathbf{A}_\nu \end{pmatrix}, \tag{6}$$

is the metric in configuration space $\mathcal{C}$ and

$$\mathsf{M} = \sum_{\alpha=1}^{n-1} \boldsymbol{\rho}_\alpha \otimes \boldsymbol{\rho}_\alpha - |\boldsymbol{\rho}_\alpha|^2\, \mathsf{I}, \tag{7}$$

$$\mathbf{A}_\mu = \mathsf{M}^{-1} \sum_{\alpha=1}^{n-1} \boldsymbol{\rho}_\alpha \times \frac{\partial \boldsymbol{\rho}_\alpha}{\partial q^\mu}, \tag{8}$$

$$g_{\mu\nu} = \sum_{\alpha=1}^{n-1} \frac{\partial \boldsymbol{\rho}_\alpha}{\partial q^\mu} \times \frac{\partial \boldsymbol{\rho}_\alpha}{\partial q^\nu} - \mathbf{A}_\mu \cdot \mathsf{M} \cdot \mathbf{A}_\nu, \tag{9}$$

are the inertia tensor, the Coriolis gauge potential and the metric on shape space, respectively.

A velocity vector of the form $v^a = (\boldsymbol{\omega}, 0)$ is purely rotational or *vertical* since $\dot{q}^\mu = 0$ and thus the body's shape is not changing. *A priori* there is not a canonical way to define a purely internal velocity vector, because the condition $\boldsymbol{\omega} = 0$ is not gauge independent and therefore it cannot be physically meaningful. However, we can define a vector to be purely internal or *horizontal* if it is orthogonal to all vertical vectors w.r.t. the metric G_{ab} introduced in (6). As it turns out, a vector is horizontal iff its associated angular momentum vanishes. In terms of shape and orientation coordinates, the angular momentum (referred to the body frame) can be written as:

$$\mathbf{L} = \mathsf{M} \cdot (\boldsymbol{\omega} + \mathbf{A}_\mu \dot{q}^\mu), \tag{10}$$

and thus, a vector is horizontal if $\boldsymbol{\omega} dt = -\mathbf{A}_\mu dq^\mu$. In this manner it can be observed that the gauge potential $\mathbf{A}_\mu$ maps infinitesimal changes in shape space to infinitesimal rotations under the condition of vanishing angular momentum. When the body undergoes a finite deformation $q^\mu(t)$ subjected to the condition $\mathbf{L} = 0$, the previous relation can be integrated such that the body and the space frames are related through the rotation:

$$\mathsf{R}(t) = \mathsf{R}_0 \mathcal{P}\exp\left(-\int_{q_0}^{q(t)} \mathsf{A}_\mu dq^\mu \right), \tag{11}$$

where R_0 is a rotation relating both frames at $t = 0$, $\mathcal{P}\exp$ is the path-ordered exponential and the antisymmetric matrix A_μ is associated with the gauge potential according to $(\mathsf{A}_\mu)_{ij} = -\epsilon_{ijk} A^k_\mu$. Of course, in general this rotation is gauge-dependent and not meaningful from the physical viewpoint, nevertheless for closed paths – *i.e.*, for cyclic deformations of the body's shape – it actually describes a change of the body's orientation. In fact, just as in any Yang-Mills gauge theory, it is possible to define an associated curvature form or Coriolis tensor

$$\mathbf{B}_{\mu\nu} = \partial_\mu \mathbf{A}_\nu - \partial_\nu \mathbf{A}_\mu - \mathbf{A}_\mu \times \mathbf{A}_\nu, \tag{12}$$

such that a cyclic deformation in shape space under conditions of vanishing angular momentum, represented by the infinitesimal parallelogram spanned by the vectors y^μ and z^μ, produces the gauge-invariant infinitesimal rotation $\boldsymbol{\omega} dt = -\mathbf{B}_{\mu\nu} y^\mu z^\nu$. We then notice that in order to get a non trivial rotation, we need both the enclosed area by the body's deformation path in shape space and the curvature tensor to be different from zero.

The dynamics of the system can also be described by means of the gauge-covariant Hamiltonian:

$$H = \frac{1}{2}\mathbf{L} \cdot \mathsf{M} \cdot \mathbf{L} + \frac{1}{2}(p_\mu - \mathbf{A}_\mu \cdot \mathbf{L}) g^{\mu\nu} (p_\nu - \mathbf{A}_\nu \cdot \mathbf{L}) + V(q), \tag{13}$$

Journal of Physics: Conference Series **624** (2015) 012007 doi:10.1088/1742-6596/624/1/012007

where $p_\mu = g_{\mu\nu}\dot{q}^\nu + \mathbf{A}_\mu \cdot \mathbf{L}$ is the momentum conjugate to shape coordinate q^μ. The first and second terms in previous expression correspond to the vertical and horizontal contributions to the kinetic energy, respectively. From the Hamiltonian (13) we immediately note that if $\mathbf{L} = 0$, then the shape and orientation degrees of freedom decouple and we can independently solve for $q^\mu = q^\mu(t)$ and then plug it into expression (11) to obtain the orientation trajectory in $SO(3)$. Alternatively, one could assume that the body deformation is controlled externally so that by conveniently tailoring a cyclic path $q^\mu(t)$ in shape space, the body would get rotated by (11) as desired. Even though the latter scenario may be interesting for applications in quantum control theory, the system in that case is not longer isolated as we initially assumed.

Clearly the condition $\mathbf{L} = 0$ can be imposed in the classical context in a straightforward way, but in the quantum realm such a condition is too restrictive since it would imply that the system's state is the $l = 0$ angular momentum's eigenstate, in which the dispersion of the orientation angle operators is infinite. Thus, a more sensible condition in the quantum regime would be that the expectation value of $\mathbf{L}$ vanished. Therefore, if we are interested in describing a net rotation of a quantum deformable body produced by a cyclic change of its shape, we need to find the eigenvectors of the corresponding Hamiltonian operator at some point, but since M, $\mathbf{A}_\mu$ and $g^{\mu\nu}$ are operators themselves, the problem may be quite difficult. Indeed, since (13) commutes with $\mathbf{L}^2$ and L_{sz}, we know that its eigenfunctions are also eigenfunctions of $\mathbf{L}^2$ and L_{sz} with eigenvalues l and m, *i.e.*:

$$\psi_{lm}(\mathsf{R}, q^\mu) = \sum_{k=-l}^{l} \chi_k^l(q) D_{mk}^l(\mathsf{R})^*, \tag{14}$$

where $D_{km}^l(\mathsf{R})$ are the Wigner functions corresponding to the $(2l+1) \times (2l+1)$ irreducible matrix representation of $\mathsf{R} \in SO(3)$, and $\chi_k^l(q)$ is the gauge-dependent wavefuntion in the section $\mathcal{S}$. The previous expression tells us that for eigenstates of the system, its wavefunction for any orientation R can be obtained from the wavefunction in the section $\chi_k^l(q^\mu)$ through a rotation, and then in some sense it is the quantum version of relation (3). In this way, to solve the eigenvalue problem accounts to diagonalize an $(2l+1) \times (2l+1)$ matrix of operators in shape space, which is rather complicated in general. To simplify things a bit, we will consider quasi-rigid deformable bodies.

Let us now assume that the body consists of a system of n coupled harmonic oscillators and let us restrict ourselves to the analysis of small vibrations around a non-collinear equilibrium configuration q_0^μ, so that $q^\mu = q_0^\mu + \lambda x^\mu$, where λ is a small ordering parameter representing the ratio of deformation length scale to the equilibrium configuration length scale. We further choose x^μ to be both Riemann normal coordinates and normal modes of the potential, so that the potential can be written as $V(x) = \omega_{(\mu)}^2 x^\mu x_\mu / \lambda^2$, where we have implicitly assumed that $\partial_\mu^2 V = \omega_{(\mu)}^2/\lambda^4$. On the other hand, to further simplify things we use Poincaré gauge, $\mathbf{A}_\mu x^\mu = 0$, that is, in this gauge it holds $\mathbf{A}_\mu(0) = 0$. As it has been previously showed by Littlejhon and Mitchell [10], Poincaré gauge can be geometrically interpreted as choosing the gauge so that the section is the horizontal lift of radial lines emerging from the equilibrium configuration, which of course depends on the coordinates used in shape space. If Riemann coordinates are chosen, then the section is flat and it coincides with the Eckart frame. Finally, we scale $p_\mu \to p_\mu/\lambda$, $H \to \lambda^2 H$, $\mathsf{M} \to \lambda^2 \mathsf{M}$, $\mathbf{A}_\mu \to \lambda \mathbf{A}_\mu$ and $\mathbf{L} \to \mathbf{L}$, and write $R_{\mu\alpha\beta\nu}$ in terms of M and $\mathbf{B}_{\mu\nu}$[1], so that the Hamiltonian (13) up to $O(\lambda^2)$ can be expanded as

$$H = \frac{1}{2}\left(p^\mu p_\mu + \omega_{(\mu)}^2 x^\mu x_\mu\right) + \frac{1}{2}\lambda^2 (\mathbf{L} - \mathbf{S}) \cdot \mathsf{M}^{-1} \cdot (\mathbf{L} - \mathbf{S}), \tag{15}$$

[1] Namely, $2R_{\mu\nu\sigma\tau} = \mathbf{B}_{\mu\nu} \cdot \mathsf{M} \cdot \mathbf{B}_{\sigma\tau} + \mathbf{B}_{\mu[\tau} \cdot \mathsf{M} \cdot \mathbf{B}_{\sigma]\nu}$ *cf.* relation (5.61) from [9].

where M^{-1} in the above expression is evaluated at q_0^μ and we have introduced the "internal angular momentum" $\mathbf{S}$ and "shape angular momentum" S:

$$\mathbf{S} = \frac{1}{4}\mathsf{M}\cdot\mathbf{B}_{\mu\nu}S^{\mu\nu}, \quad S^{\mu\nu} = x^\mu p^\nu - x^\nu p^\mu. \tag{16}$$

The only term in the Hamiltonian (15) that couples the shape and orientation degrees of freedom is the crossed one, which is of the form $\mathbf{L}\cdot\mathsf{M}^{-1}\cdot\mathbf{S}$, and since $\mathbf{L}$ is the generator of rotations, when acting on the orientation sector of the wavefunction, it can be interpreted as an infinitesimal rotation about the unit vector $\widehat{\mathsf{M}^{-1}\cdot\mathbf{S}}$ by an angle $|\mathsf{M}^{-1}\cdot\mathbf{S}|$, which are both operators in shape space. At this point we are ready to show that a change of its shape induces a rotation of a body as the dynamical process generated by the Hamiltonian (15).

3. Three-body case

As it turns out, to "see" the cat rotating while it changes its shape, it is enough to model it as a three-body problem. In this case the shape space has dimension $3n-6=3$ and we can define $n-1=2$ Jacobi vectors as:

$$\boldsymbol{\rho}_{s1} = \sqrt{\mu_1}\left(\mathbf{r}_{s1} - \mathbf{r}_{s3}\right), \tag{17}$$
$$\boldsymbol{\rho}_{s2} = \sqrt{\mu_2}\left(\mathbf{r}_{s2} - \mathbf{R}_{s,13}\right), \tag{18}$$

where $\mu_1 = m_1m_3/m_{13}$, $\mu_2 = m_2m_{13}/M$ and $\mathbf{R}_{s,13}$ is the center of mass of subsystem $m_{13} = (m_1+m_3)$. A convenient set of coordinates in shape space is:

$$\begin{aligned} q_1 &= |\boldsymbol{\rho}_{s1}|^2 - |\boldsymbol{\rho}_{s2}|^2, \\ q_2 &= 2\boldsymbol{\rho}_{s1}\cdot\boldsymbol{\rho}_{s2}, \\ q_3 &= 2|\boldsymbol{\rho}_{s1}\times\boldsymbol{\rho}_{s2}| \geq 0, \end{aligned} \tag{19}$$

which makes it clear that the shape space is one half of $\mathbb{R}^3$. The q_3 coordinate measures the area of the body, and the plane $q_3=0$ then corresponds to collinear shapes (which have zero area) while the q_3 axis contains symmetric shapes, *i.e.*, the shapes whose inertia tensor is degenerate in the body xy plane.

For the sake of concreteness, let us adopt now the north regular gauge [9, 11] which is also an Eckart (and Poincaré) gauge for the equilibrium shape $q_0^\mu = (0,0,1)$, which corresponds to the configuration given by one mass located at each vertex of an equilateral triangle with edges of unit length. If, moreover, we consider small oscillations around equilibrium q_0^μ, *i.e.*, $q^\mu = q_0^\mu + \lambda x^\mu$, such that x^μ are Riemann and normal modes of the potential, then we can apply the the quasi-rigid bodies results obtained in the previous Section 2. Normal modes x^μ are related to the coordinates q^μ as defined in (19) according to:

$$\begin{aligned} q_1 &= -\lambda(x_1 + \sqrt{2}x_2), \\ q_2 &= -\lambda(x_2 - \sqrt{2}x_1), \\ q_3 &= (1 - 2\lambda x_3), \end{aligned} \tag{20}$$

where x_3 corresponds to the breathing mode of higher frequency $\omega_3 = \sqrt{2}$ (in natural units), while x_1 and x_2 are degenerate orthogonal modes of frequency $\omega_1 = \omega_2 = 1$.

The moment of inertia tensor for the equilibrium configuration is

$$M_{ij} = \begin{pmatrix} 1/2 & 0 & 0 \\ 0 & 1/2 & 0 \\ 0 & 0 & 1 \end{pmatrix}, \tag{21}$$

the only non trivial component of Coriolis tensor is $\mathbf{B}_{12} = 2\hat{\mathbf{z}}$ and therefore the "internal angular momentum" reduces to $\mathbf{S} = S_{12}\hat{\mathbf{z}}$, where $S_{12} = x_1 p_2 - x_2 p_1$ is the angular momentum in the plane $x_1 x_2$ of shape space. That is, the fibre bundle in this case is curved only on the plane $x_1 x_2$, which from the physical point of view can be understood because only an out-of-phase combined excitation of modes x_1 and x_2 can generate internal angular momentum.

Then, the Hamitonian takes the form:

$$H = \frac{1}{2}\left(p_1^2 + p_2^2 + x_1^2 + x_2^2\right) + \frac{1}{2}\left(p_3^2 + 2x_3^2\right) + \lambda^2\left(\mathbf{L}^2 - \frac{1}{2}L_z^2 - L_z S_{12} + \frac{1}{2}S_{12}^2\right). \tag{22}$$

As claimed, in the classical case the condition of vanishing angular momentum makes it possible to find the solutions of the equations of motion for shape coordinates independently of the orientation ones. From the previous Hamiltonian (22), it can be realized that $\dot{S}_{12} = 0$ and so S_{12} is a constant of motion. In this case $x^\mu = x^\mu(t)$ correspond to harmonic oscillations to leading order in λ:

$$x^\mu(t) = x_0^\mu \cos\omega_\mu t + \frac{p_0^\mu}{\omega_\mu}\sin\omega_\mu t, \tag{23}$$

with frequency $\omega_\mu = (1, 1, \sqrt{2})$ and $p^\mu(t) = \dot{x}^\mu(t)$. Since S_{12} is a constant of motion and the gauge field $\mathbf{A}_\mu$ is pseudo Abelian in this case, relation (11) yields:

$$R_{ij}(t) = \begin{pmatrix} \cos\left(\lambda^2 S_{12} t\right) & \sin\left(\lambda^2 S_{12} t\right) & 0 \\ -\sin\left(\lambda^2 S_{12} t\right) & \cos\left(\lambda^2 S_{12} t\right) & 0 \\ 0 & 0 & 1 \end{pmatrix}, \tag{24}$$

which represents a rotation about $\hat{\mathbf{z}}$ by an angle $\alpha(t) = -\lambda^2 S_{12} t$, such that a cyclic change of shape (23) with $x_3(0) = 0 = p_3(0)$ gives rise to $\alpha(2\pi) = -2\pi\lambda^2 S_{12}$. In Fig. 1 we show the cyclic deformation sequence that produces the rotation (24), with $S_{12}(0) = 2$ for three periods $0 \leq t \leq 3T = 3(2\pi/\omega_1)$. Clearly, if initial conditions are chosen such that $S_{12}(0)$'s sign is opposite, then the angle's sign of the resulting rotation is also opposite.

Coming back to the quantum case, it is convenient to define the annihilation and creation operators in shape space as usual $a_\mu = (x_\mu + ip_\mu)/\sqrt{2}$ and $a_\mu^\dagger$ (along this work we will work in units such that $\hbar = 1$), and analogously the annihilation and creation operators associated with angular excitations in the $x_1 x_2$ plane as $a_\pm = (a_1 \mp ia_2)/\sqrt{2}$ and $a_\pm^\dagger = (a_1^\dagger \pm ia_2^\dagger)/\sqrt{2}$, where $a_+(a_+^\dagger)$ annihilates (creates) one positive angular excitation and $a_-(a_-^\dagger)$ annihilates (creates) one negative angular excitation [13]. We also define the corresponding number operators $N_\mu = a_\mu^\dagger a_\mu$ and $N_\pm = a_\pm^\dagger a_\pm$, so that the Hamiltonian (22) can be written as:

$$H = (N+1) + \omega_3\left(N_3 + \frac{1}{2}\right) + \lambda^2\left(\mathbf{L}^2 - \frac{L_z^2}{2} - L_z S + \frac{S^2}{2}\right), \tag{25}$$

with $N = N_+ + N_-$ and $S = N_+ - N_-$. The previous Hamiltonian[2] is defined in the Hilbert space $L^2(\mathcal{C}, d^3q\ d\mathsf{R})$, with $d\mathsf{R}$ being the Haar normalized measure in $SO(3)$. It is not difficult to see that a complete set of mutually compatible operators commuting with the Hamiltonian is $\{N, S, N_3, \mathbf{L}^2, L_z, L_{sz}\}$, see [14], and then it is clear that Hamiltonian eigenstates $|n, s, n_3, l, m, k\rangle$ are labelled by the corresponding quantum numbers $n_3, n, l \in \{0, 1, 2, \ldots\}$, $m, k = -l, -l+1, \ldots, l-1, l$ and $s = -n, -n+2, \ldots, n-2, n$. Now, the vanishing angular momentum condition implies that the state of the system has to be such that $l = 0$ and thus

[2] We should include the geometric potential $V_2 = D^{-1/4}\partial_\mu(g^{\mu\nu}\partial_\nu D^{1/4})/2$ with $D = \det(\mathsf{M})\det(g_{\mu\nu})$ in the quantum Hamiltonian, however in the small oscillations regime it reduces to a constant that we neglect.

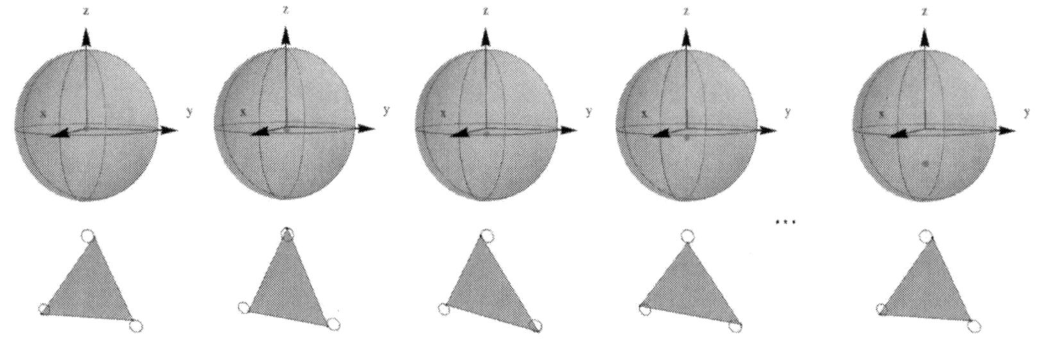

Figure 1. Sequence of the orientation change produced by a cyclic deformation in the plane x_1x_2 in shape space, for the three body problem. The figure at the top represents $SO(3)$ as the 3-ball of radius π with identified antipodes at the boundary $S^2(\pi)$, any point in this space corresponds to a rotation in the axis-angle representation, the vector joining the point to the origin corresponds to the axis of rotation while its magnitude corresponds to the angle of rotation. We present different configurations for the different values of time $t = 2\pi n/4$, with $n = 0, 1, 2, 3, \ldots, 12$, $S_{12}(0) = 2$ and $\lambda = 0.2$: at the top we show the (red) dot that represents the orientation configuration in the fibre $SO(3)$ and at the bottom the corresponding body's shape (for illustrative purposes the body's deformation was magnified by a factor of four).

$m = 0$, but since the term in the Hamiltonian that produces the rotation is $\mathbf{L}\cdot\mathsf{M}^{-1}\cdot\mathbf{S} = -\lambda^2 L_z S$, then when acting on a state satisfying the condition the only rotation it can generate is the identity. In order to see a more general rotation we shall impose the *weak* vanishing angular momentum condition $\langle\mathbf{L}\rangle = 0$.

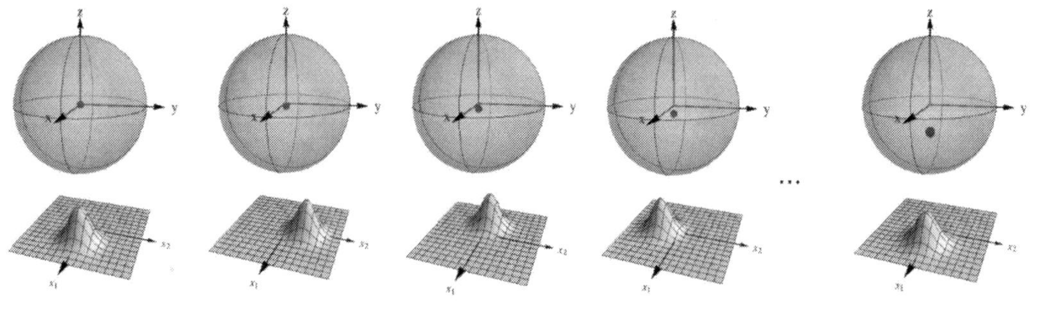

Figure 2. Sequence of the orientation change produced by a cyclic deformation in the plane x_1x_2 in shape space for the three body problem, in the quantum regime. The figure at the top represents $SO(3)$ as in Fig. 1. We present the time-evolution of the probability density associated with the initial wavefunction $\Psi(\mathsf{R}, q^\mu)$ given by (26) for the different values of time $t = 2\pi n/4$, with $n = 0, 1, 2, 3, \ldots, 12$, $\lambda = 0.2$ and $\alpha_{(+,-)} = (\sqrt{2}, 0)$. At the top we show the surface of constant (in orientation coordinates) orientation reduced probability density $P_t^O(\mathsf{R}) = .99 \max(P_t^O)$ that represents the orientation configuration in the fibre $SO(3)$ and at the bottom the reduced probability in the shape space $P_t^S(x_1, x_2)$.

Let us thus consider a more general state $|\Psi\rangle$ satisfying the weak angular momentum condition

$\langle\Psi|\mathbf{L}|\Psi\rangle = 0$, with Ψ given by:

$$\Psi(\mathsf{R}, q^{\mu}) = \frac{1}{\sqrt{\pi}}\psi_{\alpha_+,\alpha_-}(x_1, x_2)\psi_0(x_3)\Phi(\phi), \tag{26}$$

where $\psi_{\alpha_+,\alpha_-}(x_1, x_2) = \prod_{i=1}^{2} \mathrm{e}^{-\frac{1}{2}(x_i-\langle x_i\rangle)^2}\,\mathrm{e}^{i\langle p_i\rangle\left(x_i-\frac{1}{2}\langle x_i\rangle\right)}$ is a coherent state with $\langle x_i\rangle = \sqrt{2}\,\mathrm{Re}(\alpha_i)$ and $\langle p_i\rangle = \sqrt{2}\,\mathrm{Im}(\alpha_i)$ as usual, with $\alpha_1 = (\alpha_+ + \alpha_-)/\sqrt{2}$ and $\alpha_2 = i(\alpha_+ - \alpha_-)/\sqrt{2}$ for $\alpha_\pm \in \mathbb{C}$, $\psi_0(x_3)$ is the breathing mode's ground state, and $\Phi(\phi) = \Phi(\mathsf{R}_{\hat{\mathbf{n}}}(\phi)) = N_\Phi\,\mathrm{e}^{\cos\phi/2}$ is a wavefunction in $SO(3)$ in the axis-angle representation $(\hat{\mathbf{n}}, \phi)$. The state $\Phi(\phi)$ describes a system that is most probably oriented in coincidence with the body frame.

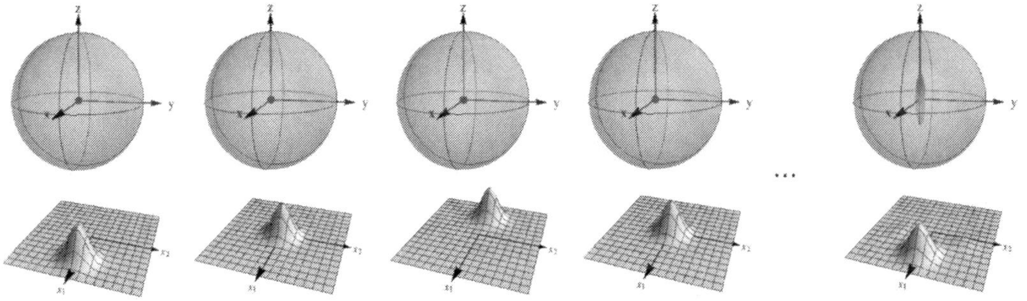

Figure 3. Sequence of the orientation change produced by a cyclic deformation in the plane x_1x_2 in the shape space for the three body problem, in the quantum regime. The figure at the top represents $SO(3)$ as in Fig. 1. We present the time-evolution of the probability density associated with the initial wavefunction $\Psi(\mathsf{R}, q^\mu)$ given by (26) for the different values of time $t = 2\pi n/4$, with $n = 0, 1, 2, 3, \ldots, 20$, $\lambda = 0.2$ and $\alpha_{(+,-)} = (\sqrt{2}, -\sqrt{2})$. At the top we show the surface of constant (in orientation coordinates) orientation reduced probability density $P_t^O(\mathsf{R}) = .99\max(P_t^O)$ that represents the orientation configuration in the fibre $SO(3)$ and at the bottom the reduced probability in the shape space $P_t^S(x_1, x_2)$.

The time evolution of the wavefunction (26) can be computed directly, though no completely trivially. The Hamiltonian (15) can be separated into three terms, $H_S = (N+1) + \omega_3(N_3 + 1/2) + \lambda^2 S^2/2$ that only depends and acts on the shape degrees of freedom, $H_O = \lambda^2(\mathbf{L}^2 - L_z^2/2)$ that is basically the kinetic energy of the rotor, and $H_I = -\lambda^2 L_z S$ that couples R and q^μ. All three terms commute with each other, and thus the complete time evolution can be decomposed into the action of the three corresponding unitary transformations. The first and the second evolve the shape and orientation sectors of the wavefunction, respectively and in an independent way. The action of $\mathrm{e}^{-i\lambda^2 H_O t}$ can be computed by expanding the function $\Phi(\phi)$ in the H_O's eigenvectors $\{D^l_{km}\}$, while that of $\mathrm{e}^{-iH_S t}$ can be expressed in two steps: one corresponding to the λ-independent term which just shifts the coherent state label to $\alpha_\pm(t) = \alpha_\pm\,\mathrm{e}^{-it}$, and one associated with the λ-dependent term of H_S which does not give rise to a coherent state, but can be approximated by $\mathrm{e}^{-i\lambda^2 S^2 t/2} \approx 1 - i\lambda^2 S^2 t/2$, since $\lambda^2\langle S^2\rangle T/2$ is small. The third coupling term on the other hand is the most interesting one, when it is applied to the wavefunction it produces a coherent state labelled by $\alpha^k_\pm(t) = \alpha_\pm(t)\,\mathrm{e}^{\pm i\lambda^2 kt}$, where k is the eigenvalue of L_z, *i.e.*,

$$\Psi_t(\mathsf{R}, q^\mu) \approx \frac{1}{\sqrt{\pi}}\,\mathrm{e}^{-i\frac{\omega_3 t}{2}}\,\psi_0(x_3)\sum_{l,k} c^l_k \varphi^k_t(x_1, x_2)\,\mathrm{e}^{-i\lambda^2\left(l(l+1)-\frac{k^2}{2}\right)t}\,D^l_{kk}(\mathsf{R}), \tag{27}$$

where $\varphi^k_t(x_1, x_2) = (1 - i\lambda^2 S^2 t/2)\psi_{\alpha^k_+(t),\alpha^k_-(t)}(x_1, x_2)$ and $c^l_{k,m} = \int D^{l\,*}_{k,m}(\mathsf{R})\Phi(\mathsf{R})d\mathsf{R} = c^l_k\delta_{km}$ which turns out to vanish for $m \neq k$. To illustrate the system's change in orientation as its shape changes due to time evolution, we project the probability density $P_t(\mathsf{R}, q^\mu) = |\Psi(\mathsf{R}, q^\mu)|^2$ onto the section $\mathcal{S}$ and onto the fibre $SO(3)$, and thus we define the reduced probability density in shape space by integrating out the orientation degrees of freedom, $P_t^S(q^\mu) = \int P_t(\mathsf{R}, q^\mu) d\mathsf{R}$ and the orientation reduced probability density $P_t^O(\mathsf{R}) = \int P_t(\mathsf{R}, q^\mu)\sqrt{g} d^{3n-6}q$. In Fig. 2 we show the quantum version of the cyclic deformation and the corresponding rotation sequence that is produced by the Hamiltonian as time increases with $\alpha_{(+,-)} = (\sqrt{2}, 0)$, which implies that $\langle S \rangle = 2$, just as in the classical situation shown in Fig. 1. It can be noted that the system rotates clockwise as its shape changes in a similar fashion than its classical counterpart, however the angle of the rotation after three complete periods in SS seems to be slightly smaller than in the classical case. We can notice as well that the surface of most probable orientation in this case corresponds to an sphere whose radius grows as time increases. Both effects are related to the dispersion of the wavefunction in orientation space. And also just as in the classical case we could choose (26) such that $\alpha_{(+,-)} = (0, \sqrt{2})$, so that $\langle S \rangle = -2$ and thus the body would rotate counter clockwise. Furthermore, we could also fix as our initial state a wavefunction like (26) with $\alpha_{(+,-)} = (\sqrt{2}, \sqrt{2})$ or equivalently with $\alpha_{(1,2)} = (2, 0)$ which corresponds to a coherent state associated with a linear excitation along x_1 and therefore $\langle S \rangle = 0$. In contrast to the classical case, this system's state can produce a non vanishing probability of changing its orientation in spite of enclosing a null area, see Fig. 3. The cigar-shaped probability distribution shown in the latter figure describes a cat whose orientation is rotated, with respect to the initial one, by all angles (approximately) within the interval $[-\pi/2, \pi/2]$ about the z-axis with equal probability. So that if the cat is initially oriented, say horizontally, it would end landing on any part of its body located between its feet and its head, continuously and with equal probability, becoming a “continuous Schrödinger cat”.

4. Conclusions

We presented a quantum description of the free-falling cat problem in terms of geometrical methods applied to the n-body problem. In concrete, we answered affirmatively to our initial motivating question: can the quantum cat control its muscles so that it lands on its feet as its classical analogue? The specific scenario which more closely resembles the corresponding classical analogue takes place when the cat's wavefunction is a coherent state in both, the shape and orientation sectors; furthermore, we showed that more exotic purely quantum scenarios may happen, including one describing a “continuous Schrödinger cat”. This result might remind us that applying quantum mechanics without inquiring on its interpretation and foundations may have dangerous, even fatal, consequences, a message that is commonly found in professor Mielnik's work.

Acknowledgments

The authors wish to acknowledge partial financial support from DGAPA-UNAM grant IN114712 and CONACyT grant 103486. HHC thanks Nora Bretón, David Fernández and Piotr Kielanowski for the invitation to present this work at the “International Conference Quantum Control, Exact or Perturbative, Linear or Nonlinear to celebrate 50 years of scientific path of Professor Bogdan Mielnik”, and specially Bogdan Mielnik for interesting and inspiring discussions.

References

[1] C. Frohlich. *Scientific American*, **263**, 155-64, 1980.
[2] T. R. Kane and M. P. Scher. *J. Solid Struct.* **5**, 663-667, 1969.
[3] C. Frohlich. *Am. J. Phys.* **47**, 583-92, 1979.
[4] A. Guichardet. *Ann. Inst. H. Poincaré, Phys. Theor.*, 40/3:329-342, 1984.
[5] B. Mielnik, *Rep. Math. Phys.*, 12, 331, 1977.
[6] B. Mielnik, *J. Math. Phys.*, 27, 2290, 1986.

Journal of Physics: Conference Series **624** (2015) 012007 doi:10.1088/1742-6596/624/1/012007

[7] D.J. Fernandez and B. Mielnik, *J. Math. Phys.* 35, 2083, 1994.
[8] B. Milenik. *J. Phys. Conf. Ser.*, 512, 012035, 2014.
[9] R. G. Littlejhon and M. Reinsch. *Reviews Mod. Phys.* Vol. 69, No. 1, 1997.
[10] R. G. Littlejhon and K. A. Mitchell. Gauge theory of small vibrations in polyatomic molecules, *Geometry, Mechanics, and Dynamics* ed Newton P, Holmes P and Weinstein A (Springer New York) pp 407428 ISBN 978-0-387-95518-6 URL http://dx.doi.org/10.1007/0-387-21791-6 13.
[11] R. G. Littlejohn, K. A. Mitchell, M. Reinsch, V. Aquilanti and S. Cavalli. *Phys. Rev. A*, Vol 58, No. 5, 3705-3717, 1998.
[12] R. G. Littlejohn, K. A. Mitchell, M. Reinsch, V. Aquilanti and S. Cavalli. *Phys. Rev. A*, Vol 58, No. 5, 3718-3738, 1998.
[13] Messiah, *Quantum Mechanics*, Ed. Dover.
[14] L. D. Landau and L. M. Lifshitz, Quantum Mechanics, Third Edition: Non-Relativistic Theory (Volume 3), Pergamon Press, 1981.

Diffractive Physics with ALICE at the LHC: the control of quantum collisions

G Herrera Corral
Physics Department, CINVESTAV
PO Box 14 740, Mexico D.F. 07360, Mexico
E-mail: gherrera@fis.cinvestav.mx

Abstract. We report on the status of the construction and installation of a new detector in the forward region of the ALICE experiment at the LHC. This detector will allow the study of processes with gaps at larger rapidity than those presently covered. A setup of two stations called AD (stands for ALICE Diffractive) on the right and on the left of the interaction point enhances significantly the efficiency to study diffractive physics and photon induced processes.

1. Introduction: The Large Hadron Collider

The Large Hadron Collider (LHC) [1] accelerates protons as well as heavy ions in a 27 km long ring located in the European Center for Nuclear Research (CERN) at Geneva, Switzerland. The production and acceleration of protons and ions starts in LINAC 2 and LINAC 3, respectively. The protons accelerated in LINAC 2 are injected with an energy of 50 MeV into a Proton Synchrotron Booster where they reach an energy of 1.4 GeV. The last step for the beam before LHC is the Super Proton Synchrotron (SPS) that was modified to deliver a high brightness proton beam. It takes 26 GeV protons from the PS and brings them to an energy of 450 GeV before they are transferred to the LHC. The accelerators complex at CERN is shown in Fig. 1.

The LINAC 3 produces 4.2 MeV/u lead ions. It uses the Low Energy Injector Ring (LEIR) as a storage and cooler unit which in turn, provides the ions to the Proton Synchrotron with an energy of 72 MeV/nucleon. Ions are further accelerated by the Proton Synchrotron and the Super Proton Synchrotron before they are injected into the LHC where they reach an energy of 2.76 TeV/nucleon.

The total cross section of proton-proton interaction at 7 TeV could be inferred from hadronic cross section measurements at lower energies [2]. It would be around 110 mbarn and it corresponds to about 60 mbarn of inelastic scattering processes. The LHC reaches a luminosity of 10^{34} sec^{-1}cm^{-1}, which means that the inelastic interaction rate is

$$\text{rate} = 10^{34} \times 60 \times 10^{-3}\,\text{barn} \times 10^{-24}\,\frac{\text{cm}^2}{\text{barn}}\frac{\text{collisions}}{\text{s}}.$$

Out of these events, approximately 150×10^6 are diffractive processes. It is therefore natural that the ALICE collaboration becomes interested in collecting most of the hundredths of millions of diffractive collisions per second that carry important information about non perturbative QCD phenomena.

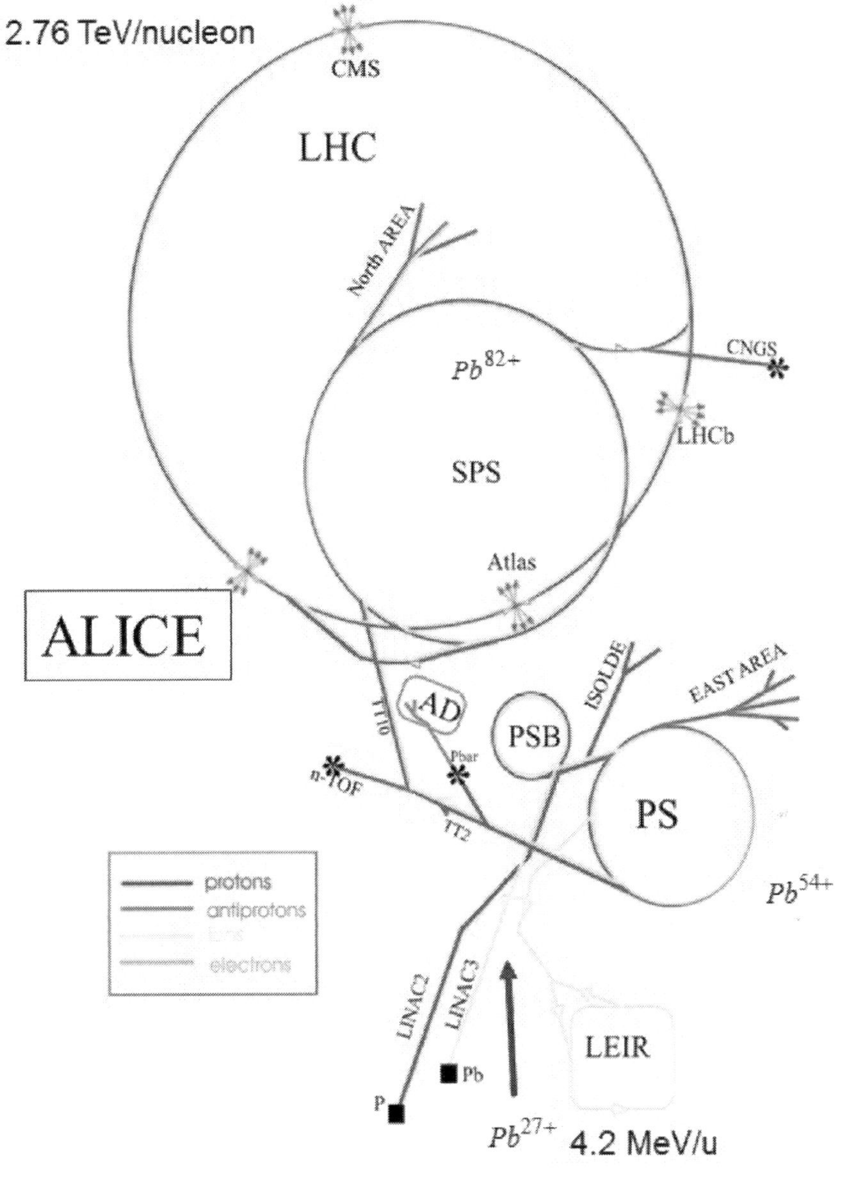

Figure 1. Accelerators at CERN, © CERN.

During the fall 2009 bunches of protons were injected into the LHC ring. During the start-up phase, first collisions with protons at 900 GeV were observed. In 2010 the beam was ramped to 3.5 TeV, it remained at this energy in 2011 while the number of bunches and their intensity was increased. By the middle of 2011 the maximum number of bunches, i.e. 1380, was achieved for a spacing of 50 ns between bunches. The performance of the LHC was then increased by reducing the emittances of the beams that were delivered by the injectors. In 2012 the beam energy was increased to 4 TeV and the bunch spacing was made to stay at 50 ns. A remarkable fact of the LHC performance during 2011 and 2012 was the high bunch intensity and lower than nominal emittance. The bunch intensity obtained was 150% of nominal. In 2012 the protons per bunch reached 1.7×10^{11} while the nominal design value is 1.15×10^{11}. The normalized emittance in the collision was 67% of the nominal values, which represents a significant improvement too. The LHC proved to be able to handle these brighter beams in terms of beam-beam effects. For the experiments this meant higher pile-up, an issue that imposes a challenge for the readout, analysis and storage speeds. The nominal mean number of events per bunch crossing at design values is 19 but, in 2012, with the parameters above, this number became 37. All of these was successfully managed by the experiments that announced the Higgs discovery in July 2012.

ALICE recorded data on proton-proton interactions during 2012 at a luminosity of $5 \times 10^{30}\,\mathrm{cm}^{-2}\mathrm{s}^{-1}$ with collisions between enhanced satellites bunches and the main bunches. Since the beginning of the LHC operation there have been two runs of heavy ion beams, one in 2010 for commissioning and analysis and the second one in 2011 with an instant luminosity above the nominal. These two heavy ion runs provided an integrated luminosity of 9.7 $\mu\mathrm{b}^{-1}$ (2010) and 166 $\mu\mathrm{b}^{-1}$ (2011) with a centre-of-mass energy of $\sqrt{S_{NN}} = 2.76$ TeV. In addition, in 2013 a proton-lead run at $\sqrt{S_{NN}} = 5.02$ TeV, provided a 30 nb^{-1} integrated luminosity. The data collected in all these runs is being analyzed and has already produced very interesting physics results. In 2015 the LHC will operate at a higher energy. The goal is to run the physics programme at 13 TeV in the centre of mass. The accelerator recommissioning will start in March 2015 with a 50 ns bunch spacing to reduce it to 25 ns later on during the year. In November 2015 a 4 week long heavy ions run is foreseen. ALICE will be collecting the data of lead-lead interactions at the highest energy ever seen.

2. The ALICE experiment

ALICE is a multipurpose detector that exploits the unique potential of lead-lead collisions at the LHC energies [3]. ALICE also studies proton-proton interactions in research topics where it is competitive with other LHC experiments. Most of all, the study of proton-proton reactions gives the possibility to compare a number of phenomena that occur in lead-lead interactions. The ALICE detector provides identification of hadrons, electrons, muons and photons produced in the collisions. It measures also their trajectories, momentum and energy in specific regions covered by various sub detectors (Fig. 2).

One of the main differences of ALICE with the other experiments at the LHC is particle identification. ALICE uses all known techniques for particle identification:

- The Inner Tracking System is formed by six cylindrical layers of silicon detectors (Silicon Pixel, Silicon Strip and Silicon Drift) that provides not only a precise determination of the main and secondary vertices of unstable particles but also energy loss measurements that separates particle species in the central pseudo-rapidity region.
- The Time Projection Chamber is used as the main tracking detector for charged particles. It is the biggest Time Projection Chamber ever built and consists of 88 m^3 cylinder filled with gas and divided in two drift regions by a central electrode. The gas ionization of charged particles traversing the volume provides energy loss measurements for particle identification.
- The Time-Of-Flight system combines time measurement with the momentum and track length measured in the tracking detectors to identify particles. It separates pions and kaons with momenta lower than 2 GeV/c and protons with momenta lower than 4 GeV/c. It consists of a multigap resistive plate chamber with a double stack of 0.5 mm glass.
- The Transition Radiation Detector is the main electron detector. It identifies electrons of high momenta ($p_t > 1\,\mathrm{GeV/c}$). The detector consists of several modules, with radiator and a drift chamber which is operated with a mixture of Xe and CO_2 , that are surrounding the interaction point in the barrel region.
- The High Momentum Particle Identification detectors is a ring imaging Cerenkov detector. It consists of seven modules 1.5×1.5 meter each mounted in an independent support cradle.

It provides particle identification in the 1 to 5 GeV/c momentum range complementing the barrel detectors.

In the forward direction on one side of the interaction point, a set of tracking chambers inside a dipole magnet measures muons. An absorber stops all the products of the interaction except for the muons which travel across and reach the tracking chambers. The whole structure is called muon arm. The barrel region of ALICE is located inside a big solenoid that provides a magnetic field of 0.5 T. The central tracking and particle identification systems cover $-0.9 < \eta < 0.9$. Electrons and photons are measured in the central part: photons are measured in PHOS which is a high resolution calorimeter located 4.4 meters below the interaction point. PHOS is made of lead-tungstate (PWO) crystals which have a high light output. It covers $-0.12 < \eta < 0.12$ in pseudo-rapidity and 100 degrees in azimuth for a total area of 8 m^2. An Electromagnetic Calorimeter covers the central barrel. EMCAL is made of plastic scintillators and lead plates in a sandwich array which is read out with wavelength-shifting optical fibers. Avalanche photodiode sensors are used to convert light into an electric signal. A Forward Multiplicity Detector (FMD) consists of several stations of silicon strip detectors on both sides of the interaction point. A Zero Degree Calorimeter (ZDC) covers the very forward region providing information on the charge multiplicity and energy flow. It is a spaghetti calorimeter made of a stack of heavy metal plates with grooves that host quartz fibers. A honeycomb proportional counter for photon multiplicity measurements (PMD Photon Multiplicity Detector) is located in the forward direction on one side of the ALICE detector. The trigger system is complemented by a high level trigger (HLT) system which makes use of a computer farm to select events after read-out. In addition, the HLT system provides a data quality monitoring.

A Mexican group from CINVESTAV and the National Autonomous University of Mexico designed and built the V0A detector which is one of the two detectors that form the V0 system. The V0 system is used as the main interaction trigger.

The V0A is installed at a distance of 3.3 meters from the interaction point. It is an array of 32 cells of plastic scintillators, distributed in 4 rings forming a disc with 8 sectors. A detailed description of the V0 system can be found in [4]. The V0 system is part of the baseline of ALICE. It provides level 0 trigger signals, beam gas suppression, a default centrality measurement, and the event reaction plane in the ion-ion collisions as well as the on-line luminosity measurement.

A second group of researchers from CINVESTAV, the University of Puebla, the National Autonomous University of Mexico and University of Sinaloa designed and built the Cosmic Ray Detector (ACORDE). It is an array of 60 scintillator counters located in the upper part of the ALICE magnet [5]. The Cosmic Ray Detector generates a single muon trigger to calibrate the Time Projection Chamber and other components of ALICE. It also generates a multi-muon trigger to study cosmic rays with the help of tracking systems like the ITS and the TPC.

After several months shutdown in 2013 and 2014, the ALICE experiment will take data with both protons and lead ions in 2015. Thereafter, in 2016-17, the experiment will collect more data on Pb-Pb interactions at $\sqrt{S_{NN}} = 5.5$ TeV, just before the long break in 2018.

ALICE would complete the initial program on Pb-Pb before the shutdown in year 2022. After year 2022 an upgraded detector will be able to take data at a minimum bias rate of 50 kHz for Pb Pb collisions. The upgraded detector will have a better determination of the main vertex and a tracking system for low transverse momentum. The upgraded ALICE experiment will preserve the particle identification capabilities.

The upgrade of ALICE includes a smaller radius beam pipe, a new inner tracking system as well as a high rate read-out for the Time Projection Chamber, Transition Radiation Detector, Time-of-Flight, Electromagnetic Calorimeter, Muon Arm and Trigger Detectors. The V0 detector will be replaced with a new design to collect more light in order to define more precisely the nominal region of interaction and to achieve a time resolution of the order of 100-200 picoseconds. The new detector is planned to be installed in 2018 during the long shutdown.

In order to collect 10 nb^{-1}, several years of data taking after 2022 are needed.

In addition to the research on heavy ion interactions and proton-proton collisions, ALICE pursues a diffraction physics program by extending the detector efficiency to smaller diffractive masses for single and double diffraction events. A new detector system increases the detection of gaps in the pseudo-rapidity distribution of particles produced in the collision. This improves the capabilities of ALICE to study central diffractive processes.

3. Diffractive Physics

Quantum Chromo-Dynamics is the theory of strong interactions. Though it has been successful to describe many processes the strong coupling constant limits the success of the theory for nonperturbative scattering. In the strong interaction gluons carry the colour charge. However, gluons may group in a way that their colour charges add up to neutralize. A neutral object like this would behave in a different way than single gluons with colour charge. The dynamics of such colourless object is at present poorly understood.

Diffractive reactions are driven by the strong force. The high energy proton collisions of this kind are characterized by the exchange of a neutral state of gluons, called pomeron. It leaves a gap in the rapidity distribution of the particles produced. The interaction has the quantum numbers of the vacuum. In such reactions, the colliding protons deflect only a little and continue their trajectory very close to the original direction.

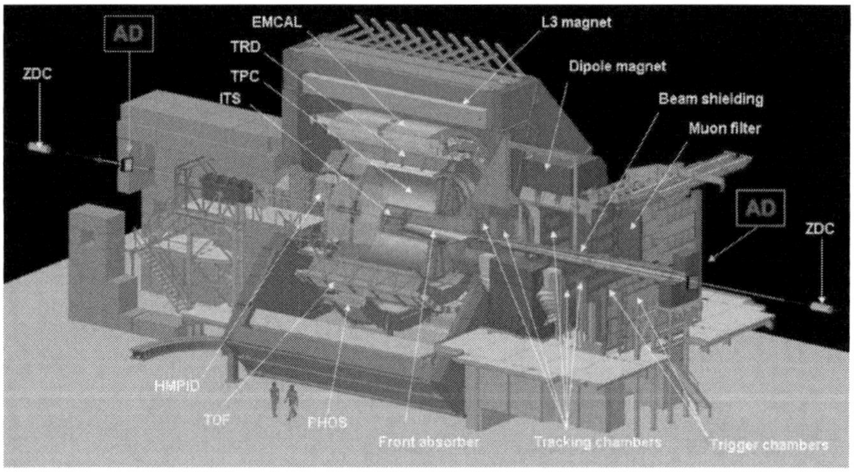

Figure 2. The ALICE detector systems and the new AD device, © CERN-ALICE.

The total cross section for proton-proton collisions is the sum of elastic, non-diffractive and diffractive interactions (Fig. 3). The cross section at the energy in the centre of mass at which the LHC will run in 2015 is unknown. The scaling with the energy of the cross sections for the different processes is model dependent.

ALICE has a forward muon spectrometer at a rapidity region $-4.0 < \eta < -2.5$ which is not covered by the experiments CMS or ATLAS. The muon arm of ALICE may give us the possibility to study J/ψ production in single diffractive events. A measurement of J/ψ production in ultra-peripheral Pb Pb collisions at $\sqrt{S}_{NN} = 2.76$ TeV center-of-mass energy has been published [6]. A measurement of inelastic, single- and double-diffraction cross section in p-p [7], has been reported too. The experiments ATLAS and CMS at the LHC have also made proposals for installing forward detectors for diffractive studies [8]. These experiments, however, are not suited to access a very low transverse momentum of centrally produced tracks. ALICE

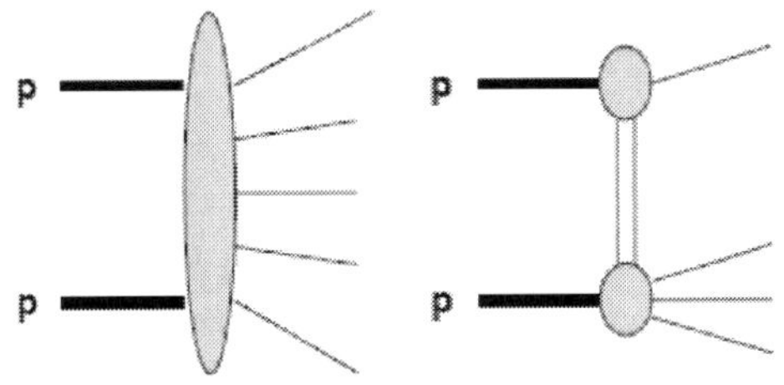

Figure 3. Left: non diffractive collision with colour exchange. Right: Single diffractive interaction with no colour exchange.

has excellent particle identification capabilities and can resolve low transverse momentum tracks, i.e., ALICE is in a unique position to study soft and hard diffractive events at the LHC. We have already installed two additional detectors around the interaction point in order to improve the efficiency to tag diffractive events [9]. The detector system could be easily extended to four stations of scintillation counters. In order to cover the spectra of small diffractive masses, as well as to enhance the sensitivity of the detector to tag diffractive events, a pseudo-rapidity coverage beyond 5 units is needed.

In addition, the high flux of photons associated with electromagnetic fields of lead ions will produce high luminosity photon-photon collisions. The physics of photon interactions spans a big variety of QED and QCD processes like lepton pair and jet production. ALICE would be able to undertake a review of resonance production in Central Diffractive events. In previous experiments [10], unexpected structures appeared but they were not fully understood. The absence of the ρ^0 resonance may be taken as indication of double Pomeron exchange. It would be very interesting to further study production mechanisms in this channel. By analysing vector meson production, Odderon signatures can be studied [11], since they might be the result of Photon-Odderon and/or Pomeron-Odderon exchange. Some theoretical developments based on exclusive production, predict new physics [12]. ALICE is well suited to explore these predictions.

The experimental study of diffractive interactions is very challenging since most of the time the protons remain in the beam pipe after a soft collision. The very low transverse momentum of the products of such collisions gives rise to technical difficulties to measure and to characterize the reaction. The incorporation of new detectors placed far from the interaction point will enhance significantly the efficiency to tag diffractive and low transverse momentum events.

4. ALICE Forward Hodoscopes - AD

During Run 1 in 2012 a couple of pad stations on both sides of the interaction point were installed. They were successfully used as a beam radiation monitor. The system was capable of detecting minimum ionizing particles. It measured background and provided an online determination of luminosity. The asynchronous read out of charge deposited in the detector worked rather well. For Run 2 we decided to integrate the detectors to the read-out of ALICE. The array was completely redesigned and the final position changed. One of the old stations will remain in the previous position at a distance of 8 meters from the interaction point on the A side. It will continue working as part of the beam diagnostic system.

For the data taking in Run 2 of the LHC, the ALICE collaboration installed, in the forward region, two small detectors made of scintillation plastic. One of the two stations is located on the C side, at 20 meters from the interaction point. This station is inside the LHC tunnel. The second station is on the opposite side (A side), at 18 meters from the interaction point.

The AD device on the A side covers a pseudo rapidity interval: $4.8 < \eta < 6.3$ while the C-side detector covers: $-7.0 < \eta < -4.9$. With this new system, the ALICE sensitivity to diffractive masses will be close to the diffraction threshold of 1.08 GeV/c^2 which corresponds to the sum of proton and pion masses.

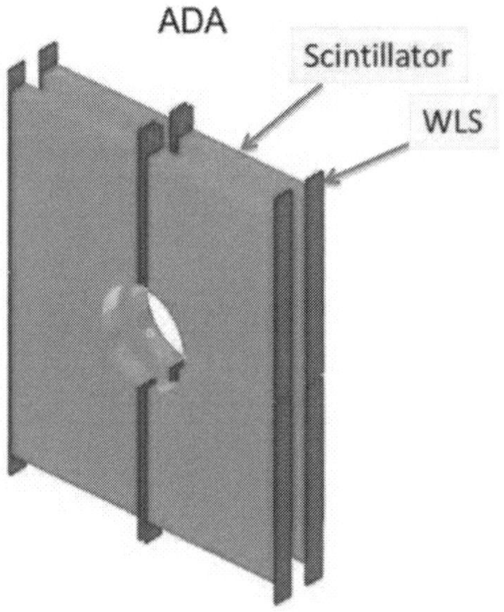

Figure 4. Design of the ALICE Diffractive Detector (AD). The WLS bars transfer the light to transparent optical fibers connected to photo-mulitplier tubes, © CERN.

Figure 5. Proto 0 of the ALICE Diffractive Detector: AD. The prototype was tested in a pion beam of 7 GeV provided by Proton Synchrotron of CERN in October 2014.

Each AD detector consists of 8 cells of scintillation plastic of about 22×22 cm and 2.5 cm thick each, arranged around the beam pipe in two layers (Fig. 4). The light produced by BC404 plastic is collected by Wave Length Shifting bars attached, but not glued, on two sides of each cell. The WLS bars transfer the collected light to a bunch of 96 transparent optical fibres per bar, that route the light to the PMTs (outside the tunnel for the C-side). The light is converted into an electric pulse by a fine mesh PMT from Hamamatsu R5946 (hybrid assembly H6153-70). The final prototype of the detector can be seen in Fig. 5. The signal from the PMT is sent to an electronic card which delivers two signals: one, amplified by a factor of 10 and clamped at about 300 mV which is used for timing measurement, the second, direct signal, is used for charge integration. The shoe box with this electronics are several meters away from the detector close to the Front End Readout electronics. The Front End Electronics provides signals for the level 0 trigger of ALICE. The readout electronics is the one presently used in the V0 detector [4]. The trigger signals of the AD counters will expand the pseudo-rapidity coverage of the present Minimum Bias trigger which is used in multiplicity studies. Moreover it will be possible to trigger on charge deposition in the two AD detectors providing an extended centrality trigger in both Pb-Pb and proton-Pb collision studies.

Acknowledgements

I want to thank the organizers of this International Conference in honour of Bogdan Mielnik for inviting me to deliver this talk. A Conference like this gives me the opportunity to express my gratitude to him for all what I have learned being close to him during many years. Having Bogdan as advisor of my postgraduate studies left a mark in my education as a physicist but having him as a friend marked my life. In particular, the celebration of 50 years of Bogdan's scientific career is an occasion for me to say that it was an honour to work with one of the most distinguished physicists in our country. I feel privileged to meet one original mind and a free spirit.

References

[1] L. Evans, Eur. Phys. J. C34 (2004)57.
[2] Review of Particle Physics, Particle Data Group, Chinese Physics C 38(2014)090001.
[3] ALICE Collaboration, J. of Phys. Nucl. Part. Phys. G30 (2004)1517-1763 J. of Phys. Nucl. Part. Phys. G32 (2006)1295-2040.
[4] Y. Zoccarato. W. Tromeur, S. Aguilar, R. Alfaro, E. Almaraz Avina, A. Anzo, E. Belmont, C. Cheshkov, B. Cheynis, C. Combaret, G. Contreras, E. Cuautle, L. Ducroux, L. Gonzlez Trueba, V. Grabski, J.Y. Grossiord, G. Herrera Corral, A. Martinez, L. Massacrier, A. Menchaca-Rocha, L.M. Montao, G. Paic, C. Prez, A. Sandoval, R. Tieulent, L. Valencia Paolomo, S. Vanzetto Nucl. Instrum. Meth. A626-627(2011)90-96.
[5] A. Fernandez, E. Gmez, G. Herrera, R. Lpez, I. Len-Monzn, M.I. Martnez, C. Pagliarone, G. Paic, S. Romn, G. Tejeda, M.A. Vargas, S. Vergara, L. Villaseor, A. Zepeda Nucl. Instr. and Meth. A572 (2007)102-103.
[6] ALICE Collaboration, (B. Abelev et al.), Phys. Lett. B718 (2013) 1273-1283.
[7] ALICE Collaboration, (B. Abelev et al.), Eur. Phys. J. C73 (2013) 6, 2456.
[8] C. Royon, arXiv:1008.3207 hep-ex, 18th International Workshop on Deep Inelastic Scattering and Related Subjects (DIS 2010), Florence Italy, 19-23 Apr. 2010.
[9] M. Bombara, E. Calvo, V. Canoa, Y. Delgado, D. Evans, A. Fernandez, A. Gago, E. Garcia, R. Gomez Jimenez, P. Gonzalez Zamora, G. Herrera Corral, E. Kryshen, I. Kralik, C. Lazzeroni, P. Ladron de Guevara, I. Leon Monzon, R. Lietava, M. I. Martinez, Z.L. Matthews, A. Menchaca, S. Navin, V. Nikulin, J. Nystrand, C. Perez, M. Poghosyan, R. Schicker, J. Schukraft, J.D. Tapia Takaki, O. Villalobos Baillie, M. Zhalov, ALICE-INT-2010-014 version 1.0
[10] T. Akesson, et al. (Axial Field Spectrometer Collaboration), Phys. Lett. B 133 (1983) 268.
[11] A. Schafer, L. Mankiewicz and O. Nachtmann, Phys. Lett. B 272 (1991) 419; A. Bzdak, L. Motyka, L. Szymanowski and J.R. Cudell, Phys. Rev. D 75 (2007)094023.
[12] L. A. Harland-Lang, V. A. Khoze, M. G. Ryskin and W. J. Stirling, Eur. Phys. J. C65 (2010) 433 -448; R.S. Pasechnik, A. Szczurek and O. V. Teryaev, Phys. Rev. D 78 (2008) 014007.

Journal of Physics: Conference Series **624** (2015) 012009 doi:10.1088/1742-6596/624/1/012009

Quantum control using genetic algorithms in quantum communication: superdense coding

Francisco Domínguez-Serna[1], **Fernando Rojas**[2]

[1] Posgrado en Física de Materiales, Centro de Investigación Científica y de Educación Superior de Ensenada, Ensenada 22890, Baja California, México

[2]Centro de Nanociencias y Nanotecnología Departamento de Física Teórica, Universidad Nacional Autónoma de México, Ensenada, Baja California, 22860 México

E-mail: `francisco.ds@gmail.com`

Abstract. We present a physical example model of how Quantum Control with genetic algorithms is applied to implement the quantum superdense code protocol. We studied a model consisting of two quantum dots with an electron with spin, including spin-orbit interaction. The electron and the spin get hybridized with the site acquiring two degrees of freedom, spin and charge. The system has tunneling and site energies as time dependent control parameters that are optimized by means of genetic algorithms to prepare a hybrid Bell-like state used as a transmission channel. This state is transformed to obtain any state of the four Bell basis as required by superdende protocol to transmit two bits of classical information. The control process protocol is equivalent to implement one of the quantum gates $\hat{X}$, $\hat{Z}$ $\hat{Y}$ in the charge subsystem. Fidelities larger than 99.5 % are achieved for the hybrid entangled state preparation and the superdense operations.

1. Introduction

Quantum control (QC) aims to systematically manipulate one or more control functions to drive the quantum dynamics of a system towards a specific set of desired final states and produce the quantum information processing task [1, 2]. QC has been extensively used to control dynamics of physical systems applied to quantum computation and communications [3–6]. Genetic Algorithms(GA), on the other hand, are a very active area of QC [7] and optimization resource have proven convergence without restriction of any range of possible control designs [8]. GA are an analogy to the evolution of living organisms [9] of mutation.

Quantum superdense code (QSC) is one of the most well known applications of quantum information as proposed by Bennet et. al [10], in which the information of two classical bits $\{00, 01, 10, 11\}$ in Alice's possession can be transmitted to Bob with a single qubit. This code requires the use of maximally entangled state which is possible to manipulate in order to transmit the classical two bits.

In this paper we consider a model of a hybrid system with a double quantum dot with one single electron with spin and Dresselhaus spin-orbit coupling [11, 12]. The electron spin and the occupancy form a hybrid basis, in the sense that the entangled state is formed by a qubit of charge (the occupancy) and a qubit of spin and this is the basis for the entangled channel required on the QSC protocol. Hybrid entangled state (HES) preparation will be studied first, and then the four unitary transformations needed for the QSC to transmit the two bit state will be addressed.

The QSC is treated as a control problem, where the required final states are optimized through a fidelity measure by means of genetic algorithms (GA) [13, 14]. Tunneling and site energies are time dependent tuning parameters used to design a control pulse.

1.1. Genetic Algorithms

The evolution analogy of GA consists in forming chromosomes which are different possible solutions of a problem, and the genes of the chromosomes will be the values of every variable of each solution, i.e., a chromosome is a set of parameters while the genes are each of the parameters individually. We used the GA available in the PIKAIA suite [13, 15], where the solutions evolve with genetic recombination occurring at breeding and random probabilistic mutations at breeding events. GA can be understood in a quantum mechanics point of view as follows: given that a quantum state at a given time can be obtained after the actuation of a time evolution operator parameterized with a vector $\vec{b}_i$ (control parameter, electric pulse width and amplitude etc) on initial states as $|\psi(t)\rangle = \hat{U}(\vec{b}_i, t)\,|\psi(0)\rangle$. GA generate an initial population of i different $\hat{U}(\vec{b}_i, t)$ operators that will act on $|\psi(0)\rangle$. GA will test an objective function called *fitness function* for all i that will be constructed as

$$f_{\text{fit}}(\vec{b}_i) = |\,\langle\psi_{\text{fit}}|\,\psi(t = t_m)\rangle|^2 = |\,\langle\psi_{\text{fit}}|\,\hat{U}(\vec{b}_i, t)\,|\psi(0)\rangle\,|^2, \tag{1}$$

where $|\psi_{\text{fit}}\rangle$ is the objective or ideal quantum state that one wants to obtain to perform the the quantum processing task. Internally, PIKAIA maximizes the objective function $f_{\text{fit}}(\vec{b}_i)$ in a bounded n-dimensional space B : $\vec{b}_i \equiv (b_1, b_2, ..., b_n)$ with $b_k \in [0.0, 1.0]\forall k$. The algorithm will pick up a set of better *fitted* solutions as the parents for the next offspring. For instance, in a 2 dimensional B space where the parents form the two operators $\hat{U}(\mathbf{0.105}714, \mathbf{0.542}219, t)$, $\hat{U}(0.542\mathbf{774}, 0.674\mathbf{219}, t)$, the "children" solutions could be $\hat{U}(0.105\mathbf{774}, 0.542\mathbf{219}, t)$ and $\hat{U}(0.542714, 0.674\mathbf{219}, t)$ (the breeding). Mutation could happen at breeding, this makes any digit of a variable equally probable to be chosen and randomly modified, for example in the variable $0.5_1 4_2 2_3 2_4 1_5 9_6$ any digit position (from 1 to 6) could be selected, once a position is selected by a *roulette* algorithm, that digit has a probability to be substituted for any number (including the same number) For example the children solutions could *mute* to $\hat{U}(0.10\mathbf{3}774, 0.\mathbf{5}42219)$ and $\hat{U}(0.5427\mathbf{2}4, 0.\mathbf{6}74219)$, where the bold numbers have changed in some of them. The whole process generates the new offspring that will once again pass the same process of parents selection and breeding, repeated for an explicit amount of *generations* or until a *fitness* condition is achieved.

1.2. Physical Model and Quantum Superdense Coding

The studied system consists of a double lateral quantum dot shown in Fig. 1, where tunneling $t(t)$ and the site energies $E_{L,R}(t)$ are time dependent control parameters that can be tailored to accomplish QSC, and the tunneling with change of spin is given by the Dresselhaus spin-orbit (SO) interaction t_{SO} which is considered constant. At $t = 0$ a spin up electron is placed in the left QD. The four dimensional computational basis, for the two qubit system is formed for the charge qubit with the presence of the electron in sites left $|\Phi_L\rangle \equiv |0\rangle$ and right $|\Phi_R\rangle \equiv |1\rangle$, and for the spin qubit with electron spin up $|\uparrow\rangle \equiv |0\rangle$ and down $|\downarrow\rangle \equiv |1\rangle$ that together form the hybrid basis $\{|00\rangle, |01\rangle, |10\rangle, |11\rangle\} \equiv \{|\Phi_L \uparrow\rangle, |\Phi_L \downarrow\rangle, |\Phi_R \uparrow\rangle, |\Phi_R \downarrow\rangle\}$. The base of this problem is occupancy and spin as follows

$$|\Psi(t)\rangle = \alpha_1\,|\Phi_L \uparrow\rangle + \alpha_2\,|\Phi_R \uparrow\rangle + \alpha_3\,|\Phi_L \downarrow\rangle + \alpha_4\,|\Phi_R \downarrow\rangle\,. \tag{2}$$

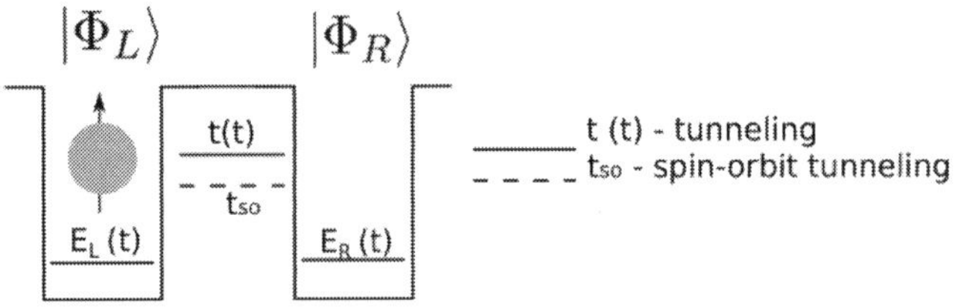

Figure 1. Quantum dot model for the QSD at $t = 0$ with a spin up electron in left site. Site energies and tunnelings are time dependent control parameters.

As described before, the QSC target is to transmit from Alice to Bob two classical bits of information with single qubit operations. First, Alice and Bob must share a maximally entangled state formed by two qubits (each in possession of one qubit). Let us say Alice has the charge qubit and Bob the spin qubit. QSC is achieved by unitary operations on Alice's qubit which transforms the shared state in one of the four distinguishable Bell-like states at Bob's side. In this work we consider as communication channel required in the QDC protocol, the hybrid spin-site Bell-like family of states given by

$$\begin{aligned} |\Phi^{\pm}_{\rm LR}\rangle &= \frac{1}{\sqrt{2}}(|\Phi_L \uparrow\rangle \pm e^{i\pi/4}\,|\Phi_R \downarrow\rangle), \\ |\Psi^{\pm}_{\rm LR}\rangle &= \frac{1}{\sqrt{2}}(|\Phi_L \downarrow\rangle \pm e^{i\pi/4}\,|\Phi_R \uparrow\rangle). \end{aligned} \tag{3}$$

QSC requires the ability to achieve each of the states in (3) starting with an initial state, for instance, $|\Phi^{+}_{LR}\rangle$. Given that all operations act on Alice's qubit, the 4 transformations needed to send the two classical bits $(a1, a2), a1 = \{0, 1\} a2 = \{0, 1\}$ are summarized in Table 1, where the unitary operations represent the transformation that the initial state $|\Phi^{+}_{\rm LR}\rangle$ experiences to become any of the final Bell-like family states, as well as the required fitness functions to achieve that at a measurement time t_m. With $|\Phi^{+}_{LR}\rangle$ as the initial state if $a1 = 1$ then a Z gate is applied to the charge qubit, similarly if $a2 = 1$ then a X gate is applied. These transformations are the Pauli matrices [16] as $X = |0\rangle\langle 1| + |1\rangle\langle 0|$ and $Z = |0\rangle\langle 0| - |1\rangle\langle 1|$ with $Y = ZX$ on the site subsystem. These gates are achieved with the GA control applied to our model using pulse, position, width and amplitudes as chromosomes and basis of the control parameter space. Once the transformation has been applied, Alice's qubit is sent to Bob. Bob can recover the two classical bits of information with the inverse process of entangling to decode the Bell-like state. This is achieved with the applications of a $CNOT$ and a Haddamard H gates. The $CNOT = |00\rangle\langle 00| + |01\rangle\langle 01| + |11\rangle\langle 10| + |10\rangle\langle 11|$ is a two qubit gate, and $H = (1/\sqrt{2})(|0\rangle\langle 0| + |0\rangle\langle 1| + |1\rangle\langle 0| - |1\rangle\langle 0|)$ creates superpositions of single qubit states.

The fitness function is composed with the fidelity of each desired state plus the concurrence C added as a measure of the degree of entanglement. The fidelity is simply $|\langle\varphi|\Psi(t = t_m)\rangle|^2$, where $|\varphi\rangle$ is any of the hybrid Bell-like states. Concurrence is used as a measure of entanglement for two qubits systems [17]. In (2) basis, the concurrence can be calculated as $C = 2|\alpha_1\alpha_4 - \alpha_2\alpha_3|$. This quantity has been added to the fitness function to ensure that the final states be maximally entangled.

2. Quantum Control with GA

To implement QCS in the model shown in Fig. 1 all the states in (3) must be accessible. Different state preparations are needed, one for the HES, and the other four for the QSC protocol transformations. The quantum control procedure to prepare all the required states is described as 4 steps:

Table 1. Quantum superdense coding to send two classical bit $(a1, a2)$, $a1 = \{0, 1\}$, $a2 = \{0, 1\}$, the unitary operations and fitness functions f required for each two classical bits.

(a1, a2)	Unitary operations	Final state	Fitness Function
00	$I_A \otimes I_B$	$\|\Phi^+_{LR}\rangle$	$f_{\Phi_{LR}} = \|\langle\Phi^+_{LR}\|\Psi(t_m)\rangle\|^2 + C$
01	$X_A \otimes I_B$	$\|\Psi^+_{LR}\rangle$	$f = \|\langle\Psi^+_{LR}\|\Psi(t_m)\rangle\|^2 + C$
10	$Z_A \otimes I_B$	$\|\Psi^-_{LR}\rangle$	$f = \|\langle\Psi^-_{LR}\|\Psi(t_m)\rangle\|^2 + C$
11	$Y_A \otimes I_B$	$\|\Phi^-_{LR}\rangle$	$f = \|\langle\Phi^-_{LR}\|\Psi(t_m)\rangle\|^2 + C$

(i) The system and the control Hamiltonians that contain the space of parameters (chromosomes) to be optimized are defined. This two level system is simply described as,

$$H(t) = H_0 + H_S + H_C(t, b), \tag{4}$$

where H_0 represents the sites Hamiltonian, containing site energies and tunneling independent of spin, H_S is the spin Hamiltonian with the spin-orbit effect and $H_C(t, \mathbf{b})$ is the Control Hamiltonian, which includes the $\mathbf{b}$ *chromosomes* described above as the arguments of the control functions of the pulse structure. Each Hamiltonian has the following structure,

$$H_0 = \begin{bmatrix} E_L & t_0 & 0 & 0 \\ t_0 & E_R & 0 & 0 \\ 0 & 0 & E_L & t_0 \\ 0 & 0 & t_0 & E_R \end{bmatrix}, \tag{5}$$

where E_L and E_R are left and right site energies respectively, and t_0 is a spin-independent tunneling constant,

$$H_S = \begin{bmatrix} 0 & 0 & \gamma|k|E_L e^{i\phi} & \gamma|k|t_0 e^{i\phi} \\ 0 & 0 & \gamma|k|t_0 e^{i\phi} & \gamma|k|E_R e^{i\phi} \\ \gamma|k|E_L e^{-i\phi} & \gamma|k|t_0 e^{-i\phi} & 0 & 0 \\ \gamma|k|t_0 e^{-i\phi} & \gamma|k|E_R e^{-i\phi} & 0 & 0 \end{bmatrix}, \tag{6}$$

where γ is a material-dependent constant given by Dresselhaus spin-orbit effect and $|k|e^{i\phi}$ is the electron wave-vector as $k = k_x + ik_y$.

$$H_C(t, b) = \begin{bmatrix} \sigma^+_{LR}(t, b) & t_{LR}(t, b) & 0 & 0 \\ t^*_{LR}(t, b) & \sigma^-_{LR}(t, b) & 0 & 0 \\ 0 & 0 & \sigma^+_{LR}(t, b) & t_{LR}(t, b) \\ 0 & 0 & t^*_{LR}(t, b) & \sigma^-_{LR}(t, b) \end{bmatrix}. \tag{7}$$

$H_C(t, b)$ is formed by the control parameters defined as the time dependent pulses

$$\begin{aligned} t_{LR}(t, b) &= b_{LR} e^{-\frac{(t - t_{LR})^2}{2\tau^2_{LR}}} - t_0 \tanh \frac{(t - t_{LR})^2}{2\tau^2_{LR}}, \\ \sigma^+_{LR}(t, b) &= b\sigma_{LR} e^{-\frac{(t - t\sigma_{LR})^2}{2\tau\sigma^2_{LR}}} - t_0 \left(\tanh \frac{(t - t\sigma_{LR})^2}{2\tau\sigma^2_{LR}} - 1 \right), \\ \sigma^-_{LR}(t, b) &= -\sigma^+_{LR}(t, b). \end{aligned} \tag{8}$$

where the widths, positions and amplitudes are $\mathbf{b} = (b_{LR}, b\sigma_{LR}, t_{LR}, t\sigma_{LR}, \tau_{LR}, \tau\sigma_{LR})$ These parameters are constructed in a way that $t_{LR}(t, b)$ and $\sigma^+_{LR}(t, b)$ can change the spin independent tunneling and site energies, respectively. The control parameters that are varied by GA is the vector $\mathbf{b}$. The control pulses correspond to charge subsystem that can be controlled with electric fields directly applied to the QD configuration. This is because we are not considering magnetic fields, needed if the spin subsystem was manipulated. $t_{LR}(t,b)$ is a Gaussian-shaped tunneling control pulse that includes a hyperbolic function to cancel all transition probabilities out of control pulse action time. $\sigma^{+}_{LR}(t,b)$ is a Gaussian-shaped pulse too, where the term t_0 has been added just for consistency, this pulse changes the energy in sites left and right anti-symmetrically.

(ii) The fitness function, depends on the final desired state and will be explicitly defined for every step of the QSC as shown in Table 1. All the objetive or fitness function is to maximize the probability of finding the system in every hybrid Bell basis state which by redundancy has maximum concurrence,

(iii) Integration of the time dependent wave equation up to time of measurement t_m and obtain $|\Psi(t_m)\rangle|$.

(iv) Genetic algorithm application to find the optimal control parameters. Some assumptions are made for the calculations: a $\gamma = 24eV\mathring{A}$ is used corresponding to GaAs [12]. Tunneling and spin-orbit parameters are defined as $t_0 = 1meV, |k| = 0.34, \phi = \pi/5, \hbar = 1$ what makes time to have arbitrary units (au in the text), in what follows all energy units are considered in mili-electron volt (meV) and is used $\langle k_z^2\rangle = (\pi/\omega)2 = 24.6nm^{-2}$ [18], considering the quantum dot wide as $\omega = 20nm$. PIKAIA was set to 50 generations and default settings, which include 100 individuals, crossover probability of 0.85 and mutation variable between 0.005 and 0.25.

3. Results

3.1. Hybrid Entangled State preparation

As a first step, the hybrid entangled state needs to be created as an information channel. The final desired state is,

$$|\Phi^{+}_{LR}\rangle = \frac{1}{\sqrt{2}}(|\Phi_L \uparrow\rangle + e^{i\pi/4}|\Phi_R \downarrow\rangle), \tag{9}$$

i.e., for the state preparation, as described in Table 1 the fitness function is $f_{\text{fit}} = |\langle\Phi^{+}_{LR}|\,\psi(t=t_m)\rangle|^2 + C$. Considering the initial state $|\psi(t=0)\rangle = |\Phi_L \uparrow\rangle$, a possible solution can be found with GA, given the defined control parameters is $\mathbf{b} = (7.209, 7.652, 310, 63, 64, 250)$ for (8) with a fidelity of 0.9972. This process can be easily understood from Fig. 2. In this figure, the time evolution of the Bell-like states probabilities is shown, together with the shape of the time dependent control pulses (inset), with the parameters described above, applied to obtain the final state (9).

Once the maximum probability of being in $|\Phi^{+}_{LR}\rangle$ is reached, we se that it is maintained, and all the others states probabilities drop to zero. By analyzing the dynamics of the probability plots, we obtain an equivalent quantum circuit of this process, which is illustrated in Fig. 3. The X and Z rotations are generated by the action of both the tunneling and site energy $t_{LR}(t,b)$ and $\sigma^{\pm}_{LR}(t,b)$, respectively. The $CNOT$ gate is accomplished by the combination of SO effects with $t_{LR}(t,b)$ together. Corresponding to the control pulses, the action of a partial Z gate is first seen by promoting a unbalance of site energies $\sigma^{\pm}_{LR}(t,b)$. In a similar way, the tunneling pulse starts at about 100 time units, and the action of an X gate with a phase could be deduced. The joint effect of $\mathrm{e}^{-i\frac{\pi Z}{4}}\,\mathrm{e}^{-i\frac{\pi X}{2}}$ is equivalent to a Hadamard gate with a phase. After that, the action of a two qubit interaction(the SO term) is depicted when control pulses are going down. All these actions represent a synthesis of the probabilities and complex amplitudes of each Bell-like state and can not be clearly seen because of the joint action of different interactions with evolving rations between them. However, the quantum circuit shows a generalization of Bell

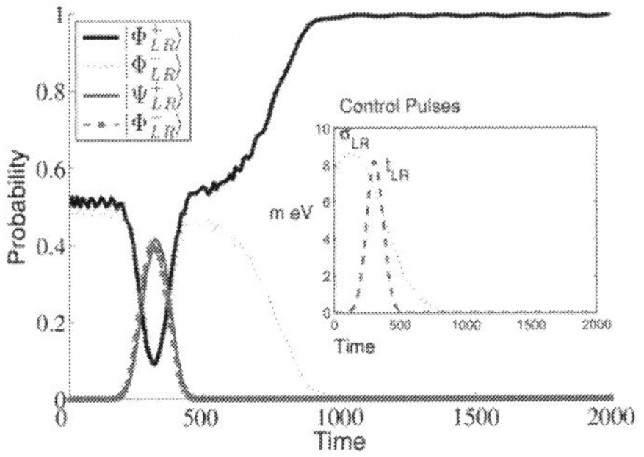

Figure 2. Bell-like probability evolutions for the hybrid entangled state $|\Phi^+_{LR}\rangle$ preparation with a fidelity of 0.997. The control pulses are presented in the inset.

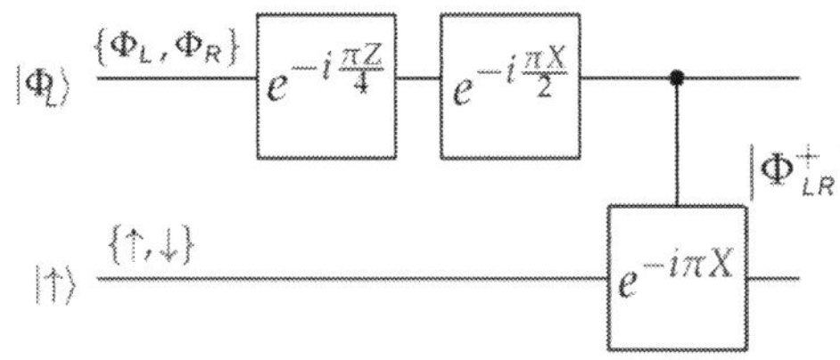

Figure 3. Quantum circuit for the HES $|\Phi^+_{LR}\rangle$ preparation with the initial condition $|\Phi_L \uparrow\rangle$

state creation operations that require only a Hadamard and a CNOT [16] but with different phases that arise because of SO effect.

3.2. Quantum Superdense Coding operations

We describe now, the physical process for the protocol for the QSC. After the preparation of the HES (9), the system requires the ability to achieve the 4 Bell hybrid basis states to accomplish QSDC of sending the four classical two bits. QSDC is implemented as shown in Fig. 4. This figure summarizes the operations in Table 1. The center of this figure represents the initial Bell-like HES that was previously prepared. From this initial state the four Bell-like HES can be easily accessed by proper pulse shapes. The four figures show the Bell-like time evolution probabilities through the application of the control pulses shown in the corresponding insets. Each operation is accompanied with the corresponding quantum circuit obtained from the time evolution parameters. One set of parameters is found to accomplish each of the states and are described in Table 1 by the control pulses (8). The structure and parameters of the control pulses for each two bit state to be sent are detailed as follows.

To send the two bit 00 no control pulses are needed, because the system is already in the final desired state and the quantum circuit is only an identity operation. This is shown in Fig. 4(a), as no alteration is made, fidelity is 1. To send 01, the control pulse obtained with GA is shown in Fig. 4(b). The corresponding control parameters are $\mathbf{b} = (0.7851, 0, 745, 0, 388, 0)$, and the quantum circuit shows the X gate is applied with a global phase π, this is because only the tunneling control pulse is present. The fidelity achieved is of 0.998; to send 10 the control parameters obtained are $\mathbf{b} = (1.2039, 0.5766, 20, 992, 101, 496)$ as shown in Fig. 4(c) with a fidelity of 0.998. In this case the quantum circuit has a dominant part of the Z gate, the control pulses show the tunneling present for a small amount of time, which promotes the phase change needed to accomplish the final state in a more complex way that only the Z gate, while the site energies are dominant in the control pulses.

Finally, to send 11, the parameters of the control pulses of Fig. 4(d) are $\mathbf{b} = (1.7116, 1.9260, 997, 222, 249, 284)$ with a fidelity of 0.9838. The quantum circuit implemented shows Z and X rotations together that are equivalent to a Y rotation, these are present because both the tunneling and the site energies are present for a time interval as seen in the control pulses in the figure. The two gates operate at different times, when site energies decrease tunneling increases in the order shown in the quantum circuit.

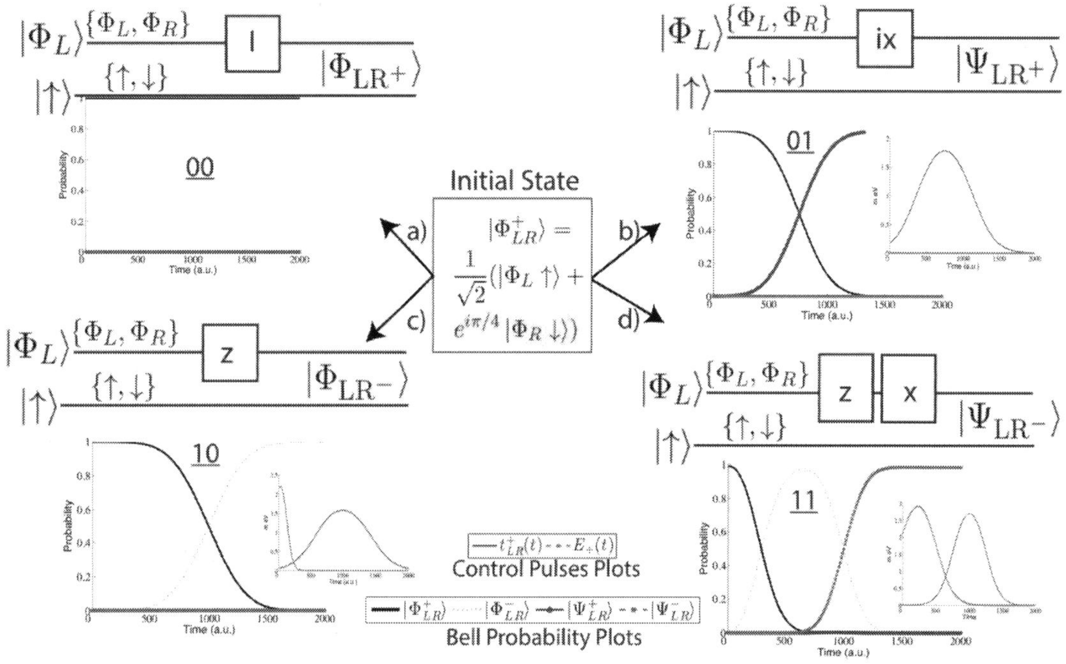

Figure 4. Quantum Superdense Coding scheme with the hybrid spin-charge system and control pulses with the initial $|\Phi^+_{LR}\rangle$. Each subfigure shows how to transmit the two bit basis (transform to the final hybrid Bell-like state) (a) 00 (Φ^+_{LR}), only the identity operation is needed, (b) 01 ($|\Psi^+_{LR}\rangle$) with an X gate with phase, obtained with the tunneling pulse shown, (c) 10 (Φ^-_{LR}) the Z gate is emulated with site energies disbalance and phase adjusted with tunneling, (d) 11 ($|\Psi^-_{LR}\rangle$) Z and X gates are needed, achieved by the action of site energies unbalance and tunneling.

4. Conclusions

We have presented an example of how GA can be applied in quantum information processing. We implemented in a physical model the QSC protocol, with a double QD system using a spin-orbit entangled stated as a communication channel. Given the nature of the QD model with only one electron, hybrid states between charge and spin states arose naturally.

A maximally entangled hybrid Bell-like state was generated from an initial state, controlling the dynamics, using the GA approach.

The control parameters of the system were Gaussian shaped time dependent site energies and transition probabilities between sites.

The four unitary operations required to send the two bits of classical information $\{00, 01, 10, 11\}$, which transform a given Bell-like state to the other three, were performed with large fidelities or fitness. The GA parameters used allowed a fast convergence to the solution within just two minutes of calculation time. This work illustrates how straightforward a physical quantum system can be controlled using the GA approach to carry out a quantum information task.

Acknowledgments

The work was supported by DGAPA Project No. IN112012 and by CONACyT Phd Schollaship(FDS)

[1] Dong D and Petersen I R 2009 (*Preprint* 0910.2350) URL `http://arxiv.org/abs/0910.2350`

[2] D'Alessandro D 2007 *Introduction to Quantum Control and Dynamics* (CRC Press) ISBN 1584888830
[3] Xue D, Zou J, Li J G, Chen W and Shano B 2010 *J. Phys. B* **43** 45503
[4] Delgado F 2010 *Phys. Rev. A* **81** 42317
[5] Branczyk A, Mendonça P, Gilchrist A, Doherty A and Bartlett S 2007 *Phys. Rev. A* **75** 12329
[6] Krause J, Reitze D, Sanders G, Kuznetsov A and Stanton C 1998 *Phys. Rev. B* **57** 9024–9034
[7] Giraldi G A, Portugal R and Thess R N 2004 (*Preprint* 0403003) URL `http://arxiv.org/abs/cs/0403003`
[8] Zingg D W, Nemec M and Pulliam T H 2008 *Revue européenne de mécanique numérique* **17** 103–126 ISSN 17797179 URL `http://www.tandfonline.com/doi/abs/10.3166/remn.17.103-126`
[9] Holland J C 1975 *Adaptation in Natural and Artificial Systems* primera ed (Michigan, US: The U. of Michigan Press)
[10] Bennett C H and Wiesner S J 1992 *Phys. Rev. Lett.* **69** 2881–2884
[11] Dresselhaus G 1955 *Phys. Rev* **100** 580–586
[12] Perel V I, Tarasenko S A, YAssievich I N, Ganichev S A, Bel'kov V V and Prettl W 2003 *Phys. Rev. B* **67** 201304
[13] Carbonneau P 1995 *The Astrophysical Journal Supplement Series* **101** 309–334
[14] Navarro-Muñoz J C, Rosu H C and López-Sandoval R 2006 *Physical Review A* **74** 52308
[15] Carbonneau P and Knapp B 1995 *A user's guide to PIKAIA 1.0* 1st ed (Boulder, Colorado, US: High Altitude Observatory.National Center for Atmospheric Research)
[16] Nielsen M and Chuang I L 2000 *Quantum Computation and Quantum Information* 1st ed (Cambridge University Press)
[17] Hill S and Wootters K 1997 *Phys. Rev. Lett.* **78** 5022–5025
[18] Studer M, Walser M P, Baer S, Rusterholz H, Schön S, Schuh D, Wagscheider W, Enssin K and Salis G 2010 *Cond. Mat. Hall* 1–7

Mielnik's evolution loop hexagon in new light

Kurt Bernardo Wolf
Instituto de Ciencias Físicas, Universidad Nacional Autónoma de México,
Av. Universidad s/n, Cuernavaca, Morelos 62210, México
E-mail: bwolf@fis.unam.mx

Abstract. The problem of controlling quantum wavefunctions by means of potential jolts and periods of free evolution was broached by Bogdan Mielnik in 1977. This *quantum control* became a subject of great interest for the preparation of atomic and particle systems. We point out here that these manipulations are also realized in paraxial geometric and wave optics, with lenses and free spaces, and even more transparently than with matter.

PACS: 02.20.Qs, 02.30.Yy, 03.65.Vf, 42.15.Eq
Keywords: Hamiltonian systems, Paraxial optics.

1. Introduction

In his 1977 paper, Bogdan Mielnik [1] showed that the free evolution of quantum wavefunctions can be reversed in time by applying an appropriate succession of harmonic oscillator potential jolts, within the Schrödinger formalism of quantum mechanics. This he reformulated in 1986 [2] as the existence of evolution loops, namely products of operators which are equivalent to *a* (or *the*) unit transformation. Mielnik's *evolution loop hexagon*, shown in Fig. 1, appeared in this paper and was the logo for the 2014 conference on Quantum control, exact or perturbative, linear or nonlinear, held at CINVESTAV and topic of these Proceedings. Much of the context for Mielnik's hexagon was doubtless supported by a previous fundamental article of Białynicki-Birula, Mielnik and Plebański [3] that analyzes the Baker-Campbell-Hausdorff problem of exchanging the exponentials of non-commuting operators.

The purpose of this article is to present the same problems (and solutions) when the system is optical, and classical, to be compared with the quantum system. The idealized potential jolts are replaced by rather real refracting surfaces, while the time periods τ_i when the system is free become, naturally enough, spaces z_i of free propagation along the optical axis — which by tradition points horizontally to the right. Moreover, as I shall contend, optical systems are richer than the Schrödinger's model of quantum mechanics. This is so because these refracting surfaces can mimick harmonic oscillator jolts only in the *paraxial* régime, where ray or beam angles are small and we look near to the optical axis. Beyond the paraxial régime of geometric and wave optics extends the *metaxial* régime, which constitutes a series expansion involving the nonlinearities called aberrations. Finally, there is the *global* régime, where the true nature of optical systems requires rays or beams in all directions on the sphere — or circle, or hypersphere, according to the dimension of the model.

In Section 2 we present three mathematical realizations of the symplectic algebra of 2×2 matrices that describe quadratic systems: the quantum oscillator jolts and free evolution, and also the paraxial lenses and empty spaces given in Section 3. In Section 4 we point out the

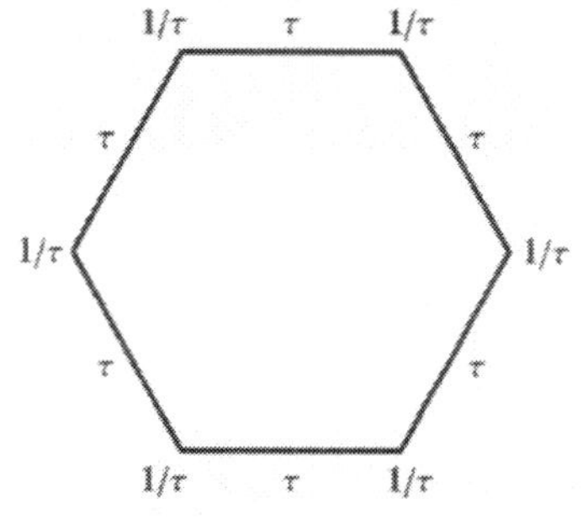

$$\left[\exp\left(-\frac{i}{\hbar}\frac{mq^2}{2}\frac{1}{\tau}\right)\exp\left(-\frac{i}{\hbar}\frac{\hat{p}^2}{2m}\tau\right)\right]^6 \equiv \mathbb{I}$$

Figure 1. The Mielnik hexagon illustrates the application of six harmonic oscillator potential jolts of strength $1/\tau$, separated by six time periods τ of free evolution. This was the logotype of the conference on Quantum control, exact or perturbative, linear or nonlinear, held at CINVESTAV in October 2014.

subtle but important difference between classical and wave/quantum systems contained in the symplectic *versus* metaplectic units by means of the Bargmann parametrization of the universal cover of the symplectic groups, built out of products of lower- and upper-triangular matrices that realize lenses and free spaces. In Section 5 we collect some conclusions and point to extensions.

2. Three models under one algebra

Consider the following basis of three linearly independent functions $j_i(q,p)$ that are quadratic in q (position), and p (momentum),

$$j_+ := \tfrac{1}{2}p^2, \quad j_- := \tfrac{1}{2}q^2, \quad j_0 = \tfrac{1}{2}pq. \tag{1}$$

Corresponding to these we have three linear operators, built with Poisson brackets $\bar{\jmath}_i := \{j_i, \circ\}$ with the classical coordinates of position and momentum,

$$\bar{\jmath}_+ = -p\partial_q, \quad \bar{\jmath}_- = q\partial_p, \quad \bar{\jmath}_0 = \tfrac{1}{2}(p\partial_p - q\partial_q), \tag{2}$$

which act on *beam density* functions $\rho(q,p)$. The usual and unique Schrödinger quantization of (1) from the basic operators of position and momentum, $q \mapsto \hat{Q} = q\,\cdot$ and $p \mapsto \hat{P} = -\mathrm{i}\partial_q$ yields the following three up-to-second degree operators $\hat{J}_i$,

$$\hat{J}_+ = -\tfrac{1}{2}\partial_q^2, \quad \hat{J}_- = \tfrac{1}{2}q^2\,\cdot, \quad \hat{J}_0 = -\tfrac{1}{4}\mathrm{i}(q\partial_q + \partial_q q), \tag{3}$$

that are hermitian on the Hilbert space $\mathcal{L}^2(\mathsf{R})$ in the single variable $q \in \mathsf{R}$. Finally, we have the basis of three 2×2 real traceless matrices $\mathbf{J}_i$, given by

$$\mathbf{J}_+ = \begin{pmatrix} 0 & 1 \\ 0 & 0 \end{pmatrix}, \quad \mathbf{J}_- = \begin{pmatrix} 0 & 0 \\ -1 & 0 \end{pmatrix}, \quad \mathbf{J}_0 = \tfrac{1}{2}\begin{pmatrix} 1 & 0 \\ 0 & -1 \end{pmatrix}. \tag{4}$$

These matrices *present* the algebra $\mathsf{sp}(2,\mathsf{R})$ and its similar ones, acting on the phase space coordinates that we can fit into a two-vector $\binom{q}{p}$.

The common property of the four triads of $\mathcal{J}$'s (the notation stands for j, $\jmath$, $\hat{J}$ or $\mathbf{J}$) is to realize the Lie algebra $\mathsf{so}(2,1) = \mathsf{su}(1,1) = \mathsf{sp}(2,\mathsf{R})$, whose defining set of Lie brackets can be written as

$$\{\![\mathcal{J}_0, \mathcal{J}_+]\!\} = +\iota\mathcal{J}_+, \quad \{\![\mathcal{J}_0, \mathcal{J}_-]\!\} = -\iota\mathcal{J}_-, \quad \{\![\mathcal{J}_+, \mathcal{J}_-]\!\} = -2\iota\mathcal{J}_0. \tag{5}$$

For the functions j_i in (1), the Lie bracket is the classical Poisson bracket $\{\![j_a, j_b]\!\} \equiv \{j_a, j_b\} := (\partial_q j_a)(\partial_p j_b) - (\partial_p j_a)(\partial_q j_b)$ with $\iota = 1$. The phase space operators (2) obey (5) with the commutator Lie bracket $\{\![\jmath_a, \jmath_b]\!\} \equiv [\jmath_a, \jmath_b] := \jmath_a\jmath_b - \jmath_b\jmath_a$ also with $\iota = 1$, as do the matrices in the realization (4). Only for the quantum-mechanical operators (3) the commutation relations (5) carry $\iota = \mathrm{i}$ to preserve hermiticity in $\mathcal{L}^2(\mathsf{R})$.

Although all four sets of objects (1)–(4) realize the same Lie algebra, the exponentiation of the two operator sets, the classical (2) and the quantum (3) to Lie groups do not *quite* give the same group. While the matrix realization (4) clearly indicates that the group is that of 2×2 real matrices of unit determinant $\mathsf{SL}(2,\mathsf{R}) = \mathsf{Sp}(2,\mathsf{R})$, the quantum (and paraxial wave-optical) provides an *integral transform* realization of its two-fold cover, the *metaplectic* group $\mathsf{Mp}(2,\mathsf{R})$, as we shall elaborate below.

3. Free propagation, jolts and lenses

The formalism introduced in the previous section refers to one-dimensional models, but provides the gist of the matter. It can be made N-dimensional with the symplectic algebras $\mathsf{sp}(2N,\mathsf{R})$; if the systems are isotropic, a *radial* $\mathsf{sp}(2,\mathsf{R})$ algebra will arise, also satisfying (5), where the generator $\mathcal{J}_+$ will receive an extra 'centrifugal' term $\sim \mu/q^2$ that will lie, for $\mu \geq -\frac{1}{4}$, among the lower-bound discrete series D^k_+ of $\mathsf{sp}(2,\mathsf{R})$. Yet, the 2×2 matrix *presentation* of this algebra in (4) is unchanged.

Rigorous mathematics are not lacking, but let me appeal to the common understanding of Schrödinger quantum mechanics, while drawing a parallel sketch of geometric and wave paraxial optics. In this context we look first at the phase space variables as operators. The spectrum of the position operator $\mathcal{Q}$ is the set of observation points on a line, assumed to be continuous $q \in \mathsf{R}$. Here the quantum wave function is 'known' at a given time τ; in geometric optics q is the coordinate of the point where a line of light crosses a one-dimensional screen that is perpendicular to the optical axis at a given distance z; in wave optics z plays the same role as τ in quantum mechanics, the measurement being the (complex) elongation of the wavefield on a screen placed at z.

The momentum operator $\mathcal{P}$ is more amenable to discussion, since in all models it relates the evolution of $\mathcal{Q}$ within the screen under infinitesimal propagation in τ or z; thence the quantum $\hat{P} = -\mathrm{i}\partial_q$. In geometric optics, the matter is more geometrical: see Fig. 2. A ray of position q, and inclined by an angle θ to the optical axis will, upon propagation by infinitesimal dz, go to position $q + dq$. Now, dq on the screen is perpendicular to dz on the axis, and both form a right triangle with ds, the *optical* distance (along which a wavefield would oscillate); this triangle is similar to that formed by the quantity

$$p = n \sin\theta, \tag{6}$$

which is the *optical momentum* on the screen; then, there is a quantity $p_z \equiv -h(q,p)$ that turns out to be (minus) the geometric *optical Hamiltonian*; the hypotenuse of this triangle is the refractive index n of the medium at q. This similarity leads to the Hamilton equations [4, Ch. 2]. In the *global* régime of optics, where $\theta \in \mathsf{S}^1$ (the circle), $|p| \leq n(q)$.

Indeed, the optical Hamiltonian is

$$h(q,p) = -\sqrt{n(q)^2 - p^2} \approx \frac{p^2}{2n(q)} - n(q) + \frac{p^4}{8n(q)^3} + \frac{p^6}{16n(q)^5} + \cdots \tag{7}$$

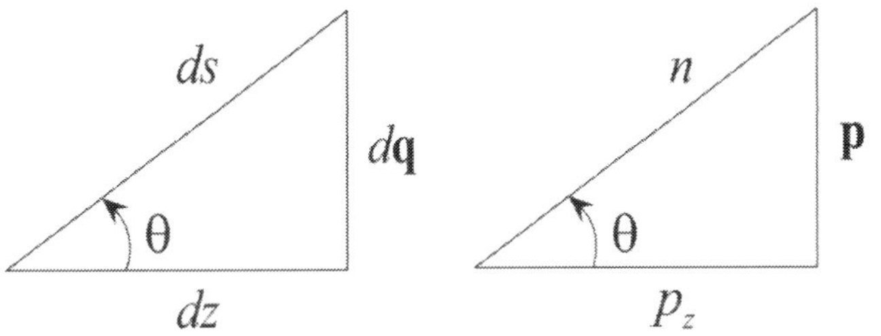

Figure 2. The left triangle, formed by infinitesimal displacements in position on the screen $d\mathbf{q}$ (in any number of dimensions) and evolution distance dz has for hypotenuse the optical distance ds. This is similar to the triangle formed by optical momentum $\mathbf{p}$, the (minus) Hamiltonian p_z, and the refractive index n at the point $\mathbf{q}$.

The paraxial approximation consists in assuming that the angle θ is near to the optical axis, so $\theta^2 \approx 0$, and so momentum (6) becomes $p = n\theta$ while the Hamiltonian (7) looses all p-terms beyond p^2, and comes to resemble a mechanical system of variable mass [5]. But then, the *paraxial* model assumes that, notwithstanding θ is small, the momentum p can be extended to all of R together with all the simplified formalism of quadratic expressions when also $n(q) \approx n_0 - \frac{1}{2}n_2\, q^2$ for isotropic media and $q^2p^2 \approx 0$; then the Hamiltonian (7) becomes $p^2/2n_0 - n_0 + \frac{1}{2}n_2q^2$. Note that the refractive index corresponds to the *negative* of a quantum potential: $n_0 - \frac{1}{2}n_2q^2 \sim -V(q)$, so rays 'fall into' regions of highest n, corresponding to the potential minimum.

With what has been said we can now broach the purely geometric free displacement (or propagation) by finite z or τ —indicate it by ζ—, generated by $\mathcal{J}_+ \sim \frac{1}{2}p^2$ and present it with the 2×2 matrices in (4), as

$$\mathcal{D}(\zeta) = \exp(-\iota\zeta\mathcal{J}_+) \quad \leftrightarrow \quad \mathcal{D}(\zeta) : \begin{pmatrix} q \\ p \end{pmatrix} = \begin{pmatrix} 1 & \zeta \\ 0 & 1 \end{pmatrix} \begin{pmatrix} q \\ p \end{pmatrix} = \begin{pmatrix} q+\zeta p \\ p \end{pmatrix}, \tag{8}$$

where ι is 1 or i as before. The classical map is linear; it skews the phase plane and obviously conserves $p = n\theta$, the direction of light propagation. We recall that to respect the order of multiplication for group elements $\mathcal{G}$ and their matrix presentation $\mathbf{M}$ on phase space two-vectors, the group elements act with the *inverse* matrix, $\mathcal{G}(\mathbf{M})\binom{q}{p} = \mathbf{M}^{-1}\binom{q}{p}$; because only then $\mathcal{G}(\mathbf{M}_1)\,\mathcal{G}(\mathbf{M}_2) = \mathcal{G}(\mathbf{M}_1\mathbf{M}_2)$.

The dynamical elements of the system are generated by $\mathcal{J}_- \sim \frac{1}{2}q^2$. In geometric optics it is clear that the effect of a lens of power g will be

$$\mathcal{L}(g) = \exp(-\iota g\mathcal{J}_-) \quad \leftrightarrow \quad \mathcal{L}(g) : \begin{pmatrix} q \\ p \end{pmatrix} = \begin{pmatrix} 1 & 0 \\ -g & 1 \end{pmatrix} \begin{pmatrix} q \\ p \end{pmatrix} = \begin{pmatrix} q \\ p - gq \end{pmatrix}. \tag{9}$$

This is the *lens* transformation on classical phase space; when $g > 0$ the rays are *refracted* (i.e., broken) towards the optical axis, and a beam of rays parallel to this axis will all meet at the *focal distance* $f := 1/g$ of a convex lens; when $g < 0$, the lens is concave. In wave optics, $\mathcal{L}(g)$ acts on wavefields multiplying them by the phase $\exp(-\mathrm{i}\frac{1}{2}gq^2)$, as a flat Fresnel lens such as we often see in pocket magnifying glasses. On the other hand, the quantum mechanical analogue of the flat lens (8) is a potential jolt. As built in [1], the evolution in the presence of a potential $V(q)/\kappa$ during a time κ is produced by the operator $\exp[-\mathrm{i}\kappa(-\frac{1}{2}\partial_q^2 + V(q)/\kappa)]$ acting on the wavefunction. When $\kappa \to 0$ the time window narrows and the potential strength increases, so there remains the phase $\exp[-\mathrm{i}V(q)]$. For the harmonic oscillator this is $V(q) = \frac{1}{2}\omega q^2$, where the oscillator strength is ω, which corresponds with the lens power g that we introduced above.

The problem is thus set: to produce, with free displacements and lenses, *all* linear (and *canonical*) transformations of phase space; correspondingly, with periods of free evolution interspersed with harmonic oscillator jolts, to produce all linear (and *unitary*) maps of wavefunctions in $\mathcal{L}^2(\mathsf{R})$. Of particular interest is *negative* free evolution, and thus the unit transformation that returns the beam or wavefunction to its original condition before it entered the optical setup or other such apparatus. In view of (8) and (9), the classical problem reduces to write any (and every) such transformation as a 2×2 matrix, and decompose this into products of lower- and upper-triangular matrices. Perhaps this is an unusual factorization for $\mathsf{Sp}(2,\mathsf{R})$, which is better known through its Iwasawa, Bargmann, or Euler-type angle and boost decompositions.

Let us compose a lens (or jolt) of strength g with a free space (or time) ζ; multiplying their matrix representatives; this 'elementary subsystem' we denote by

$$\mathbf{M} = \begin{pmatrix} 1 & 0 \\ -g & 1 \end{pmatrix} \begin{pmatrix} 1 & \zeta \\ 0 & 1 \end{pmatrix} = \begin{pmatrix} 1 & \zeta \\ -g & 1-g\zeta \end{pmatrix}. \tag{10}$$

Two such arrangements in a row are represented by their product,

$$\mathbf{M}^2 = \begin{pmatrix} 1-g\zeta & \zeta(2-g\zeta) \\ -g(2-g\zeta) & (1-g\zeta)^2 - g\zeta \end{pmatrix}, \tag{11}$$

where we note that if $g\zeta = 2$, the transformation becomes a rotation of phase space by π, namely $\begin{pmatrix} -1 & 0 \\ 0 & -1 \end{pmatrix}$. Three such elements, depicted in Fig. 3, produce

$$\mathbf{M}^3 = \begin{pmatrix} (1-g\zeta)^2 - g\zeta & \zeta(1-g\zeta)(3-g\zeta) \\ -g(1-g\zeta)(3-g\zeta) & (1-g\zeta)^3 - g\zeta(3-2g\zeta) \end{pmatrix}. \tag{12}$$

When $g\zeta = 1$ we obtain the π rotation $\begin{pmatrix} -1 & 0 \\ 0 & -1 \end{pmatrix}$, while when $g\zeta = 3$ the result is $\begin{pmatrix} 1 & 0 \\ 0 & 1 \end{pmatrix}$ — the identity matrix. This last 'identity system' was already recognized in [2, Eq. (3.25)] to yield an output which is the input function with an overall minus sign.

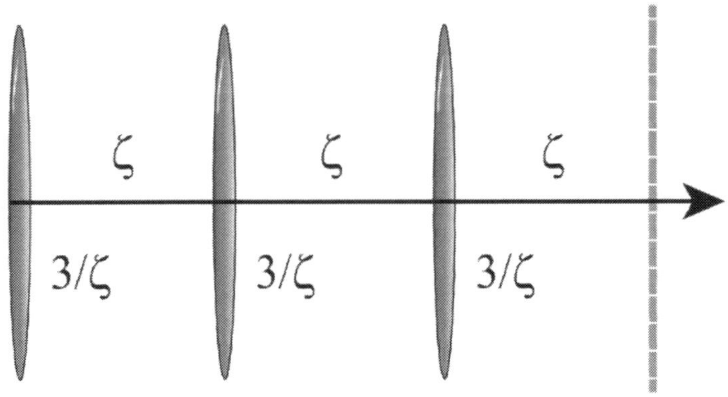

Figure 3. In geometric optics, three lenses of power $g = 3/\zeta$ and three free flights by ζ yield the identity transformation between ingoing and outgoing rays.

The above argument with matrices, which would appear to indeed simplify the Mielnik hexagon into a triangle, underscores the fact that, because $\mathsf{Sp}(2,\mathsf{R})$ is *multiply connected*, the geometric-optical realization of the symplectic identity group element $\begin{pmatrix} 1 & 0 \\ 0 & 1 \end{pmatrix}$ in (12), corresponds to *two* group elements in its two-fold cover $\mathsf{Mp}(2,\mathsf{R})$. This holds in the same way that a rotation by 2π around any axis is the identity transformation in the three-dimensional rotation group $\mathsf{SO}(3)$, but corresponds to *two* 'rotations', by 2π and by 4π, in its two-fold cover by the spin group $\mathsf{SU}(2)$. In the next section we shall see that $\mathsf{Sp}(2,\mathsf{R})$, as the circle S^1, is *infinitely* connected; the metaplectic group $\mathsf{Mp}(2,\mathsf{R})$ is only its two-fold cover.

It is of interest to note, in the context of geometric optics, that the problem of realizing any linear symplectic transformation in a (two-dimensional) paraxial optical setup composed by lenses and free flights, i.e., the 'optical' decomposition of the three-dimensional $\mathsf{Sp}(2,\mathsf{R})$ manifold, by means of upper- and lower-triangular 2×2 matrices, was solved in Ref. [6]. It turns out that two lenses and two free flights suffice to reach all such systems, *except* for a lower-dimensional submanifold that includes negative free flight; so, three lenses and two free flights suffice. (See also Ref. [4, Sec. 10.5].)

4. Symplectic and metaplectic units

The work of Valentin Bargmann on the symplectic groups, Ref. [7] for $\mathsf{Sp}(2,\mathsf{R})$ and Ref. [8] for $\mathsf{Sp}(2N,\mathsf{R})$, is as comprehensive as can be found. He uses the analogue of the characterization of complex numbers by phase and modulus, $z = \mathrm{e}^{\mathrm{i}\arg z}|z| \in \mathsf{C}$, which for matrices is their *polar* decomposition into the product of a unitary and a symmetric positive definite matrix. In the 2×2 case this is

$$\begin{pmatrix} a & b \\ c & d \end{pmatrix} = \begin{pmatrix} \cos\phi & -\sin\phi \\ \sin\phi & \cos\phi \end{pmatrix} \begin{pmatrix} \lambda + \mathrm{Re}\mu & \mathrm{Im}\mu \\ \mathrm{Im}\mu & \lambda - \mathrm{Re}\mu \end{pmatrix}, \tag{13}$$

where $ad - bc = 1$, and its *Bargmann* parameters are

$$\phi = \arg\left[a + d - \mathrm{i}(b - c)\right] \in \mathsf{R}, \tag{14}$$

$$\mu = \tfrac{1}{2}\,\mathrm{e}^{-\mathrm{i}\phi}[a - d + \mathrm{i}(b + c)] \in \mathsf{C}, \tag{15}$$

with $\lambda := +\sqrt{|\mu|^2 + 1} > |\mu|$. Note that, as in the complex number case, the 'angle' ϕ is now allowed to take values beyond the basic angular interval.

In terms of the Bargmann parameters, the product of two matrices $\mathbf{M} = \mathbf{M}_1\mathbf{M}_2$ is found to be

$$\phi = \phi_1 + \phi_2 + \arg\nu, \tag{16}$$

$$\text{where } \nu := 1 + \mathrm{e}^{-2\mathrm{i}\phi_2}\,\mu_1\mu_2^*/\lambda_1\lambda_2, \tag{17}$$

$$\mu = \mathrm{e}^{-\mathrm{i}\arg\nu}(\lambda_1\mu_2 + \mathrm{e}^{-2\mathrm{i}\phi_2}\,\mu_1\lambda_2). \tag{18}$$

It is crucial to realize that $|\nu - 1| < 1$, so $\arg\nu \in (-\frac{1}{2}\pi, \frac{1}{2}\pi)$, which implies that (16) *uniquely* defines the 'phase' ϕ beyond the basic angular interval, and thus parametrizes the universal covering group $\overline{\mathsf{Sp}}(2,\mathsf{R})$. The manifold of the symplectic group $\mathsf{Sp}(2,\mathsf{R})$ is the exterior of a one-sheeted hyperboloid, where ϕ can effect any number of turns around its waist; see for example Ref. [4, Sect. 9.4]. In the metaplectic group $\mathsf{Mp}(2,\mathsf{R})$, ϕ is counted modulo 4π.

In (10)–(12) we drew attention to three cases: $g\zeta = 1$, 2, and 3, which resulted in diagonal matrices, either $\begin{pmatrix} -1 & 0 \\ 0 & -1 \end{pmatrix}$ or $\begin{pmatrix} 1 & 0 \\ 0 & 1 \end{pmatrix}$ —which do not depend on g or ζ separately. With (14)–(15) and the products (16)–(18) we can know whether these form the 'first' or the 'second' metaplectic (and true) unit. For reference, we annotate that in the three cases of $g\zeta$, the matrices $\mathbf{M}$ in (10) correspond to the following Bargmann parameters for $\zeta = 1$:

$$\begin{aligned}
g\zeta = 1: \quad & \mathbf{M}_1 := \begin{pmatrix} 1 & 1 \\ -1 & 0 \end{pmatrix} \quad && \phi = -1.10715, \quad && \mu = 0.223607 + 0.447214\,\mathrm{i}, \\
g\zeta = 2: \quad & \mathbf{M}_2 := \begin{pmatrix} 1 & 1 \\ -2 & -1 \end{pmatrix} \quad && \phi = -\tfrac{1}{2}\pi, \quad && \mu = \tfrac{1}{2} + \mathrm{i}, \\
g\zeta = 3: \quad & \mathbf{M}_3 := \begin{matrix} 1 & 1 \\ -3 & -2 \end{matrix} \quad && \phi = -1.81577, \quad && \mu = 0.606339 + 1.697750\,\mathrm{i}.
\end{aligned} \tag{19}$$

Values of ζ close to 1 lead to values of ϕ, μ that are close to those listed above but, as we said, we are only interested in the cases when the powers of (19) are diagonal matrices, $\mathbf{1}$ or $-\mathbf{1}$, independent of ζ and g separately.

Since the hand computation of products in terms of Bargmann parameters is rather arduous, we resorted to symbolic and numeric computation with WOLFRAM MATHEMATICA which, after thorough checking, yielded the following parameter values for their powers:

$$
\begin{array}{llll}
g\zeta = 1: & \mathbf{M}_1^3 = -\mathbf{1}, & \phi = -\pi, & \mu = 0, \\
 & \mathbf{M}_1^6 = +\mathbf{1}, & \phi = -2\pi, & \mu = 0; \\
g\zeta = 2: & \mathbf{M}_2^2 = -\mathbf{1}, & \phi = -\pi, & \mu = 0, \\
 & \mathbf{M}_2^4 = +\mathbf{1}, & \phi = -2\pi, & \mu = 0, \\
 & \mathbf{M}_2^6 = -\mathbf{1}, & \phi = -3\pi, & \mu = 0; \\
g\zeta = 3: & \mathbf{M}_3^3 = +\mathbf{1}, & \phi = -2\pi, & \mu = 0, \\
 & \mathbf{M}_3^6 = +\mathbf{1}, & \phi = -4\pi, & \mu = 0.
\end{array}
\tag{20}
$$

From here we can conclude that the symplectic unit can be realized with the three-lens setup $\mathbf{M}_3^3$ shown in Fig. 3 as a minimal arrangement (see Ref. [4, Subsec. 10.5.4]), but also by means of a $\mathbf{M}_2^4$ four-lens arrangement with $g\zeta = 2$, or with six lenses where $g\zeta = 1$, as shown in Figs. 1 and 4. The metaplectic unit appears when we double the setups, and thus appears only in $\mathbf{M}_3^6$, with $g = 3/\zeta$.

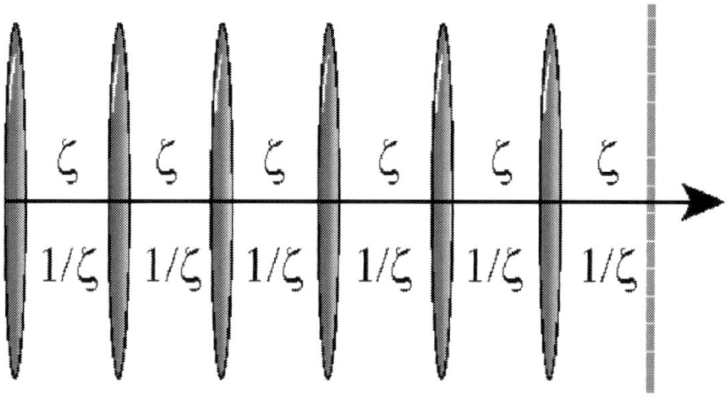

Figure 4. Six lenses of power $g = 1/\zeta$ and six free flights by ζ yield the symplectic identity transformation between ingoing and outgoing rays. The metaplectic identity is obtained by concatenating two symplectic identities each realized by the setup in the preceding Fig. 3.

5. Conclusions

When Moshinsky and Quesne defined linear canonical transforms [9] as a representation of the group of symplectic transforms that preserve the Heisenberg-Weyl algebra generated by the quantum operators $\hat{Q}$, $\hat{P}$ and $\hat{1}$, realized on the space $\mathcal{L}^2(\mathsf{R})$ of quantum wavefunctions $\psi(q)$, they wrote it as an integral transform kernel [10, Ch. 9],

$$C_{\left(\begin{smallmatrix} a & b \\ c & d \end{smallmatrix}\right)}(q, q') = \frac{\mathrm{e}^{-\mathrm{i}\pi/4}}{\sqrt{2\pi b}} \exp\left(\mathrm{i}\frac{aq'^2 - 2q'q + dq^2}{2b}\right), \tag{21}$$

where $ad - bc = 1$. The square root of b in the denominator already announces a double valuation. The product of two transformations was investigated in that same article, leading to a sign that depends on the b-elements of the two factors and the product matrices. The symplectic unit would multiply $\psi(q)$ by a minus sign while the metaplectic unit would be the true unit transform.

Although we have often written the metaplectic product law using the Bargmann parameters (13) in symbolic form, I cannot recall any paper where they have been explicitly computed for a given optical or quantum setup. In performing the exercise of analyzing the three- and six-lens optical systems in Figs. 3 and 4, as well as the product of any number of optical elements, we can also describe their quantum counterpart systems.

In the Introduction we stated that optical models are richer than quantum-mechanical ones. This is true in the metaxial (i.e., beyond the paraxial) régime because of the *aberrations* due to the non-flat nature of lenses, and because now the momentum (6) is no longer $p = n\theta$, so that now $q \mapsto q + zn\sin\theta$, and the terms $\sim p^{r+1}$ in (7) represent r^{th} order aberrations, for $r \in \{3, 5, 7, \ldots\}$. As Figs. 3 and 4 suggest, as we move the screen along the z-axis, a non-flat convex lens represents a potential jolt along the two endpoints of a segment (the border of a disk in two-dimensional screens) that grows along the z (or τ) axis from zero to the size of the lens, and then decreases to a point and vanishes.

Generally metaxial systems require functions $p^u q^v$, with u, v integers and $u + v > 2$, and their corresponding Poisson or quantum operators, the latter following the Weyl correspondence rule to ensure hermiticity. The groups built in this way are $\mathsf{A}_r \triangleleft \mathsf{Sp(2,R)}$, where A_r is the normal subgroup of rth order aberrations, generated by all monomials $p^u q^v$ with $2 < u + v \le r$, plus the assumption that $p^u q^v \approx 0$ for $u + v > r$ [4, Part 4]. In this régime the primary interest is to minimize the aberrations through the best choice of the surface shape [11] and of the refractive index profile $n(q)$. If we only have constructible physical lenses and free flights, it is unclear that these optical transformations can be bent back to unity, and it is most likely that they may not.

Acknowledgments

I thank Guillermo Krötzsch for assistance with the figures and Alejandro R. Urzúa for his interest in this work, as well as the PAPIIT-UNAM support of the Óptica Matemática projects PAPIIT IN101011 and IN101115 in this period.

References

[1] B. Mielnik, Global mobility of Schrödinger's particle, *Rep. Math. Phys.* **12**, 331–339 (1977).
[2] B. Mielnik, Evolution loops, *J. Math. Phys.* **27**, 2290–2306 (1986).
[3] I. Białynicki-Birula, B. Mielnik, and J. Plebański, Explicit solution of the continuous Baker-Campbell-Hausdorff problem and a new expression for the phase operator, *Ann. Phys.* **51**, 187–200 (1969).
[4] K.B. Wolf, *Geometric Optics on Phase Space* (Springer-Verlag, Heidelberg, 2004).
[5] S. Cruz y Cruz and O. Rosas-Ortiz, Lagrange equations and spectrum generating algebras of mechanical systems with position-dependent mass. In: Symmetry, Integrability and Geometry: Methods and Applications, *SIGMA* **9**, 004 (2013), 21 pp.
[6] K.B. Wolf and G. Krötzsch, El problema de las tres lentes, *Rev. Mex. Fís.* **47**, 291–298 (2001)
[7] V. Bargmann, Irreducible unitary representations of the Lorentz group, *Ann. Math.* **48**, 568–642 (1947).
[8] V. Bargmann, Group representations in Hilbert spaces of analytic functions. In: *Analytical Methods in Mathematical Physics*, Ed. by P. Gilbert and R.G. Newton (Gordon & Breach, New York, 1970), pp. 27–63.
[9] M. Moshinsky and C. Quesne, Linear canonical transformations and their unitary representation, *J. Math. Phys.* **12**, 1772–1780 (1971).
[10] K.B. Wolf, *Integral Transforms in Science and Engineering* (Plenum Publ. Corp., New York, 1979).
[11] K.B. Wolf and G. Krötzsch, Metaxial correction of fractional Fourier transformers, *J. Opt. Soc. Am. A* **16**, 821–830 (1999).

Quantum control and the challenge of non-Hermitian model-building

Miloslav Znojil
Nuclear Physics Institute ASCR, 250 68 Řež, Czech Republic
E-mail: `znojil@ujf.cas.cz`

Abstract. In a way inspired by the brief 2002 note "The challenge of nonhermitian structures in physics" by Ramirez and Mielnik (with the text most easily available via arXiv: quant-ph/0211048) the situation in the theory is briefly summarized here as it looks twelve years later. Our text has three parts. In the first one we briefly mention the pre-history (dating back to the Freeman Dyson's proposal of the non-Hermitian-Hamiltonian method in 1956 and to its subsequent successful "interacting boson model" applications in nuclear physics) and, first of all, the amazing recent progress reached, in the stationary case, using, in essence, an inversion of the Dyson's approach. The impact on the latter idea upon abstract quantum physics is sampled, first of all, by the reference to papers by Bender et al. (who made the non-Hermitian model-building popular under the nickname of parity-times-time-reflection-symmetric *alias* PT-symmetric quantum mechanics) and by Mostafazadeh (who reinterpreted PT-symmetry as P-pseudo-Hermiticity). In the second part of our review the emphasis is shifted to the newest, non-stationary upgrade of the formalism which we proposed in the year 2009 and which is characterized by the simultaneous participation of a triplet of Hilbert spaces $\mathcal{H}$ in the representation of a single quantum system. In the third part of the review we finally emphasize that the majority of applications of our three-Hilbert-space (THS) recipe is still ahead of us because the enhancement of the flexibility is necessarily accompanied by an enhancement of the technical difficulties. An escape out of the technical trap is proposed to be sought in a restriction of attention to quantum models living in finite-dimensional Hilbert spaces $\mathcal{H}$. As long as the use of such spaces is so typical for the quantum-control considerations, we conclude with conjecture that the THS formalism should start searching for implementations in the field of quantum control.

1. Introduction and summary

1.1. Schrödinger equation and quantum control

In accord with the standard textbooks on quantum theory [1] the evolution of a pure state of a closed quantum system is most comfortably determined by solving Schrödinger equation

$$i\hbar\frac{\partial}{\partial t}|\psi(t)\rangle^{(P)} = \mathfrak{h}(t)|\psi(t)\rangle^{(P)} \qquad (1)$$

in which one assumes that the states of the system in question are represented by elements $|\psi(t)\rangle^{(P)}$ of a properly selected physical Hilbert space $\mathcal{H}^{(P)}$ and in which the generator of evolution (called Hamiltonian) is self-adjoint, $\mathfrak{h}(t) = \mathfrak{h}^\dagger(t)$.

Using the notation conventions of Ref. [2] and abbreviating $|\psi(t)\rangle^{(P)} \equiv |\psi(t)\succ$ one traditionally assumes that at $t = 0$ the system is prepared in an initial state $|\psi(0)\succ = |\psi_i\succ$

and that it is detected, after some time $T > 0$, in a final state $|\psi(T)\succ = |\psi_f\succ$. Thus, in the most conventional approach one knows $\mathfrak{h}(t)$ and constructs the final state $|\psi_f\succ \in \mathcal{H}^{(P)}$.

An entirely different task is typical for the so called quantum-control (QC) setup. In its most elementary specification (often called bilinear model – see, e.g., review paper [3] for details) one is given just a desirable final target state $|\psi(T)\succ = |\psi_f\succ$. For the purpose, one has to select a suitable "realization Hamiltonian" $\mathfrak{h}(t) = \mathfrak{h}^\dagger(t)$ and specify the necessary "realization time" $T > 0$.

For technical reasons people often restrict their attention to finite-dimensional quantum systems living in an N-dimensional complex Hilbert spaces $\mathcal{H}^{(P)} = \ell^{(N)}$ in which the admissible self-adjoint QC Hamiltonians have the form of a superposition

$$\mathfrak{h}(t) = \mathfrak{h}_0 + \sum_{k=1}^{K} u_k(t)\mathfrak{h}_k\,. \tag{2}$$

The $(K+1)$–plet of auxiliary operators $\{\mathfrak{h}_0, \mathfrak{h}_1, \ldots, \mathfrak{h}_K\}$ is assumed time-independent. Moreover, this multiplet of operators is often chosen as a set of generators of a Lie algebra $\mathcal{L}_0$ such that the desired evolution of the system towards a given target state may be proved to exist (one speaks about a "controllability" [4]). In such a setting the target $|\psi_f\succ$ is to be reached solely via the selection of the real coefficients $u_k(t) \in \mathbb{R}$ called *control functions*.

1.2. The plan and summary of the paper

In our present paper we intend to expose the standard, above-outlined formulation of the quantum control problem to a modification. It will be inspired by the recent developments in quantum theory in which one complements the *ad hoc* choice of dynamics (i.e., of the Hamiltonian) by the possibility of an *independent* alteration of the Hilbert space itself.

The essence of the latter developments will be explained in sections 2, 3 and 4. Firstly, in section 2 we shall follow, for pedagogical reasons, older reviews [5, 6] and introduce the amendment $\mathfrak{h} \to G \neq G^\dagger$ of the generator of evolution in its simplified, time-independent version with $\mathfrak{h} \neq \mathfrak{h}(t)$ replaced by $G \neq G(t)$ in Schrödinger Eq. (1). Marginally, let us add that from the present perspective, sections 2 and 3 should be read as a mere contextual introduction. They offer a review of recent updates of quantum theory which became widely known as $\mathcal{PT}$–symmetric Quantum Mechanics (PTSQM, cf. the Bender's review paper [5]) or, in a slightly more general form, as Pseudo-Hermitian representation of Quantum Mechanics (PHRQM, cf. the Mostafazadeh's review paper [6]). As long as in both of these approaches the operators of observables must remain stationary (or, at best, quasi-stationary [7]) none of these formalism is directly applicable in the QC context.

Subsequently, section 4 will outline the upgraded and generalized (a.k.a. "three-Hilbert-space", THS) formalism of Refs. [2,8]. Our discussion will cover the case in which the manifest time dependence of $\mathfrak{h} = \mathfrak{h}(t)$ and of $G = G(t)$ is permitted. Although some of the preceding ideas remain unchanged, it will be necessary to shift the emphasis. Indeed, only a change of perspective will enable us to address the QC-related conceptual questions. In this sense, the next section 5 should be read as a more technical addendum reviewing a few aspects of necessary mathematics. The key message is that the most general time-dependent Hamiltonian-like operators may still be required to generate the standard *unitary* evolution of a given quantum system in time. We shall also explain why, in contrast to the PTSQM or PHRQM scenarios, the spectra of our present Hamiltonian-like generators $G(t) \neq G^\dagger(t)$ are, in general, complex.

In the key part and climax of our message in section 6 we shall return to the problems of quantum control, outlining briefly the possible use of the whole THS machinery for an enhancement of the flexibility and efficiency of the specific QC tasks. Preliminarily, our proposal may be summarized as opening a new approach to quantum control in which one extends the model-building freedom via a transfer of Schrödinger equation from its representation (1) in the "primary" Hilbert space $\mathcal{H}^{(P)}$ (which is assumed to appear, for any reason, unfriendly) to some of its alternative though, by assumption, equivalent and technically friendlier forms.

A few complementary comments on such a possibility will be finally formulated in our last section 7. We shall emphasize that the THS-representation-mediated introduction of the manifestly time-dependent non-Hermitian generators of evolution $G(t)$ is in fact necessary in the QC context. We believe that our considerations will offer a sufficiently strong encouragement for a more concrete model-building activity in the nearest future.

2. Time-independent non-Hermitian Hamiltonians in Quantum Mechanics

2.1. Modified Schrödinger equation

It is well known [1] that in principle, the constructive solution of Schrödinger Eq. (1) is particularly straightforward for Hamiltonians which are Hermitian, diagonalized and *not* time-dependent. Even in these cases, there exist quantum systems (like, for example, heavy atomic nuclei) for which even the brute-force numerical diagonalization of a given Hermitian $\mathfrak{h} \neq \mathfrak{h}(t)$ yields, typically, very poorly convergent results.

In the year 1956, one of the most unexpected ways out of similar difficulties has been proposed by Dyson [9]. He proposed a reparametrization $|\psi(t)\rangle^{(P)} = \Omega\,|\psi(t)\rangle^{(F)}$ of wave functions in which a "friendlier" ket $|\psi(t)\rangle^{(F)}$ was assumed to belong to a "friendlier" Hilbert space $\mathcal{H}^{(F)}$. Moreover, the time-independent mapping Ω was chosen, in contrast to common practice, *non-unitary*, yielding a nontrivial operator product $\Omega^\dagger\Omega \neq I$. In this way, the insertion in Eq. (1) led to a potentially friendlier Schrödinger equation defined in the new Hilbert space,

$$i\hbar\frac{\partial}{\partial t}|\psi(t)\rangle^{(F)} = H\,|\psi(t)\rangle^{(F)}\,, \qquad H = \Omega^{-1}\mathfrak{h}\Omega \neq H(t)\,. \tag{3}$$

Naturally, in the numerical setting the Dyson's trick and Hilbert-space invertible mapping $\Omega : \mathcal{H}^{(F)} \to \mathcal{H}^{(P)}$ only made sense if it led to an accelerated convergence but its enormous success may be found confirmed, e.g., in the recent nuclear-physics-devoted review paper [10].

A not entirely pleasant consequence of the non-Hermiticity of the Dyson's mapping may be seen in the emergence of a manifest non-Hermiticity of H. Indeed, in the new language the old Hermiticity rule reads

$$\mathfrak{h}^\dagger = [\Omega^{-1}]^\dagger\, H^\dagger\, \Omega^\dagger = \Omega\, H\, \Omega^{-1} \tag{4}$$

and may be re-written in a more compact form

$$H^\dagger\,\Theta = \Theta\, H\,, \qquad \Theta = \Omega^\dagger\,\Omega \neq I\,. \tag{5}$$

Thus, the new Hamiltonian may only be declared "quasi-Hermitian" [11].

2.2. The coexistence and mutual relations of the triplet of simultaneous representation Hilbert spaces

A new life of the same old trick has been conceived in 1998 when Bender with Boettcher [12] proposed the use of certain non-Hermitian H with real spectrum in the role of a standard quantum energy observable. Subsequently, the consistent PTSQM formalism (with its physics-inspired emphasis on the additional feature of parity-times-time-reversal symmetry) has been born, in its final form, in the year 2004 [5, 13]. In parallel, also the more general, less restrictive PHRQM version of the formalism as already known to nuclear physicists before 1992 [10] was given new life and popularity by Mostafazadeh (see his numerous publications and/or their summary in his comprehensive review paper [6]).

For our present purposes the PHRQM relations between the above-mentioned P- and F-superscripted Hilbert spaces (and between these two spaces and the third, S-superscripted space which only differs from the F-space by the use of the metric-mediated, i.e., $\Theta-$mediated inner product) may be summarized using the following diagram,

P
unitary evolution generated by
prohibitively complicated $\mathfrak{h}$
physics as in traditional textbooks
calculations = practically **impossible**

simplification Ω^{-1} ↙ ↘↖ **equivalence**

F		**S**
inner product = **trivial**	**hermitization** ⟶	inner product = **nontrivial**
Hilbert space = **friendly**		$H = H^{\ddagger} = \Theta^{-1}H^{\dagger}\Theta$ = **simple**
all physics = **falsified**		interpretation = **standard**
calculations = **feasible**		$\Theta = \Omega^{\dagger}\Omega$ = **sophisticated**

3. A note on the history and applications

3.1. The birth and the resolution of the puzzle.

Although the very compact review-like 2002 note "The challenge of non-Hermitian structures in physics" by A. Ramirez and B. Mielnik [14] is merely twelve years old, the subject and its applications in the various branches of physics developed, in between, so quickly that one should (and, in what follows, we are going to) update some of their conclusions. *Pars pro toto*, today, the Ramirez's and Mielnik's citation of the 2001 note [15] offering a vague indication of the non-Hermitian Hamiltonian's having "link with pseudo-euclidean structures" [14] would have to be complemented by the reference to the subsequent 2004 paper [16]. In the latter text the authors considered the same illustrative non-Hermitian square-well Hamiltonian but they already were able to explain its full compatibility with the first principles of conventional quantum theory. In this manner the latter authors provided a virtually exhaustive resolution of all of the related apparent paradoxes. Thus, in brief, one can only repeat that after the year 2004, the "consistent interpretation" of non-Hermitian quantum Hamiltonians $H \neq H^{\dagger}$ of Bender with Boettcher [12] could not have been declared "missing" by the authors of Ref. [14] anymore.

3.2. The current state of art and the continuing emergence of new puzzles

A brief recollection of the developments in the field during the last twelve years reveals that the related research activities did not stop after 2004. Naturally, an understanding of the basic idea was already available but multiple open questions survived. Many of them were already asked around the end of the millennium, i.e., immediately after the publication of the inspiring letter [12]. During a few years, many non-Hermitian Hamiltonians $H^{(\mathrm{NH})}$ with real spectra were then analyzed by many authors. Still, using the words of *loc. cit.*, "in all of these designs" the proper "statistical interpretation [was] still missing" [14].

Fortunately, as we already mentioned, the progress was quick. Around the year 2004, virtually all of the essential connections between the exotic-looking $H^{(\mathrm{NH})}$ and the conventional quantum theory seem to have been already established. Still, new ideas kept emerging even after the year 2004. Typically, the ambiguity problems concerning the assignment of a metric Θ to a given Hamiltonian H were never completely abandoned. Between the years 2007 - 2010 people also re-opened [17] and solved [18] the puzzling unitarity/non-unitarity conflict in the scattering arrangement. Similarly, due to the apparent failure of the semi-classical approximations, a new crisis emerged very recently [19]. Last though not least, even the lasting conflict between the intuitive and rigorous quantum-theoretical perception of the concept of locality did also hit the PTSQM theory in the past [20] as well as very recently [21]. Nevertheless, in a way paralleling these fluctuations, various versions of the general THS theory may be now declared to have acquired, under several sophisticated technical assumptions, a more or less closed and final-looking form.

4. The challenge of time-dependent non-Hermitian Hamiltonians in Quantum Mechanics

4.1. A remark on terminology

In the next-to-perfect Mostafazadeh's review paper [6] the physicists read, with satisfaction, that the "time-dependent quasi-Hermitian Hamiltonians arise naturally in the application of pseudo-Hermitian quantum mechanics in quantum cosmology". At the same time the mathematicians could feel puzzled when reading there that "in pseudo-Hermitian quantum mechanics we are bound to use quasi-stationary Hamiltonians" defined as "admitting a time-independent metric". Puzzling as a comparison of these two statements may sound (cf. also the unpublished discussion of this topic in arXiv [22]), it in fact merely reflects the Mostafazadeh's unexplained decision of working, exclusively, with the *observable* generators of the quantum time evolution. In other words, the scope of Mostafazadeh's PHRQM formulation remains restricted to the above-mentioned Schrödinger Eq. (3) in which the generators of evolution H remain compatible with the Dieudonne's quasi-Hermiticity constraint (5). Then, together with the standard requirements of the unitarity of the theory this would really imply that we must have $\Theta \neq \Theta(t)$, indeed.

In this sense, the THS time-dependent-metric representation of a quantum system as proposed in Refs. [2, 8] may be perceived as a further nontrivial generalization of the Bender's time-independent-metric PTSQM frame as well as of the Mostafazadeh's time-independent-metric formalism of PHRQM.

4.2. The challenge of time-dependent metrics

Once we admit that the crypto-Hermitian time-evolution time-independent-metric law (3) may be further generalized, our constructive considerations become straightforward (cf. [2, 8] for details). First of all, admitting the explicit time-variability of the Dyson's map $\Omega = \Omega(t)$ of the ket-vector spaces $\mathcal{H}^{(F)} \to \mathcal{H}^{(P)}$, i.e., postulating the relation

$$|\psi(t)\rangle^{(P)} = \Omega(t)\,|\psi(t)\rangle^{(F)} \tag{6}$$

an elementary insertion of this ansatz in the original Schrödinger Eq. (1) immediately yields the properly modified form of its equivalent representation in the friendlier Hilbert space $\mathcal{H}^{(F)}$,

$$i\hbar\frac{\partial}{\partial t}|\psi(t)\rangle^{(F)} = G(t)\,|\psi(t)\rangle^{(F)}\,, \qquad G(t) = H(t) - \Sigma(t)\,. \tag{7}$$

In this evolution equation the new, time-dependent isospectral image

$$H(t) = \Omega^{-1}(t)\,\mathfrak{h}(t)\,\Omega(t) = H^{\ddagger}(t) = \Theta^{-1}(t)\,H^{\dagger}(t)\,\Theta(t) \tag{8}$$

of the original Hamiltonian enters the time-dependent generator of quantum evolution in combination with the so called [23] quantum Coriolis force

$$\Sigma(t) = \mathrm{i}\Omega^{-1}(t)\,\dot{\Omega}(t)\,, \quad \dot{\Omega}(t) = \partial_t\Omega(t)\,. \tag{9}$$

We should add that the emergence of the Coriolis term $\Sigma(t)$ simply reflects the emergence of the manifest time-dependence of the inner products in the alternative physical and nontrivial-metric-endowed third Hilbert space $\mathcal{H}^{(S)}$. Secondly, we should emphasize that the latter space still coincides with $\mathcal{H}^{(F)}$ up to the metric, i.e., as a topological vector space [6]. Thirdly, we may return now to the first paragraph of this section and see that separately, *both* the "virtual-force" Coriolis operator $\Sigma(t)$ *and* the related time-dependent generator $G(t) = H(t) - \Sigma(t)$ become, in general, *unobservable*. In other words, the requirement of the manifest time-dependence of the generator in our most general but still *unitarity-guaranteeing* Schrödinger Eq. (7) implies that the spectrum of such an operator $G(t)$ *may cease to be real*.

In this sense, the standard textbook Quantum Theory admits its phenomenologically most general but still mathematically fully consistent THS representation as introduced in Ref. [8], reviewed in Ref. [2] and described by the following amended diagram

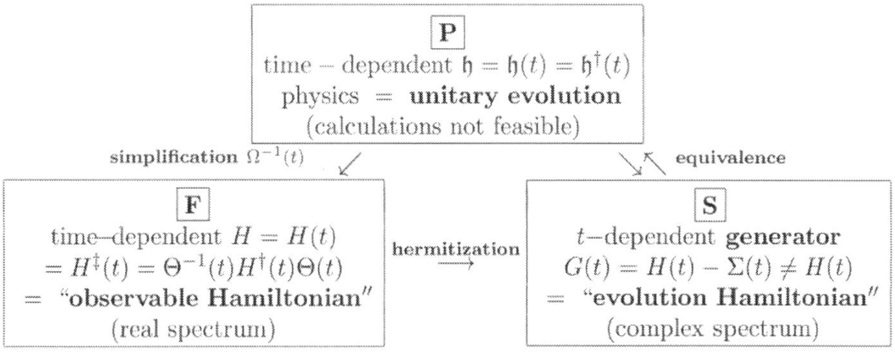

5. The properties of the sophisticated physical Hilbert space $\mathcal{H}^{(S)}$

5.1. A mixed blessing of the use of the time-dependent Dyson maps

The two-step realization $P \to F \to S$ of the unitary equivalence between the two alternative *physical* Hilbert spaces $\mathcal{H}^{(P)}$ and $\mathcal{H}^{(S)}$ has its merits (e.g., a simplification $\mathfrak{h}(t) \to H(t)$ of the observable of the instantaneous but still, in principle, measurable P−space-based energy of the system) and shortcomings (e.g., the use of a frequently rather misleading terminology). Still, the above-cited emergence of the non-Hermitian plus time-dependent forms of the hiddenly unitary quantum evolution law (7) "in ... quantum cosmology" [6] offers a sufficiently persuasive motivation for the mathematical study as well as for new proposals of phenomenological applications of the THS representations of quantum systems which are made more flexible by the permission of time-dependence in the underlying Dyson's maps $\Omega(t)$.

Naturally, the price to be paid for the maximally enhanced flexibility of Eq. (7) is not too low. In particular, the original motivation of the formalism (which proved fairly persuasive in theory, plus strong in applications) gets perceivably weakened in the time-dependent THS case [23]. Moreover, the technical difficulties further increase if we decide to invert the original "Dyson's" direction $P \to F \to S$ of the construction as incorporated in the PHRQM formalism of Refs. [6, 9, 10] in which one started all considerations from a given pair of operators $\mathfrak{h} \neq \mathfrak{h}(t)$ and $\Omega \neq \Omega(t)$. Indeed, even in the perceivably more restrictive but still time-independent-metric-unsing PTSQM formalism as summarized in Ref. [5] the situation appeared complicated since the pair of operators $\mathfrak{h} \neq \mathfrak{h}(t)$ and $\Omega \neq \Omega(t)$ only had to be reconstructed at the very end of all of the constructive manipulations (cf., e.g., an exactly solvable model [24] for illustration).

All this explains why the current progress in the cosmological applications of the THS formalism (cf., e.g., their first preliminary samples in [25]) still remains so deplorably slow. At the same time, the current tradition of the use of just finite-dimensional Hilbert spaces in the context of quantum control seems to open new perspectives for applications of the time-dependent-THS Schrödinger Eq. (7). Let us, therefore, complement our preceding introductory THS outline by a few most relevant further technicalities.

5.2. Ad hoc *notation conventions*

First of all, let us remind the readers of our review paper [2] that in parallel to the above-mentioned formal coincidence of kets $|\psi(t)\rangle^{(F)} = |\psi(t)\rangle^{(S)} = |\psi(t)\rangle$ (i.e., to the formal coincidence of the two ket-vector spaces $\mathcal{H}^{(F)}$ and $\mathcal{H}^{(S)}$) one has to keep in mind that the respective conjugate dual-space elements *alias* primed-vector-space elements *alias* linear functionals (i.e., in the Dirac's terminology, the bra-vectors ${}^{(F)}\langle\psi| \in \left[\mathcal{H}^{(F)}\right]'$ and ${}^{(S)}\langle\psi| \in \left[\mathcal{H}^{(S)}\right]'$ which are assigned to the corresponding ket vectors via the respective Hermitian-conjugation antilinear operations $\mathcal{T}$) remain different,

$$ {}^{(S)}\langle\psi| \equiv {}^{(F)}\langle\psi|\,\Theta \neq {}^{(F)}\langle\psi|\,. \qquad (10) $$

In order to emphasize this important feature of the $F \leftrightarrow S$ corrrespondence we shall use the notation of [2] and abbreviate $\langle\!\langle\psi| \equiv {}^{(S)}\langle\psi| \in \left[\mathcal{H}^{(S)}\right]'$ in what follows.

In order to suppress confusion we shall also accept another convention that all of the eligible Hermitian conjugations $\mathcal{T} : |\bullet\rangle \to \langle\bullet|$ will occur without superscripts, i.e., they will always be understood as performed solely in the trivial-metric spaces, i.e., just in our ${}^{P}-$ or ${}^{F}-$superscripted Hilbert spaces. This means that in our present paper we shall never employ the abbreviated and metric-dependent Hermitian-conjugation operation $\mathcal{T}^{(S)}$. Thus, for example, the inverse conjugation $\mathcal{T}^{-1} : \langle\!\langle\bullet| \to |\bullet\rangle\!\rangle$ will be always understood as performed just in the friendly, ${}^{F}-$superscripted Hilbert space, etc.

The use of such notation conventions enables us to characterize the unitary equivalence between our ${}^{P}-$ and ${}^{S}-$superscripted Hilbert space in an extremely compact manner, viz., via the following coincidence of the respective inner products,

$$ \prec\psi_1|\psi_2\succ \;\; (\equiv {}^{(P)}\langle\psi_1|\psi_2\rangle^{(P)}) = \langle\!\langle\psi_1|\psi_2\rangle \; (\equiv {}^{(S)}\langle\psi_1|\psi_2\rangle^{(S)})\,. \qquad (11) $$

Moreover, the conventional textbook use of an orthonormalized basis $\{\,|n\succ\,\}$ in $\mathcal{H}^{(P)}$ may be immediately paralleled by its ${}^{S}-$superscripted-Hilbert-space (bi)orthonormal-basis descendant with the respective kets $|n\rangle$ and bras $\langle\!\langle n|$, etc. On these grounds one characterizes a *physical* state of a given quantum system *either* by the kets $|\psi(t)\succ$ and bras $\prec\psi(t)|$ in the ${}^{P}-$superscripted representation *or, alternatively*, by the "simpler" kets $|\psi(t)\rangle$ and bras $\langle\!\langle\psi(t)|$ in their preferable but formally strictly unitarily equivalent ${}^{S}-$superscripted representation.

5.3. *Quantum time-evolution and its control by the* two *Schrödinger equations*

In the constructive mathematical perspective a decisive THS-representation advantage is that one never has to leave the auxiliary friendly space, treating the structure-reflecting concepts and symbols like, e.g., $\langle\!\langle\psi(t)|$ or $H^{\ddagger}$ as the mere metric-containing abbreviations. Moreover, as we already mentioned, a key benefit of our conventions is that after an ultimate return to the friendly Hilbert space we have got rid of all of the superscripts. In particular, the fully general non-Hermitian THS quantum evolution process as described in Ref. [2] may be now perceived as initiated, at time $t = 0$ and in its friendly $\mathcal{H}^{(F)}$ representation, by the choice of *two* initial ket-vectors $|\psi(0)\rangle$ and $|\psi(0)\rangle\!\rangle$, with the latter one being formally expressible, in the cases when we know the metric, as the metric-multiple $\Theta(0)\,|\psi(0)\rangle$. Next, in the Dyson-inspired direct $P \to F \to S$ recipe one has to know the generator $G(t)$ and, as long as $G(t) \neq G^{\dagger}(t)$, one must solve the *two* time-evolution Schrödinger equations,

$$ i\hbar\frac{\partial}{\partial t}|\psi(t)\rangle = G(t)\,|\psi(t)\rangle \qquad (12) $$

$$ i\hbar\frac{\partial}{\partial t}|\psi(t)\rangle\!\rangle = G^{\dagger}(t)\,|\psi(t)\rangle\!\rangle \qquad (13) $$

(incidentally, notice an unfortunate misprint in [2]). One can also find another benefit of our notation in the subsequent elementary re-derivation of formula

$$\partial_t \langle\!\langle \psi(t)|\psi(t)\rangle = 0\,, \tag{14}$$

i.e., in a reconfirmation of conservation law for the norm of state $\psi(t)$ when considered in its amended physical $^{(S)}$–superscripted representation.

6. Time-dependent Dyson maps in quantum control setup

6.1. A sample of the realization of the project of generalized non-Hermitian quantum control

In the THS generalization the traditional QC superposition ansatz (2) may be made less restrictive in several directions involving, first of all, several alternative real-control-function assumptions. Thus, the traditional Hermitian QC-related postulate (2) may be replaced, say, by its analogues describing the "evolution Hamiltonian" with a complex spectrum

$$G(t) = G_0 + \sum_{k=1}^{K_G} u_k(t) G_k \tag{15}$$

and/or the "observable Hamiltonian" with the (time-varying but, in principle, measurable) instantaneous-energy real spectrum,

$$H(t) = H_0 + \sum_{m=1}^{K_H} z_m(t) H_m \tag{16}$$

etc. Naturally, the most fundamental innovation may be expected to result from the highly nontrivial nature of the non-unitary, manifestly time-dependent Dyson's maps, say, of the same multinomial form

$$\Omega(t) = \Omega_0 + \sum_{n=1}^{K_\Omega} v_n(t)\Omega_n\,. \tag{17}$$

Obviously, as long as the knowledge of $\Omega(t)$ implies the knowledge of the Coriolis term $\Sigma(t)$, the role of assumption (17) seems fundamental. Only when we choose $K_\Omega = 1$ and set $\Omega_0 = 0$ we still obtain a transparent multinomial-operator toy model with metric $\Theta(t) = v^2(t)\Theta_1$ and with a diagonal-matrix Coriolis operator $\Sigma(t) = \mathrm{i}\dot{v}(t)/v(t)\, I = -\mathrm{i}w(t)\, I$.

7. The ultimate reconstruction challenge

In the context of our preceding illustrative example we may prolong our methodical analysis and choose, say, $K_H = 1$. This will enable us to insert all ansatzs in the Dieudonné's observability requirement (5). With the real control function $z_1(t) = z(t)$, this requirement becomes time-variation-independent and it may be separated and solved elementwise, yielding two conditions

$$H_0^\dagger\,\Theta_1 = \Theta_1\, H_0\,, \qquad H_1^\dagger\,\Theta_1 = \Theta_1\, H_1\,. \tag{18}$$

If the solution Θ_1 exists we shall be already able to derive the closed form of the generator

$$G(t) = H_0 + u(t)\, H_1 + w(t)\, H_2 \tag{19}$$

with $K_G = 2$ and $H_2 = \mathrm{i}\, I$. Thus, in a way, we shall return to a more or less standard QC scenario, with the main difference and innovation resulting from our new freedom of having the generator $G(t)$ which is non-Hermitian and which is even non-quasi-Hermitian (i.e., which does have complex eigenvalues).

In the latter context let us finally recall an unpublished preprint [26] in which Hynek Bíla tried to study a few more concrete non-Hermitian toy models living in the two-dimensional, i.e., in the first nontrivial friendly complex Hilbert space $\mathcal{H}^{(F)} = \ell^{(2)}$. This study revealed that one could also avoid the reference to the THS Schrödinger equations completely. In such an approach the necessary time-dependent metric $\Theta(t)$ has to be reconstructed via direct solution of the corresponding operator evolution differential equation of the Heisenberg-representation-resembling form which follows immediately from the definition of $\Sigma(t)$,

$$\mathrm{i}\partial_t\Theta(t) = G^\dagger(t)\Theta(t) - \Theta(t)\,G(t)\,. \tag{20}$$

Unfortunately, the Bíla's preliminary results were never completed (cf. also [27, 28]). Perhaps, the project itself could still acquire a new life in the non-Hermitian QC context.

In the conclusion let us add that the metric-determining "parallel Cauchy problem" (20) could be addressed by various techniques and under a multitude of approximations but in the QC context, the use of a finite-dimensional Hilbert-space approximation seems most promising, especially because it parallels the common practice used in the standard Hermitian models [3]. Thus, we believe that in the nearest future, the traditional Hermiticity condition $H(t) = H^\dagger(t)$ need not remain obligatory and uncircumventable, anymore.

References

[1] Landau L D and Lifshitz E M 1999 *Quantum Mechanics (Non-relativistic Theory)* (Oxford: Butterworth-Heinemann)
[2] Znojil M 2009 *Symm. Integ. Geom. Meth. Appl.* SIGMA05(2009)001 (*Preprint* arXiv:0901.0700)
[3] Dong D-Y and Petersen I R 2010 *IET Control Theory & Applications* **4** 2651
[4] D'Alessandro D 2007 *Introduction to Quantum Control and Dynamics* (Boca Raton, FL: Chapman & Hall/CRC)
[5] Bender C M 2007 *Rep. Prog. Phys.* **70** 947
[6] Mostafazadeh A 2010 *Int. J. Geom. Meth. Mod. Phys.* **7** 1191
[7] Mostafazadeh A 2007 *Phys. Lett.* B **650** 208
[8] Znojil M 2008 *Phys. Rev.* D **78** 085003
[9] Dyson F J 1956 *Phys. Rev.* **102** 1217
[10] Scholtz F G, Geyer H B and Hahne F J W 1992 *Ann. Phys.* (NY) **213** 74
[11] Dieudonne J 1961 *Proc. Int. Symp. Lin. Spaces* (Oxford: Pergamon) pp 115-122.
[12] Bender C M and Boettcher S 1998 *Phys. Rev. Lett.* **80** 5243
[13] Bender C M, Brody D C and Jones H F 2004 *Phys. Rev. Lett.* **92** 119902
[14] Ramirez A and Mielnik B 2003 *Rev. Mex. Fis. S. 2* **49** 130
[15] Znojil M 2001 *Phys. Lett.* A **285** 7
[16] Mostafazadeh A and Batal A 2004 *J. Phys. A: Math. Theor.* **37** 11645
[17] Jones H F 2007 *Phys. Rev.* D **76** 125003; Cannata F, Dedonder J-P and Ventura A 2007 *Ann. Phys.* (NY) **322** 397
[18] Znojil M 2009 *Phys. Rev.* D. **80** 045009
[19] Davies B 2000 *Bull. London Math. Soc.* **32** 432; Trefethen L N and Embree M 2005 *Spectra and pseudospectra. The behavior of nonnormal matrices and operators* (Princeton: University Press); Siegl P and Krejcirik D 2012 *Phys. Rev.* D **86** 121702(R)
[20] Günther U and Samsonov B 2008 *Phys. Rev. Lett.* **101** 230404
[21] Lee Y-C, Hsieh M-H, Flammia S T and Lee R-K 2014 *Phys. Rev. Lett.* **112** 130404
[22] Mostafazadeh A 2007 Comment on "Time-dependent quasi-Hermitian Hamiltonians and the unitary quantum evolution" (*Preprint* arXiv:0711.0137v1); Znojil M 2007 Reply to Comment on "Time-dependent quasi-Hermitian Hamiltonians and the unitary quantum evolution" (*Preprint* arXiv:0711.0514); Znojil M 2007 Which operator generates time evolution in Quantum Mechanics? (*Preprint* arXiv:0711.0535)
[23] Znojil M 2013 *Int. J. Theor. Phys.* **52** 2038
[24] Znojil M 2009 *J. Math. Phys.* **50** 122105
[25] Andrianov A A, Cannata F, Kamenshchik A Y 2006 *J. Phys. A: Math. Gen.* **39** 9975
Znojil M 2012 *J. Phys.: Conf. Series* **343** 012136
[26] Bíla H 2009 Adiabatic time-dependent metrics in PT-symmetric quantum theories (*Preprint* arXiv:0902.0474)

[27] Bíla H 2008 *Non-Hermitian Operators in Quantum Physics* (Prague: Charles University; PhD thesis)
[28] Gong J-B and Wang Q-H 2010 *Phys. Rev.* A **82** 012103; Gong J-B and Wang Q-H 2012 Time-Dependent PT-Symmetric Quantum Mechanics (*Preprint* arXiv:1210.5344)

A connection between supersymmetric quantum mechanics and Painlevé V equation

D Bermudez[1,2] **and David J Fernández C**[2]

[1]Department of Physics of Complex Systems, Weizmann Institute of Science, Rehovot 76100, Israel

[2]Departamento de Física, Cinvestav, A.P. 14-740, 07000 México D.F., Mexico

E-mail: `david.bermudez@weizmann.ac.il`, `david@fis.cinvestav.mx`

Dedicated to Professor Bogdan Mielnik for his 50 years of scientific career.

Abstract. In this article we introduce the relation between supersymmetric quantum mechanics (SUSY QM) and a second-order non-linear differential equation known as Painlevé V (PV) equation. To that end, we will first make a swift examination on the SUSY QM treatment of the radial oscillator and we will revisit its relation with the polynomial Heisenberg algebras (PHA). After that, we will formulate a theorem that connects SUSY QM to a set of solutions of the PV equation through specific PHA.

1. Introduction

In this article we will introduce a link joining supersymmetric quantum mechanics (SUSY QM) for the radial oscillator to analytical solutions of the PV equation. We will apply an approach that has already been proven efficient to solve the Painlevé IV equation [1–6]. This link has already been used both in the context of dressing chains [7–9] and in SUSY QM [1, 10–12].

The Painlevé V (PV) equation can be applied in the description of several systems in solid state physics [13], electrodynamics [14] and condense matter [15], thereby leading to studies of its geometric properties [16], numerical solutions [17], Bäcklund transformations [18], q-deformations [19], discrete versions [20], and others.

The main idea is that through SUSY QM we can obtain systems ruled by third-order polynomial Heisenberg algebras (PHA), generated by natural fourth-order differential ladder operators. In this work we will prove that these systems are connected with the PV equation. Up to our knowledge, this link was first realized in 1980 by Ablowitz *et al.* [21] and Flaschka [22]. Then, both subjects were connected through first-order SUSY QM [7–9] and later on using the higher-order case [1, 6, 10–12, 23, 24].

In view of this, we will first formulate a *reduction theorem* for the SUSY partners of the radial oscillator, which contains the necessary conditions on the transformation functions to reduce the order of the natural ladder operators associated with the SUSY partners from higher to fourth order. Then, we will study the properties of these fourth-order ladder operators in order to analyze the consequences of the theorem. In particular, we will study the different types of PV solutions that can be obtained with this method.

2. Supersymmetric quantum mechanics

In the k-th order SUSY QM one starts from a given solvable Hamiltonian

$$H_0 = -\frac{1}{2}\frac{\mathrm{d}^2}{\mathrm{d}x^2} + V_0(x), \tag{1}$$

and generates a chain of intertwined Hamiltonians as follows [25, 26]

$$H_j A_j^+ = A_j^+ H_{j-1}, \quad H_{j-1} A_j^- = A_j^- H_j, \tag{2a}$$

$$H_j = -\frac{1}{2}\frac{\mathrm{d}^2}{\mathrm{d}x^2} + V_j(x), \tag{2b}$$

$$A_j^\pm = \frac{1}{\sqrt{2}}\left[\mp\frac{\mathrm{d}}{\mathrm{d}x} + \alpha_j(x,\epsilon_j)\right], \quad j = 1,\dots,k. \tag{2c}$$

Therefore, the next equations have to be fulfilled:

$$\alpha_j'(x,\epsilon_j) + \alpha_j^2(x,\epsilon_j) = 2[V_{j-1}(x) - \epsilon_j], \tag{3a}$$

$$V_j(x) = V_{j-1}(x) - \alpha_j'(x,\epsilon_j). \tag{3b}$$

The final Riccati solution $\alpha_k(x,\epsilon_k)$ is determined by k solutions u_j of the associated Schrödinger equation:

$$H_0 u_j = -\frac{1}{2}u_j'' + V_0(x)u_j = \epsilon_j u_j, \quad j = 1,\dots,k, \tag{4}$$

which are connected with the initial Riccati solutions through $\alpha_1(x,\epsilon_j) = u_j'/u_j$.

Thus, there is a pair of k-th order operators intertwining the initial H_0 and the final Hamiltonians H_k in the way

$$H_k B_k^+ = B_k^+ H_0, \quad H_0 B_k^- = B_k^- H_k, \tag{5}$$

where $B_k^+ = A_k^+ \dots A_1^+$ and $B_k^- = A_1^- \dots A_k^-$.

The normalized eigenfunctions $\psi_n^{(k)}$ of H_k, associated to the eigenvalues E_n, are proportional to the action of B_k^+ onto the corresponding ones of H_0 (ψ_n, $n = 0, 1, \dots$). Moreover, there are k additional eigenstates $\psi_{\epsilon_j}^{(k)}$ associated to the eigenvalues ϵ_j ($j = 1,\dots,k$), which are simultaneously annihilated by B_k^-. Their corresponding explicit expressions are given by [1, 27]:

$$\psi_n^{(k)} = \frac{B_k^+\psi_n}{[(E_n-\epsilon_1)\dots(E_n-\epsilon_k)]^{1/2}}, \qquad E_n, \tag{6a}$$

$$\psi_{\epsilon_j}^{(k)} \propto \frac{W(u_1,\dots,u_{j-1},u_{j+1},\dots,u_k)}{W(u_1,\dots,u_k)}, \qquad \epsilon_j. \tag{6b}$$

Furthermore, the restriction $\epsilon_j < E_0$ naturally arises in this treatment since in this way we avoid singularities in $V_k(x)$. A diagram representing a second-order transformation is shown in Fig. 1.

3. Polynomial Heisenberg algebras

Systems described by $(m-1)$-th order polynomial Heisenberg algebras (PHA) possess three generators, one of them (the Hamiltonian H) commutes with the other two ($\mathcal{L}_m^\pm$) as ladder operators do, but the commutator $[\mathcal{L}_m^-, \mathcal{L}_m^+]$ is a polynomial in H, i.e.,

$$[H, \mathcal{L}_m^\pm] = \pm\mathcal{L}_m^\pm, \tag{7a}$$

$$[\mathcal{L}_m^-, \mathcal{L}_m^+] \equiv N_m(H+1) - N_m(H) = P_{m-1}(H), \tag{7b}$$

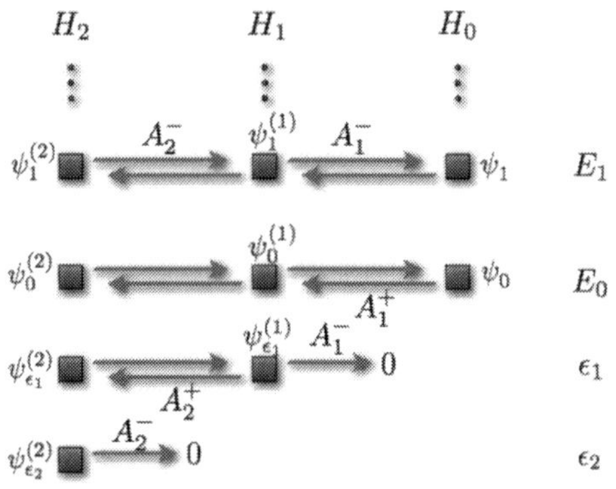

Figure 1. Second-order SUSY transformation applied to H_0 using two solutions u_1, u_2 of the initial Schrödinger equation with factorization energies $\epsilon_2 < \epsilon_1 < E_0$.

where $P_{m-1}(H)$ and $N_m(H) \equiv \mathcal{L}_m^+\mathcal{L}_m^-$ are polynomials in H of degrees $m-1$ and m, respectively. $N_m(H)$ is the analogous to the number operator for the harmonic oscillator, which can be factorized as

$$N_m(H) = \prod_{i=1}^{m}(H - \varepsilon_i), \tag{8}$$

and ε_i are the energies associated with the *extremal states*.

Note that we can realize these algebras through mth-order differential ladder operators; nevertheless, what really defines the order of the PHA is the degree $m-1$ of the polynomial $P_{m-1}(H)$ characterizing the deformation (see commutator in Eq. (7b)).

The spectrum of H can be built up by acting iteratively $\mathcal{L}_m^+$ on the extremal states of the system [10, 27, 28], which are annihilated by $\mathcal{L}_m^-$ and (in the diagonalizable case) satisfy $H\psi_{\varepsilon_i} = \varepsilon_i\psi_{\varepsilon_i}, i = 1, \ldots, m$.

Now, the general systems ruled by these PHA should be identified: for zeroth- and first-order PHA, the systems are the harmonic and radial oscillators, respectively [28–31]. On the other hand, for second- and third-order PHA, the determination of the potentials reduces to finding solutions to the Painlevé IV and V equations, denoted as PIV and PV, respectively [30, 32]. In this work, such relation will be used in reversed order, i.e., first we will find systems which are described by PHA and then we will obtain solutions to the Painlevé equations. To do that, let us establish the link between third-order PHA and PV equation.

4. Third-order PHA: fourth-order ladder operators.

Let us assume that $\mathcal{L}_4^\pm$ are fourth-order ladder operators, that can be factorized as:

$$\mathcal{L}_4^+ = A_4^+A_3^+A_2^+A_1^+ = \frac{1}{2^2}\left(\frac{\mathrm{d}}{\mathrm{d}x} - f_4\right)\left(\frac{\mathrm{d}}{\mathrm{d}x} - f_3\right)\left(\frac{\mathrm{d}}{\mathrm{d}x} - f_2\right)\left(\frac{\mathrm{d}}{\mathrm{d}x} - f_1\right), \tag{9a}$$

$$\mathcal{L}_4^- = A_1^-A_2^-A_3^-A_4^- = \frac{1}{2^2}\left(-\frac{\mathrm{d}}{\mathrm{d}x} - f_1\right)\left(-\frac{\mathrm{d}}{\mathrm{d}x} - f_2\right)\left(-\frac{\mathrm{d}}{\mathrm{d}x} - f_3\right)\left(-\frac{\mathrm{d}}{\mathrm{d}x} - f_4\right). \tag{9b}$$

Now, let us build a closed-chain, such that each pair of operators $A_j^\pm$ intertwines two Schrödinger Hamiltonians [9] as:

$$H_{j+1}A_j^+ = A_j^+H_j, \quad H_jA_j^- = A_j^-H_{j+1}, \tag{10}$$

where $j = 1, 2, 3, 4$. This leads to the following factorizations of the Hamiltonians

$$H_1 = A_1^- A_1^+ + \epsilon_1, \tag{11a}$$
$$H_2 = A_1^+ A_1^- + \epsilon_1 = A_2^- A_2^+ + \epsilon_2, \tag{11b}$$
$$H_3 = A_2^+ A_2^- + \epsilon_2 = A_3^- A_3^+ + \epsilon_3, \tag{11c}$$
$$H_4 = A_3^+ A_3^- + \epsilon_3 = A_4^- A_4^+ + \epsilon_4, \tag{11d}$$
$$H_5 = A_4^+ A_4^- + \epsilon_4, \tag{11e}$$

where the fifth Hamiltonian is not independent, but is rather given by the closure condition:

$$H_5 = H_1 - 1 \equiv H - 1. \tag{12}$$

A diagram representing the transformations and the closure relation is shown in Fig. 2.

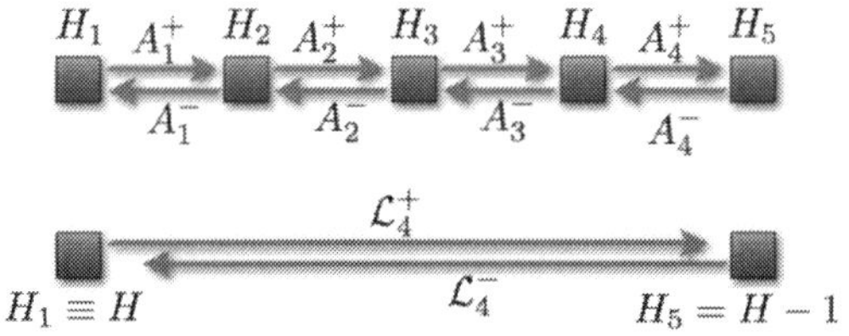

Figure 2. Diagram representing the two equivalent SUSY transformations. Above: the four-step transformation induced by the first-order operators $A_1^\pm$, $A_2^\pm$, $A_3^\pm$ and $A_4^\pm$. Below: the direct transformation achieved through the fourth-order operators $\mathcal{L}_4^\pm$.

By making the corresponding operator products of Eqs. (11), after a long calculation it turns out that all the involved functions and the potential $V(x)$ become expressed in terms of just one, $w(z)$ with $z = x^2$, which satisfies [33]:

$$\frac{\mathrm{d}^2 w}{\mathrm{d}z^2} = \left(\frac{1}{2w} + \frac{1}{w-1}\right)\left(\frac{\mathrm{d}w}{\mathrm{d}z}\right)^2 - \frac{1}{z}\frac{\mathrm{d}w}{\mathrm{d}z} + \frac{(w-1)^2}{z^2}\left(aw + \frac{b}{w}\right) + c\frac{w}{z} + d\frac{w(w+1)}{w-1}, \tag{13}$$

which is the Painlevé V equation with parameters

$$a = \frac{\alpha_1^2}{2}, \quad b = -\frac{\alpha_3^2}{2}, \quad c = \frac{\alpha_2 - \alpha_4}{2}, \quad d = -\frac{1}{8}, \tag{14}$$

where $\alpha_1 = \varepsilon_1 - \varepsilon_1$, $\alpha_2 = \varepsilon_2 - \varepsilon_3$, $\alpha_3 = \varepsilon_3 - \varepsilon_4$, $\alpha_4 = \varepsilon_4 - \varepsilon_1 + 1$.

The spectrum of H could contain four independent equidistant energy ladders starting from the extremal states [11], here we need only the expression of two of them:

$$\psi_{\varepsilon_3} \propto \exp\left[\int\left(\frac{h'}{2h} + \frac{h}{2} - \frac{\alpha_3}{h}\right)\mathrm{d}x\right], \tag{15a}$$
$$\psi_{\varepsilon_4} \propto \exp\left[\int\left(\frac{h'}{2h} + \frac{h}{2} + \frac{\alpha_3}{h}\right)\mathrm{d}x\right], \tag{15b}$$

where $h(x) = -x - g(x)$, with g being related to w through $g(x) = x/[w(x^2) - 1]$. Note that the number operator $N_4(H)$ for this system is of fourth degree, $N_4(H) = (H - \varepsilon_1)(H - \varepsilon_2)(H - \varepsilon_3)(H - \varepsilon_4)$. From Eqs. (9–12) it is straightforward to obtain the energies of the extremal states in terms of the factorization energies as $\varepsilon_j = \epsilon_j + 1$ for $j = 1, 2, 3, 4$.

On the other hand, if a system having fourth-order ladder operators is found, then it is possible to design a mechanism for generating solutions to the PV equation, similar to the one implemented for the PIV equation [3,4,6]. The key point is to identify once again the extremal states of our system. Then, from Eqs. (15) it is straightforward to show that

$$h(x) = \frac{2\alpha_3}{\left[\ln\left(\frac{\psi_{\varepsilon_4}}{\psi_{\varepsilon_3}}\right)\right]'} = \{\ln[W(\psi_{\varepsilon_3}, \psi_{\varepsilon_4})]\}'. \tag{16}$$

Furthermore, we obtain an expression for $g(x)$ given by

$$g(x) = -x - h(x) = -x - \{\ln[W(\psi_{\varepsilon_3}, \psi_{\varepsilon_4})]\}'. \tag{17}$$

Finally, $g(x)$ is related with the solution $w(z)$ of the PV equation through

$$w(z) = 1 + \frac{z^{1/2}}{g(z^{1/2})}. \tag{18}$$

Thus, we have introduced a simple recipe to generate solutions to the PV equation, based on the identification of the extremal states for systems ruled by third-order PHA, which have differential ladder operators of fourth order.

5. SUSY partners of the radial oscillator

In this section we will apply SUSY QM to the radial oscillator Hamiltonian, which is given by

$$H_\ell = -\frac{1}{2}\frac{\mathrm{d}^2}{\mathrm{d}x^2} + \frac{x^2}{8} + \frac{\ell(\ell+1)}{2x^2}, \qquad \ell \geq 0, \quad x \geq 0, \tag{19}$$

where we have added the subscript ℓ to denote the dependence of the Hamiltonian on the angular momentum index. First, let us solve the spectral problem associated to H_ℓ [11,34,35]. In order to do that, let us take $b_\ell^\pm$ such that

$$b_\ell^\pm = \frac{1}{2}\left(\frac{\mathrm{d}^2}{\mathrm{d}x^2} \mp x\frac{\mathrm{d}}{\mathrm{d}x} + \frac{x^2}{4} - \frac{\ell(\ell+1)}{x^2} \mp \frac{1}{2}\right). \tag{20}$$

Then, it is easily shown that $H_\ell b_\ell^- = b_\ell^-(H_\ell - 1)$ and $H_\ell b_\ell^+ = b_\ell^+(H_\ell + 1)$, i.e., the following commutators are obeyed $[H_\ell, b_\ell^\pm] = \pm b_\ell^\pm$. Now, we can obtain the eigenstates of H_ℓ departing from the *extremal states* $\psi_{\varepsilon_i}, i = 1, 2$, two solutions of Schrödinger equation for H_ℓ such that $b_\ell^- \psi_{\varepsilon_i} = 0$, which are given by:

$$\psi_{0\ell} \equiv \psi_{\varepsilon_1} \propto x^{\ell+1}\exp(-x^2/4), \qquad \varepsilon_1 = \frac{\ell}{2} + \frac{3}{4} \equiv E_{0\ell}, \tag{21a}$$

$$\psi_{\varepsilon_2} \propto x^{-\ell}\exp(-x^2/4), \qquad \varepsilon_2 = -\frac{\ell}{2} + \frac{1}{4} = -E_{0\ell} + 1. \tag{21b}$$

As only the first one fulfills the boundary conditions, therefore it leads to a ladder of *physical* eigenfunctions. The spectrum of the radial oscillator is therefore

$$\mathrm{Sp}(H_\ell) = \{E_{n\ell} = n + \frac{\ell}{2} + \frac{3}{4}, n = 0, 1, \ldots\}. \tag{22}$$

Note that an analogue of the number operator for the radial oscillator can be defined as

$$b_\ell^+ b_\ell^- = (H_\ell - \varepsilon_1)(H_\ell - \varepsilon_2) = (H_\ell - E_{0\ell})(H_\ell + E_{0\ell} - 1), \tag{23}$$

which is a second degree polynomial in H_ℓ, i.e., the radial oscillator is ruled by a first-order PHA.

In order to implement the SUSY transformations, we employ the general solution of the stationary Schrödinger equation for any factorization energy ϵ, which is given by [11, 36, 37]

$$u(x) = x^{-\ell} e^{-x^2/4} \left[{}_1F_1\left(\frac{1-2\ell-4\epsilon}{4}, \frac{1-2\ell}{2}; \frac{x^2}{2}\right) + \nu \frac{\Gamma\left(\frac{3+2\ell-4\epsilon}{4}\right)}{\Gamma\left(\frac{3+2\ell}{2}\right)} \left(\frac{x^2}{2}\right)^{\ell+1/2} {}_1F_1\left(\frac{3+2\ell-4\epsilon}{4}, \frac{3+2\ell}{2}; \frac{x^2}{2}\right)\right]. \tag{24}$$

Thus, the first-order SUSY partner potential of the radial oscillator becomes

$$V_1(x) = \frac{x^2}{8} + \frac{\ell(\ell+1)}{2x^2} - [\ln u(x)]''. \tag{25}$$

The conditions that must be fulfilled to produce a non-singular transformation are

$$x > 0, \quad \epsilon < E_{0\ell}, \quad \nu \geq -\frac{\Gamma\left(\frac{1-2\ell}{2}\right)}{\Gamma\left(\frac{1-2\ell-4\epsilon}{4}\right)}. \tag{26}$$

For a k-th order SUSY transformation let us take k appropriate solutions $\{u_k, \ldots, u_1\}$ in the form given in Eq. (24), for the factorization energies $\epsilon_k < \epsilon_{k-1} < \cdots < \epsilon_1 < E_{0\ell}$. The SUSY partner potential of the radial oscillator becomes now:

$$V_k(x) = \frac{x^2}{8} + \frac{\ell(\ell+1)}{2x^2} - \{\ln[W(u_1, \ldots, u_k)]\}'', \tag{27}$$

with spectrum given by $\mathrm{Sp}(H_k) = \{\epsilon_k, \ldots, \epsilon_1, E_{0\ell}, E_{1\ell}, \ldots\}$. In Fig. 3 we show examples of first (left) and second-order (right) SUSY partner potentials of the radial oscillator.

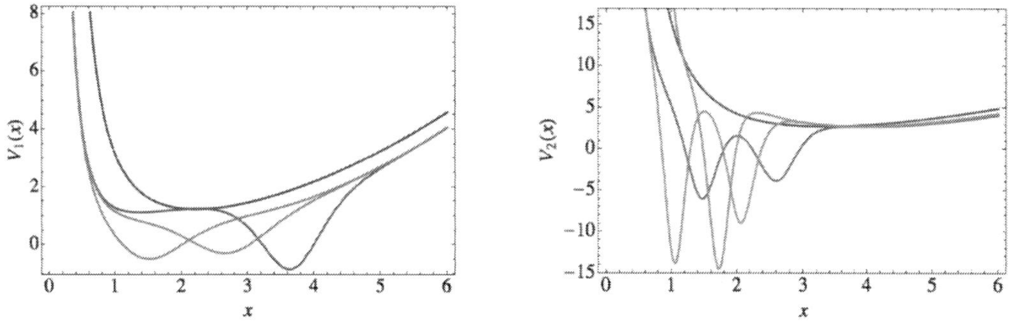

Figure 3. Radial oscillator potential (blue) and its SUSY partners for $k = 1$, $\ell = 2$, $\epsilon = 1/2$, and $\nu = \{-0.59$ (magenta), -0.4 (yellow), 1 (green)$\}$ (left plot). The right plot corresponds to $k = 2$, $\ell = 5$, $\nu_1 = 1$, and $\epsilon_1 = \{0$ (magenta), -2 (yellow), -4 (green)$\}$ with $u_2 = b_\ell^- u_1$ and $\epsilon_2 = \epsilon_1 - 1$.

Now we define a natural pair of ladder operators $L_k^\pm$ for H_k as

$$L_k^\pm = B_k^+ b_\ell^\pm B_k^-, \tag{28}$$

which are of $(2k+2)$th-order and fulfill

$$[H_k, L_k^\pm] = \pm L_k^\pm. \tag{29}$$

From the intertwining relations we can obtain the analogue of the number operator for the k-th order SUSY partners of the radial oscillator as

$$N(H_k) = L_k^+ L_k^- = (H_k - E_{0\ell})\,(H_k + E_{0\ell} - 1)\prod_{j=1}^{k}(H_k - \epsilon_j)(H_k - \epsilon_j - 1). \tag{30}$$

Since both ϵ_j and $\epsilon_j + 1$ are in the set of roots, there is a finite ladder starting and ending at ϵ_j. As the index j runs from 1 to k, this means that $\mathrm{Sp}(H_k)$ contains k one-step ladders. Note that $E_{0\ell}$ and $-E_{0\ell}+1$ are also roots of $N(H_k)$, then $\mathrm{Sp}(H_k)$ could have in principle two infinite ladders, but just the one starting from $E_{0\ell}$ is physical. We conclude that $\{H_k, L_k^-, L_k^+\}$ generates a $(2k+1)$th-order polynomial Heisenberg algebra.

6. Reduction theorem for the SUSY generated Hamiltonians H_k

In order to implement the prescription pointed out at the end of Section 4 to produce solutions to the PV equation, first of all we need to identify systems ruled by a third-order PHA, generated by fourth-order ladder operators. Note that for $k = 1$ the operator set $\{H_k, L_k^-, L_k^+\}$ of the previous section generates a third-order PHA, i.e., the first-order SUSY partners of the radial oscillator can be used directly for generating solutions to the PV equation. However, for $k > 1$ the set $\{H_k, L_k^-, L_k^+\}$ generates a PHA of order greater than three. Would it be possible to identify a subfamily of the k-th order SUSY partners of the radial oscillator Hamiltonian which, in addition of having the natural $(2k+2)$th-order ladder operators $L_k^{\pm}$ would have fourth-order ones? If so, we could generate additional solutions to the PV equation. The answer to this question turns out to be positive, and the conditions required to produce such a reduction process are contained in the following theorem [6].

Theorem. *Let H_k be the kth-order SUSY partner of the radial oscillator Hamiltonian H_0 generated by k Schrödinger seed solutions u_i which are connected by the annihilation operator b^- as*

$$u_i = (b^-)^{i-1}u_1, \quad \epsilon_i = \epsilon_1 - (i-1), \quad i = 1,\dots,k, \tag{31}$$

where $u_1(x)$ is a Schrödinger solution without zeroes, given by Eq. (24) for $\epsilon_1 < E_{0\ell} = \ell/2 + 3/4$ and

$$\nu_1 \geq -\frac{\Gamma\left(\frac{1-2\ell}{2}\right)}{\Gamma\left(\frac{1-2\ell-4\epsilon_1}{4}\right)}. \tag{32}$$

Therefore, the natural $(2k+2)$th-order ladder operator $L_k^+ = B_k^+ b^+ B_k^-$ of H_k turn out to be factorized in the form

$$L_k^+ = P_{k-1}(H_k)\,l_k^+, \tag{33}$$

where $P_{k-1}(H_k) = (H_k - \epsilon_1)\dots(H_k - \epsilon_{k-1})$ is a polynomial of degree $k-1$ in H_k and l_k^+ is a fourth-order differential ladder operator,

$$[H_k, l_k^+] = l_k^+, \tag{34}$$

such that

$$l_k^+ l_k^- = (H_k - E_{0\ell})\,(H_k + E_{0\ell} - 1)\,(H_k - \epsilon_k)(H_k - \epsilon_1 - 1). \tag{35}$$

The proof of this theorem will be given in a forthcoming paper [38]. It implies that once the factorization in Eq. (33) is fulfilled, the natural ladder operators $L_k^{\pm}$ are expressed as a product of a polynomial of degree $k-1$ in H_k times the fourth-order ladder operator $l_k^{\pm}$. This means that the $(2k+1)$th-order PHA obtained through the SUSY transformation involving k connected solutions with $\epsilon_i = \epsilon_1 - (i-1)$, $i = 1,\dots,k$, can be *reduced* to a third-order one [39].

Recall that these algebras are closely related to the PV equation. This means that when we reduce the higher-order algebras, we open the possibility of obtaining new solutions of the PV equation, similar to what happens in the case of second-order PHA and PIV equation [1–6]. In the following sections we will discuss the method to obtain solutions of the PV equation.

7. Solutions to the PV equation through SUSY QM

In order to generate solutions to the PV equation we need to find systems ruled by third-order PHA, with fourth-order ladder operators. In fact, the key point of the technique is to identify the extremal states of the system, as well as their associated energies. Then, we use Eqs. (17)-(18) to calculate the PV solution and Eq. (14) to determine the parameters, both of which are symmetric under the exchanges $\varepsilon_1 \leftrightarrow \varepsilon_2$ and $\varepsilon_3 \leftrightarrow \varepsilon_4$. Thus, from the $4! = 24$ possible permutations of the four indexes we will get six different solutions to the PV equation, some of which could have singularities. Let us apply the technique to the radial oscillator potential and to its first-order SUSY partners.

We have seen that the radial oscillator Hamiltonian has second-order differential ladder operators $b_\ell^\pm$ leading to a first-order PHA. We can construct also pairs of fourth-order differential ladder operators which will give place to a third-order PHA, for example, $L_4^\pm = a_{\ell+1}^+ b_{\ell+1}^\pm a_{\ell+1}^-$, where $\sqrt{2}\, a_{\ell+1}^\pm = \mp \mathrm{d}/\mathrm{d}x - (\ell+1)/x + x/2$. The analogue to the number operator becomes $L_4^+ L_4^- = (H_\ell + E_{0\ell} - 1)\,(H_\ell - E_{0\ell})\,(H_\ell - E_{1\ell})^2$. Its roots suggest naturally three extremal states: two of them are those associated to the first-order PHA, and another one is the first excited state $\psi_{1\ell}$ of H_ℓ with eigenvalue $E_{1\ell}$. The last one is chosen as the 'mathematical eigenstate' of H_ℓ associated to the eigenvalue $E_{1\ell}$, denoted as $\psi_{1\ell}^\perp$, which satisfies $W(\psi_{1\ell}, \psi_{1\ell}^\perp) = 1$. Now, we number them as follows:

$$\psi_{\varepsilon_1} \propto x^{\ell+1} \exp(-x^2/4), \qquad \varepsilon_1 = E_{0\ell}, \tag{36a}$$

$$\psi_{\varepsilon_2} \propto x^{-\ell} \exp(-x^2/4), \qquad \varepsilon_2 = -E_{0\ell} + 1, \tag{36b}$$

$$\psi_{\varepsilon_3} = \psi_{1\ell}, \qquad \varepsilon_3 = E_{1\ell}, \tag{36c}$$

$$\psi_{\varepsilon_4} = \psi_{1\ell}^\perp, \qquad \varepsilon_4 = E_{1\ell}. \tag{36d}$$

Using these expressions and Eqs. (14),(17),(18), it is straightforward to obtain the PV solution with parameters:

$$a = \frac{(2\ell+1)^2}{8}, \qquad b = 0, \qquad c = -\frac{(2\ell+7)}{4}, \qquad d = -\frac{1}{8}. \tag{37}$$

Now, by exploring all possible permutations of the indexes leading to different PV solutions, we obtain six combinations of parameters, which are shown in Table 1. For each one of them we obtain a solution of the PV equation, of which three non-trivial examples are:

$$w(z) = 1 + \frac{z}{2\ell+1}, \tag{38a}$$

$$w(z) = 1 - \frac{2\ell + z + 1}{2}, \tag{38b}$$

$$w(z) = 1 + \frac{z\Gamma(\ell + 1/2, 0, z/2) - 2^{\ell+1/2} \mathrm{e}^{z/2} (-z)^{\ell+1/2}}{(2\ell - 1)\Gamma(\ell + 1/2, 0, z/2)}, \tag{38c}$$

where $\Gamma(a, z_0, z_1)$ is the generalized incomplete gamma function. An illustration of the radial oscillator potential for different values of ℓ and the corresponding PV solutions $w(z)$ of equation (38c) are shown in Fig. 4.

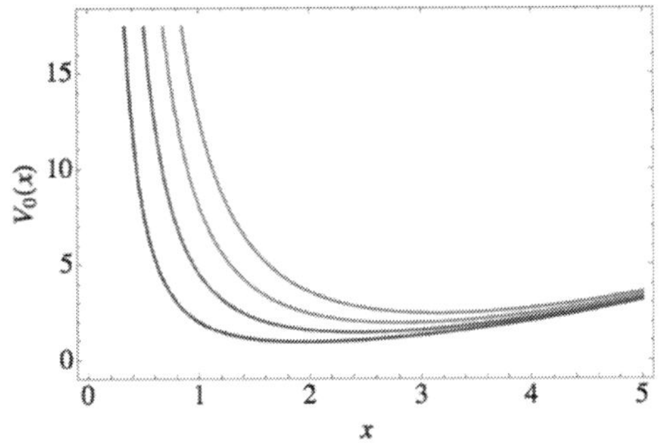

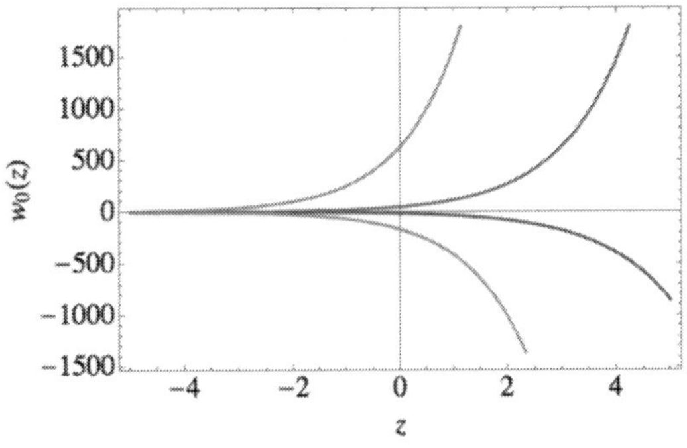

Figure 4. Radial oscillator potentials (left) and the corresponding PV solutions $w(z)$ of Eq. (38c) (right) for ℓ =3/2 (blue), 5/2 (magenta), 7/2 (yellow) and 9/2 (green).

Order	$8a$	$8b$	$4c$
1234	$(2\ell+1)^2$	0	$-2\ell-7$
1324	4	$-(2\ell+3)^2$	$2\ell-1$
1423	4	$-(2\ell+3)^2$	$2\ell-1$
2314	$(2\ell+3)^2$	-4	$-2\ell-3$
2413	$(2\ell+3)^2$	-4	$-2\ell-3$
3412	0	$-(2\ell+1)^2$	$-2\ell-3$

Table 1. Parameters of PV equation for the six permutations of the radial oscillator potential.

We must remember that, due to the radial oscillator is also described by a first-order PHA, then the third-order case must reduce to the original first-order PHA. Indeed, this is what happens since $L_4^+ = b_\ell^+(H_\ell - E_{0\ell})$ and $L_4^- = (H_\ell - E_{0\ell})b_\ell^-$.

On the other hand, the first-order SUSY partners of the radial oscillator Hamiltonian possess also a pair of natural fourth-order differential ladder operators

$$L_4^\pm = A_1^+ b_\ell^\pm A_1^-, \tag{39}$$

where $A_1^\pm$ are the first-order intertwining operators and $b_\ell^\pm$ are the second-order ladder operators of Section 5. The operators $L_4^\pm$ give place to a third-order PHA since

$$N(H_1) = L_4^+ L_4^- = (H_\ell + E_{0\ell} - 1)(H_\ell - E_{0\ell})(H_\ell - \epsilon)(H_\ell - \epsilon - 1). \tag{40}$$

The roots of $N(H_1)$ and the SUSY procedure suggest now the following extremal states: two of them are the SUSY transformed extremal states of the radial oscillator, another one is the new ground state created by the SUSY transformation. The last one will be an unphysical eigenstate of H_1 associated to $\epsilon + 1$. Let us choose now the following initial ordering:

$$\psi_{\varepsilon_1} \propto A_1^+ b_\ell^+ u, \qquad \varepsilon_1 = \epsilon + 1, \tag{41a}$$

$$\psi_{\varepsilon_2} \propto A_1^+ \left[x^{-\ell} \exp\left(-\frac{x^2}{4}\right)\right], \qquad \varepsilon_2 = -E_{0\ell} + 1, \tag{41b}$$

$$\psi_{\varepsilon_3} \propto \frac{1}{u}, \qquad \varepsilon_3 = \epsilon, \tag{41c}$$

$$\psi_{\varepsilon_4} \propto A_1^+ \left[x^{\ell+1} \exp\left(-\frac{x^2}{4}\right)\right], \qquad \varepsilon_4 = E_{0\ell}. \tag{41d}$$

Using these expressions and Eqs. (14),(17),(18), one can obtain the PV solution and its corresponding parameters:

$$a = \frac{(4\epsilon + 2\ell + 3)^2}{32}, \quad b = -\frac{(4\epsilon - 2\ell - 3)^2}{32}, \quad c = -\frac{(2\ell + 1)}{4}, \quad d = -\frac{1}{8}. \tag{42}$$

Now, by permuting the indexes of Eqs. (41) we can obtain different PV solutions. Although they can be quickly calculated with any symbolic software, they are quite large to be written down explicitly (see some plots of them in Figure 5). For that reason, we choose a simple case, using as seed solution $u(x)$ the ground state of the radial oscillator. In this case, of the six possible solutions obtained by permutations, half of them reduce to zero and the other half to rational functions. The corresponding parameters of the PV equation are given in Table 2 and the nontrivial solutions are the following:

$$w(z) = 1 + \frac{z}{2\ell - z + 1}, \tag{43a}$$

$$w(z) = 1 + \frac{z[4\ell^2 + 8\ell^3 - 15z^2 - 2\ell(2 + 5z^2)]}{16\ell(-1 + \ell + 2\ell^2)}, \tag{43b}$$

$$w(z) = 1 + \frac{z[8\ell^3 - 4\ell^2(z-1) - 2\ell(5z^2 + 2z + 2) + 5(z-3)z^2]}{-8\ell^3(z-4) + 4\ell^2(z^2 - 3z + 4) + 2\ell(5z^3 + 2z^2 - 2z - 8) - 5(z-1)z^3}. \tag{43c}$$

Order	$8a$	$8b$	$4c$
1234	$(2\ell+3)^2$	0	$-2\ell-1$
1324	4	$-(2\ell+1)^2$	$2\ell+1$
1423	4	$-(2\ell+1)^2$	$2\ell+1$
2314	$(2\ell+1)^2$	-4	$-2\ell-5$
2413	$(2\ell+1)^2$	-4	$-2\ell-5$
3412	0	$-(2\ell+3)^2$	$2\ell-3$

Table 2. Parameters of the PV equation for the six permutations of the first-order SUSY partners of the radial oscillator.

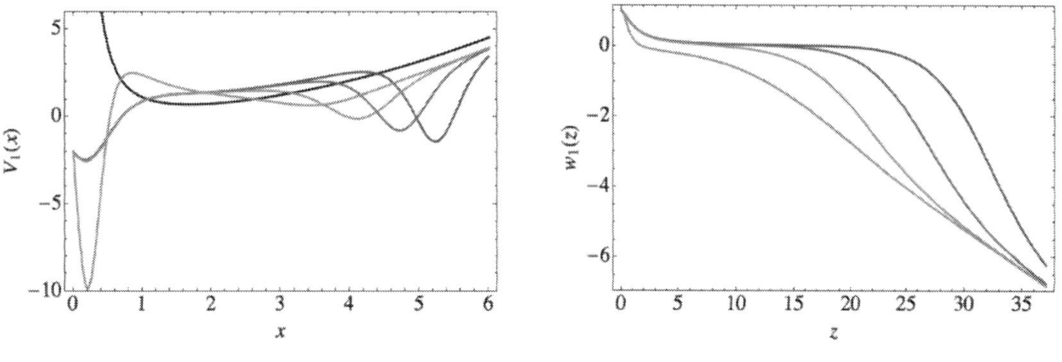

Figure 5. SUSY partner potential $V_1(x)$ of the radial oscillator (black) (left) and the solutions $w_1(z)$ to PV equation (right) for $\ell = 1$, $\epsilon_1 = 1$, and $\nu_1 = \{0.905$ (blue), 0.913 (magenta), 1 (yellow), 10 (green)$\}$.

8. Conclusions

In this paper we introduced an algebraic technique to solve the PV equation and, based on this method, we were able to obtain some of its solutions. In order to do that, we derived the SUSY partners of the radial oscillator and we obtained the natural ladder operators connected with these Hamiltonians. Then we formulated a reduction theorem, in which we point out the requirements for the natural $(2k+2)$th-order ladder operators associated to the kth-order SUSY partners of the radial oscillator to be reduced to fourth-order ones, which are related with third-order PHA and the PV equation.

Using the connection between the SUSY partners of the radial oscillator and third-order PHA, in Section 7 we were able to derive a method to obtain solutions of the PV equation. We presented the general method to get explicit solutions through the radial oscillator, and its first-order SUSY partners.

This paper further expands previous works of the same authors related with the Painlevé IV equation. The difference is that the connection with the PV equation is more elaborated, and working out explicit solutions becomes increasingly complicated, even at low orders of SUSY. We realize also that there are many other solutions that can be obtained by this method. We will continue working in this direction in the near future.

Acknowledgement

The authors acknowledge the financial support of Conacyt (Mexico) through fellowship 207672 and project 152574.

References

[1] Bermudez D and Fernández D 2011 *SIGMA* **7** 025 (14 pages)
[2] Bermudez D and Fernández D 2011 *Phys. Lett. A* **375** 2974–2978
[3] Bermudez D 2012 *SIGMA* **8** 069 (10 pages)
[4] Bermudez D and Fernández D 2012 *AIP Conf. Proc.* **1420** 47–51
[5] Bermudez D and Fernández D 2013 *Trends in Mathematics. Geometric Methods in Physics, XXX Workshop* **1** 199–209
[6] Bermudez D 2013 *Polynomial Heisenberg algebras and Painlevé equations* PhD thesis Cinvestav Mexico
[7] Veselov A and Shabat A 1993 *Funkts. Anal. Prilozh.* **27** 1
[8] Dubov S, Eleonsky V and Kulagin N 1994 *Chaos* **4** 47–53
[9] Adler V 1994 *Physica D* **73** 335
[10] Andrianov A, Cannata F, Ioffe M and Nishnianidze D 2000 *Phys. Lett. A* **266** 341
[11] Carballo J, Fernández D, Negro J and Nieto L 2004 *J. Phys. A* **37** 10349
[12] Mateo J and Negro J 2008 *J. Phys. A: Math. Theor.* **41** 045204
[13] Levi D 1992 *Painlevé transcendents, their asymptotics and physical applications* **1** 353–360
[14] Winternitz P 1992 *Painlevé transcendents, their asymptotics and physical applications* **1** 425–431
[15] Kanzieper E 2002 *Phys. Rev. Lett.* **89** 250201 (4 pages)
[16] Schief W 2002 *Anziam J.* 141–148
[17] Abramov A and Yukhno L 2013 *Comp. Math. and Math. Phys.* **53**
[18] Sakka A 2005 *Z. Naturforsch* **60**
[19] Masuda T 2003 *Nagoya Math. J.* **169**
[20] Mughisieb A 2007 *Bäcklund transformations and discrete Painlevé V equation* MSc thesis University of Gaza
[21] Ablowitz M, Ramani A and Segur H 1980 *J. Math. Phys.* **21** 1006–1015
[22] Flaschka H 1980 *J. Math. Phys.* **21** 1016–1018
[23] Filipuk G and Clarkson P 2008 *Stud. Appl. Math.* **121** 157–188
[24] Marquette I 2009 *J. Math. Phys.* **50** 012101 (23 pages)
[25] Andrianov A, Ioffe M and Spiridonov V 1993 *Phys. Lett. A* **174** 273
[26] Fernández D 2010 *AIP Conf. Proc.* **3** 36
[27] Fernández D and Hussin V 1999 *J. Phys. A: Math. Gen.* **32** 3603
[28] Dubov S, Eleonsky V and Kulagin N 1992 *Sov. Phys. JETP* **75** 446
[29] Fernández D 1984 *Development on the factorization method in quantum mechanics (in Spanish)* MSc thesis Cinvestav Mexico
[30] Adler V 1993 *Theor. Math. Phys.* **101** 1381

[31] Sukhatme U, Rasinariu C and Khare A 1997 *Phys. Lett. A* **234** 401
[32] Willox R and Hietarinta J 2003 *J. Phys. A: Math. Gen.* **36** 10615
[33] Bermudez D and Fernández D 2014 *AIP Conf. Proc.* **1575** 50–88
[34] Fernández D, Negro J and del Olmo M 1996 *Ann. Phys.* **252** 386–412
[35] Cabrera-Munguia I and Rosas-Ortiz O 2008 *J. Phys: Conf. Ser.* **128** 012042 (11 pages)
[36] Junker G and Roy P 1998 *Ann. Phys.* **270** 155–177
[37] Carballo J 2001 *SUSY QM and families of potentials almost isospectral to the radial oscillator (in Spanish)* PhD thesis ESFM-IPN Mexico
[38] Bermudez D, Fernández D and Negro J 2015 *To be published*
[39] Andrianov A and Sokolov A 2005 *Nucl. Phys. B* **660** 25–50

One dimensional Dirac-Moshinsky oscillator-like system and isospectral partners

A Contreras-Astorga

Department of Mathematics and Actuarial Science, Indiana University Northwest, 3400 Broadway, Gary IN 46408, USA

E-mail: `aloncont@iun.edu`

Abstract. Two different exactly solvable systems are constructed using the supersymmetric quantum mechanics formalism and a pseudoscalar one-dimensional version of the Dirac-Moshinsky oscillator as a departing system. One system is built using a first-order SUSY transformation. The second is obtained through the confluent supersymmetry algorithm. The two of them are explicitly designed to have the same spectrum as the departing system and pseudoscalar potentials.

Keywords: Dirac equation, Dirac-Moshinsky oscillator, pseudoscalar potential, confluent SUSY algorithm

1. Introduction

Supersymmetric quantum mechanics (SUSY) is a tool that allows to find new exactly solvable stationary Schrödinger equations. Moreover, the SUSY formalism provides a method to obtain, in principle, a quantum system with a desired spectrum. To apply it we need a solvable initial system and at least one of its solutions, as a result a new system, often referred as SUSY partner, and its solutions are generated [9, 11, 16]. This tool has been applied to a large variety of quantum interactions such as the harmonic oscillator [15], the hydrogen atom [12, 21], the Pöschl-Teller potentials [6], among many others. Moreover, it has been possible to extend the SUSY formalism so it can be used to find solutions of different kind of equations like the Fokker-Planck equation [22, 24], the time dependent Schrödinger equation [3], mass position dependent Schrödinger equation [13], Painlevé equations [4] and Dirac equations [5, 7, 8, 10, 14, 18, 20].

The purpose of this work is to apply the SUSY technique to a one dimensional Dirac Moshinsky oscillator-like system [17, 23] to obtain new isospectral families using the supersymmetric quantum mechanic formalism. In particular we will focus on two versions of this technique: the first-order SUSY quantum mechanics will be used to generate a one-parameter family of isospectral systems, this part is closely related to Bogdan Mielnik's seminal article on isospectral Hamiltonians to the harmonic oscillator, and second, a three parametric family will be generated through the confluent supersymmetry algorithm.

This paper is organized as follows. Section 2 gives a brief review of the first-order and the confluent case of supersymmetric quantum mechanics. In section 3 the confluent supersymmetry algorithm for Dirac equations is described. Section 4 is devoted to present and apply the SUSY formalism to a one dimensional version of the Dirac-Moshinsky oscillator in order to find new isospectral systems. Our conclusions will be presented in the last section.

2. Supersymmetric quantum mechanics

In this section we explain how to make a SUSY transformations focusing on two types of them. The simplest case of this technique is the first-order SUSY transformation and it permits to add a new level below the ground state energy, to delete the ground state energy or to create a new isospectral system. The second type is the confluent SUSY transformation and it allows to add an energy level in any position, to delete any existing level and also new isospectral systems can be generated.

Consider the following one dimensional stationary Schrödinger equation

$$\Psi'' + (\epsilon - U_0)\Psi = 0, \tag{1}$$

where the energy ϵ is a real-valued constant, $U_0 = U_0(x)$ is the potential energy and the prime denotes derivative with respect to the position. Through a SUSY transformation a new Schrödinger equation and its solutions can be obtained if the solutions of the original equation (1) are known.

2.1. First-order SUSY transformation

To apply a first-order SUSY transformation suppose that we know a solution u for the equation

$$u'' + (\lambda - U_0)u = 0, \tag{2}$$

where λ is a real constant known as factorization energy, then the function

$$\Phi = \frac{u'}{u}\Psi - \Psi' \tag{3}$$

is solution of the Schrödinger equation

$$\Phi'' + (\epsilon - U_1)\Phi = 0, \tag{4}$$

where

$$U_1 = U_0 - 2\,(\ln u)'' \tag{5}$$

is the transformed potential, also called SUSY partner of U_0. In the case $\epsilon = \lambda$, $\Psi = u$, the function (3) gives the trivial solution, it can be shown by direct substitution that the function

$$\Phi = \frac{1}{u} \tag{6}$$

is solution of (4) under these circumstances. When we consider boundary conditions for quantum mechanical problems with discrete spectra, the eigenvalues of the two Schrödinger equations (1) and (4) are exactly the same or at most they differ only in the lowest eigenvalue or ground state energy. If the goal is to obtain regular transformed potentials, this technique imposes the restriction $\lambda \leq \epsilon_0$ where ϵ_0 is the ground state energy of the original system.

2.2. Confluent SUSY transformation
In order to apply the second-order SUSY transformation we need two transformations functions u_1 and u_2 satisfying the following set of equations

$$u_1'' + (\lambda - U_0)u_1 = 0, \tag{7}$$
$$u_2'' + (\lambda - U_0)u_2 = u_1, \tag{8}$$

this set is referred as a Jordan chain. Once the two transformation functions are known we can construct the function

$$\Phi = \frac{W_{u_1,u_2,\Psi}}{W_{u_1,u_2}}, \tag{9}$$

where W represent the Wronskian of the functions in its index. The function Φ is a solution of the Schrödinger equation

$$\Phi'' + (\epsilon - U_2)\Phi = 0 \tag{10}$$

where the potential U_2 is given by

$$U_2 = U_0 - 2\,(\ln W_{u_1,u_2})''\,. \tag{11}$$

As in the first order transformation, a special case is presented when $\lambda = \epsilon$ and $\Psi = u_1$; direct substitution shows that

$$\Phi = \frac{u_1}{W_{u_1,u_2}} \tag{12}$$

is a non trivial solution of (10) in this situation. For quantum mechanical problems with discrete spectra, equations (1) and (10) with same boundary conditions could have the same spectrum or, in the more general case, the spectra will differ in only one eigenvalue. The main difference between the first-order and the confluent transformations is that in the last case the restriction for the values of λ disappears.

3. Dirac equation with pseudoscalar potentials and the confluent supersymmetry algorithm

The stationary Dirac equation in one spatial dimension is given by [18, 19]

$$i\,\sigma_2\,\Psi' + (V_0 - E)\,\Psi = 0, \tag{13}$$

where σ_2 stands for the second Pauli matrix, E denotes a real-valued number, and the spinor $\Psi = (\Psi_1, \Psi_2)^T$ represents the solution. Furthermore, we assume the potential V_0 to be of pseudoscalar form, that is,

$$V_0 = m\,\sigma_3 + q_0\,\sigma_1, \tag{14}$$

for a constant, positive mass m, Pauli matrices σ_1, σ_3, and a function $q_0 = q_0(x)$. In components, the Dirac equation (13) for the potential (14) can be written as

$$\Psi_1' - q_0\,\Psi_1 + (E + m)\,\Psi_2 = 0, \tag{15}$$
$$\Psi_2' + q_0\,\Psi_2 - (E - m)\,\Psi_1 = 0. \tag{16}$$

Solving the first of these equations for Ψ_2 gives the relation

$$\Psi_2 = \frac{1}{E+m}\left(q_0\,\Psi_1 - \Psi_1'\right). \tag{17}$$

Substitution of this expression into (16) leads to the equation

$$\Psi_1'' + \;E^2 - m^2 - q_0^2 - q_0'\;\;\Psi_1 = 0. \tag{18}$$

Hence, any spinor that solves the initial Dirac equation (13), is also a solution to equations (17), (18) and vice versa. Let us now rewrite (18) as follows

$$\Psi_1'' + (\epsilon - U_0)\,\Psi_1 = 0, \tag{19}$$

where $\epsilon = E^2 - m^2$ and $U_0 = q_0^2 + q_0'$. Comparing equations (17) with (3) we can see that the component Ψ_2 can be obtained as a first-order SUSY transformation where the function q_0 can be written as

$$q_0 = \frac{u'}{u}. \tag{20}$$

The equation fulfilled by u can be found substituting the relation (20) in the definition of the potential $U_0 = q_0^2 + q_0'$, the resulting equation is

$$u'' - U_0 u = 0, \tag{21}$$

i.e. the transformation function u is a solution of the Schrödinger equation (19) when $\epsilon = \lambda = 0$. Thus, in any stationary Dirac equation in one dimension with pseudoscalar potential the component Ψ_1 satisfies a Schrödinger equation and the Ψ_2 can be obtained through a first-order SUSY transformation where the factorization constant λ is equal to zero.

Now, to obtain a new exactly solvable Dirac equation with a pseudoscalar potential in the form

$$i\,\sigma_2\,\Phi' + (V_1 - E)\,\Phi = 0, \tag{22}$$

where $\Phi = (\Phi_1, \Phi_2)^T$, E a real-valued constant and the pseudoscalar potential as

$$V_1 = m\,\sigma_3 + q_1\,\sigma_1, \tag{23}$$

we can use the solutions of (13) and the second-order SUSY formalism to obtain the component Φ_1, then through a first order SUSY transformation the function q_1 and the component Φ_2 can be constructed.

Let us assume that we have found three transformation functions $u_1,\ u_2,\ u_3$, the first two are solution of the Jordan chain (7), (8) while u_3 is solution of (2) for $\lambda = 0$. Then the function Φ_1, given by (9) is a solution of the Schrödinger equation

$$\Phi_1'' + (\epsilon - U_2)\,\Phi_1 = 0. \tag{24}$$

The potential U_2 is given by the expression (11), that is,

$$U_2 = U_0 - 2\,\left[\ln\left(W_{u_1,u_2}\right)\right]''. \tag{25}$$

Analogous to (20), we can find q_1 with a solution of (24) transforming u_3 as given in (9), for instance

$$q_1 = \frac{W'_{u_1,u_2,u_3}}{W_{u_1,u_2,u_3}} - \frac{W'_{u_1,u_2}}{W_{u_1,u_2}}. \tag{26}$$

The component Φ_2 can be obtained applying a first order SUSY transformation to the functions Ψ_1 in the following way

$$\Phi_2 = \frac{1}{E+m}\left(q_1\Phi_1 - \Phi_1'\right). \tag{27}$$

In the last step it remains to convert (26) into a transformed pseudoscalar Dirac potential V_1, given by

$$V_1 = m\,\sigma_3 + q_1\,\sigma_1. \tag{28}$$

The explicit form of this potential is obtained after insertion of (26). We get

$$V_1 = m\ \sigma_3 + \left[\frac{W'_{u_1,u_2,u_3}}{W_{u_1,u_2,u_3}} - \frac{W'_{u_1,u_2}}{W_{u_1,u_2}}\right]\sigma_1. \tag{29}$$

The family of pseudoscalar potentials (29) enters in the transformed Dirac equation as follows

$$i\ \sigma_2\ \Phi' + (V_1 - E)\ \Phi = 0. \tag{30}$$

The first component Φ_1 of the solution spinor $\Phi = (\Phi_1, \Phi_2)^T$ of our equation (30) is given in (9), while the second component Φ_2 is in (27).

4. One dimensional Dirac-Moshinsky oscillator and isospectral SUSY partners

Consider the following one-dimensional Dirac-Moshinsky oscillator-like equation:

$$i\sigma_2\Psi' + (x\sigma_1 + m\sigma_3 - E)\Psi = 0, \qquad x \in (-\infty, \infty) \tag{31}$$

where $\Psi = (\Psi_1, \Psi_2)^T$ is a two component spinor and the boundary conditions to be fulfilled are

$$\lim_{|x|\to\infty} \Psi_1 = \lim_{|x|\to\infty} \Psi_2 = 0. \tag{32}$$

Comparing (13) and (14) with (31) it can be seen that for this system $q_0 = x$. Furthermore, from (18), the equation satisfied by the component Ψ_1 of the spinor Ψ is

$$\Psi_1'' + (E^2 - m^2 - x^2 - 1)\Psi_1 = 0, \tag{33}$$

and with the substitution $\epsilon = E^2 - m^2$ and $U_0 = x^2 + 1$ it has the standard form of a Schrödinger equation (see (1)). Considering the boundary conditions (32), the eigenvalues of (33) are

$$\epsilon_n = 2n + 2, \qquad n = 0,\ 1,\ 2, \ldots, \tag{34}$$

and the corresponding eigenfunctions are

$$\Psi_{1,n} = \frac{1}{\sqrt{\pi^{1/2} 2^n n!}} \exp\left(-\frac{x^2}{2}\right) H_n(x) \tag{35}$$

where H_n are the Hermite polynomials [2]. The second component can be found using the expression (17) and simplified using properties of the Hermite polynomials as

$$\Psi_{2,n} = \frac{1}{E_n + m}\left(x\Psi_{1,n} - \Psi'_{1,n}\right) = \frac{\sqrt{2(n+1)}}{E_n + m}\Psi_{1,n+1}. \tag{36}$$

The eigenvalues then are given by solving for E the relation $\epsilon = E^2 - m^2$:

$$|E_n| = \sqrt{2n + 2 + m^2}. \tag{37}$$

For this one dimensional Dirac-Moshinsky like system both components (35) and (36) of the spinor Ψ are solution of the harmonic oscillator Schrödinger equation. Figure 1 shows the function $q_0 = x$ and three normalized probability densities $P = \Psi^T\Psi$ for $n = 0$ (blue curve), $n = 1$ (purple curve) and $n = 2$ (green curve).

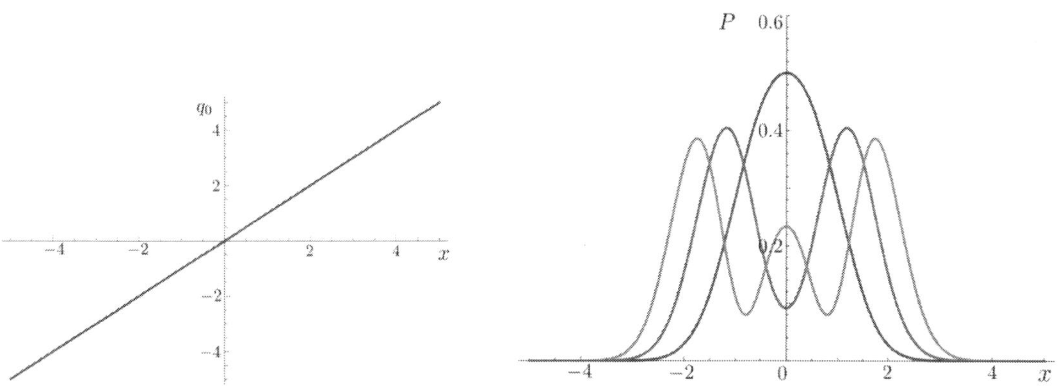

Figure 1. Function q_0 and probability densities associated to the normalized spinor Ψ for $n = 0$ (blue curve), $n = 1$ (purple curve) and $n = 2$ (green curve).

4.1. Isospectral Dirac problems using the first-order SUSY.
To obtain a new exactly solvable Dirac equation

$$i\sigma_2\widetilde{\Psi}' + (\widetilde{V} - E)\widetilde{\Psi} = 0, \qquad x \in (-\infty, \infty)$$
$$\lim_{|x|\to\infty} \widetilde{\Psi}_1 = \lim_{|x|\to\infty} \widetilde{\Psi}_2 = 0. \tag{38}$$

with the spectrum (37) using the first order transformation we can keep one component of the spinor $\widetilde{\Psi}$ as given by (35) or (36) and then build the second with a general 1-SUSY transformation. In this work we start by taking $\widetilde{\Psi}_1 = \Psi_1$. Now, in order to construct the component $\widetilde{\Psi}_2$ for the new system we need a transformation function u satisfying

$$u'' - (x^2 + 1)u = 0. \tag{39}$$

The general solution of this equation is

$$u = \exp\left(\frac{x^2}{2}\right)\left[C_1 + C_2\frac{\sqrt{\pi}}{2}\mathrm{Erf}(x)\right], \tag{40}$$

where Erf stands for the Error function [2]. The function $\widetilde{q}_0$ involved in the new pseudoscalar potential will be given by (20) as

$$\widetilde{q}_0 = \frac{u'}{u} = x + \frac{2C_2\exp\left(-\frac{x^2}{2}\right)}{2C_1 + C_2\sqrt{\pi}\ \mathrm{Erf}(x)}. \tag{41}$$

In order to avoid singularities for the new function $\tilde{q}_0$ we should set $C_1 = 1$ and $|C_2| < 2/\sqrt{\pi}$. Then the new Dirac equation is

$$i\sigma_2\widetilde{\Psi}' + \left\{m\sigma_3 + \left[x + \frac{2C_2\exp\left(-\frac{x^2}{2}\right)}{2 + C_2\sqrt{\pi}\ \mathrm{Erf}(x)}\right]\sigma_1 - E\right\}\widetilde{\Psi} = 0, \qquad x \in (-\infty, \infty) \tag{42}$$

and its boundary conditions as in (38). The second component is obtained with the relation (17) and can be expressed as

$$\widetilde{\Psi}_{2,n} = \frac{1}{E_n + m}\left[\sqrt{2(n+1)}\Psi_{1,n+1} + \frac{2C_2 \exp\left(-\frac{x^2}{2}\right)}{2 + C_2\sqrt{\pi}\,\mathrm{Erf}(x)}\Psi_{1,n}\right]. \tag{43}$$

The eigenvalues of the new Dirac equation (42) are given in (37), i.e. the system (31) and the family of equations (42) characterized by the continuous parameter C_2 are isospectral.

Figure 2 shows on the left the plot of the function $q_0 = x$ (blue curve) and the function $\widetilde{q}_0$ given by (41) when $C_1 = 1$ and $C_2 = -1.1$ (purple curve). On the right the normalized probability densities $\widetilde{P} = \widetilde{\Psi}^T\widetilde{\Psi}$ for $n = 0, 1, 2$ were plotted. A comparison between the probability densities of the original system (31) and the new one (42) shows how the deformation in the function q_0 induced by the SUSY transformations affects them even though the two systems are isospectral. For this new systems the component $\widetilde{\Psi}_1$ is solution of the harmonic oscillator Schrödinger equation while $\widetilde{\Psi}_2$ fulfills a Schrödinger equation with a Abraham-Moses-Mielnik potential [1,15].

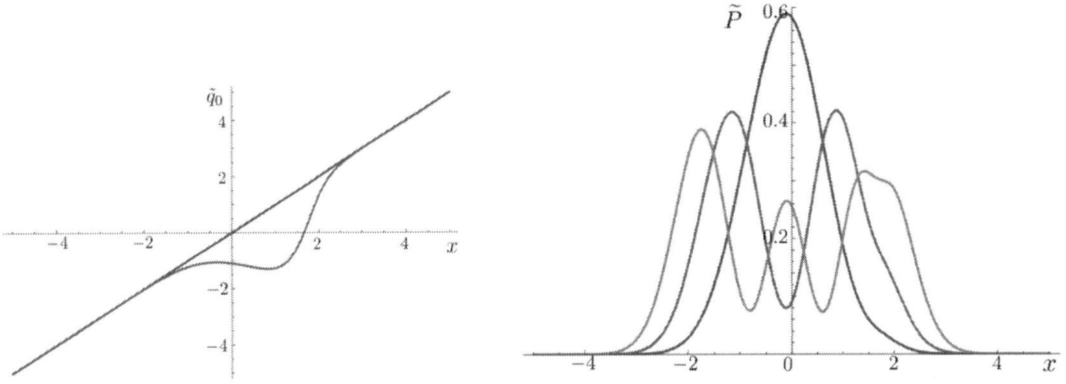

Figure 2. On the left, functions $\widetilde{q}_0$ (purple curve) and q_0 (blue curve). On the right, probability densities associated to the normalized spinor $\widetilde{\Psi}$ for $n = 0$ (blue curve), $n = 1$ (purple curve) and $n = 2$ (green curve) when $C_2 = -1.1$

4.2. Isospectral partners of the Dirac Moshinsky Oscillator using confluent SUSY transformations

To obtain new isospectral and exactly solvable Dirac problems we can use the confluent supersymmetry algorithm for Dirac equations. The new Dirac equation will have the form

$$\begin{aligned} &i\sigma_2\Phi' + (V_1 - E)\Phi = 0, \qquad x \in (-\infty, \infty) \\ &\lim_{|x|\to\infty} \Phi_1 = \lim_{|x|\to\infty} \Phi_2 = 0, \end{aligned} \tag{44}$$

where $\Phi = (\Phi_1, \Phi_2)^T$ and V_1 is a pseudoscalar potential with a parametrizing function q_1 to be found. First we can apply the second-order confluent algorithm to the Schrödinger equation (33) to obtain the equation to be fulfilled by Φ_1, then applying a first order SUSY transformation the parametrizing function q_1 involved in the new potential V_1 and the component Φ_2 of the spinorial solution can be constructed.

Journal of Physics: Conference Series **624** (2015) 012013 doi:10.1088/1742-6596/624/1/012013

To perform the confluent SUSY transformation we need two transformation functions solving the system (7)-(8) with $U_0 = x^2 + 1$ (see (33)). We will use $\lambda = \epsilon_k$, i.e. the factorization energy being one eigenvalue of (33) and

$$u_1 = \Psi_{1,k}, \qquad k = 0,\ 1,\ 2,\ \ldots, \tag{45}$$

$$u_2 = \Psi_{1,k} + C_3\ \Psi_{1,k} \int^x \frac{1}{\Psi_{1,k}^2} ds + \Psi_{1,k} \int^x \frac{\int^t \Psi_{1,k}^2 ds}{\Psi_{1,k}^2} dt \tag{46}$$

where C_3 is an arbitrary constant. We are using as transformation function u_1 a square integrable solution of the Schrödinger equation (7). The first two terms of u_2 are solution of the homogeneous equation (8) while the last term is a particular solution obtained using the variation-of-constants formula. Now, through the confluent SUSY transformation we can obtain the new Schrödinger equation

$$\Phi_1'' + (\epsilon - U_1)\Phi_1 = 0, \tag{47}$$

where the new potential U_1 is given by (11). Using the transformation functions in (45) and (46) we obtain

$$U_1 = x^2 + 1 - 2\left[\ln\left(C_3 + \int_{-\infty}^x \Psi_{1,k}^2 dx\right)\right]''. \tag{48}$$

With the indicated integration limits and since $\Psi_{1,k}$ is a normalized eigenfunction of (33) we can ensure that the potential U_1 is free of singularities when $C_3 \in (-\infty, -1) \cup (0, \infty)$. The solution of (47) can be written in terms of (35) and the transformation functions u_1 and u_2 as

$$\Phi_1 = \frac{W_{u_1,u_2,\Psi_1}}{W_{u_1,u_2}} = \left(\epsilon_k - \epsilon + \frac{\Psi_{1,k}\ \Psi_{1,k}'}{C_3 + \int_{-\infty}^x \Psi_{1,k}^2 dx}\right)\Psi_1 - \left(\frac{\Psi_{1,k}^2}{C_3 + \int_{-\infty}^x \Psi_{1,k}^2 dx}\right)\Psi_1'. \tag{49}$$

Note that the Wronskian in the numerator was simplified by Laplace expansion and (19) was used for further simplification. The previous formula is valid when $\epsilon \neq \epsilon_k$, in case the the factorization energy ϵ_k coincides with the energy ϵ we have to consider (12), then

$$\Phi_1 = \frac{\Psi_{1,k}}{C_3 + \int_{-\infty}^x \Psi_{1,k}^2 dx}. \tag{50}$$

To proceed we need the third transformation function, u_3 satisfying (2) when $\lambda = 0$, its solution is

$$u_3 = \exp\left(\frac{x^2}{2}\right)\left[1 + C_4 \frac{\sqrt{\pi}}{2} \mathrm{Erf}(x)\right], \tag{51}$$

where C_4 is an arbitrary constant. Now, using the Jordan chain (7)-(8) and (2) the Wronskian W_{u_1,u_2,u_3} and its first derivative can be expressed as

$$\begin{aligned} W_{u_1,u_2,u_3} &= -\Psi_{1,k}^2\ u_3' + \left[\epsilon_k\left(C_3 + \int_{-\infty}^x \Psi_{1,k}^2 dx\right) + \Psi_{1,k}\ \Psi_{1,k}'\right] u_3, \\ W_{u_1,u_2,u_3}' &= \left[\epsilon_k\ \left(C_3 + \int_{-\infty}^x \Psi_{1,k}^2 dx\right) - \Psi_{1,k}'\ \Psi_{1,k}\right] u_3' + \Psi_{1,k}'^2 u_3, \end{aligned} \tag{52}$$

these two expressions have to be substituted in (26) to find

$$q_1 = \frac{W_{u_1,u_2,u_3}'}{W_{u_1,u_2,u_3}} - \left(\frac{\Psi_{1,k}^2}{C_3 + \int_{-\infty}^x \Psi_{1,k}^2 dx}\right). \tag{53}$$

The component Φ_2 of the spinor Φ is given by (27) as

$$\Phi_2 = \frac{1}{E+m}\left(q_1\Phi_1 - \Phi_1'\right), \tag{54}$$

unfortunately, the explicit form of this function is very long, such that we do not show it here but the explicit expressions for the functions involved are in (49), (50) and (53).

The function q_1 is plotted in Fig 3 (purple curve). The figure shows how the parametrizing function q_0 (blue curve) is deformed after the confluent SUSY transformation. Also, on the right, you can find the first three normalized probability densities $\hat{P} = \Phi^T\Phi$. The parameters characterizing this particular transformation are $\epsilon_k = \epsilon_1$, i.e. we are using the first excited level in the Schrödinger scheme as transformation function, the constant $C_3 = 0.04$ (see (46)) and finally the third parameter appearing in (51) is $C_4 = 1$. Note that with the selected parameter the spinor component Φ_1 is given by (49) for $n = 0,\ 2,\ 3,\ 4, \ldots$ and by (50) when $n = 1$. The function q_1 was constructed using a non-standard SUSY transformation in the sense that a factorization energy ϵ_k above the ground state energy was used.

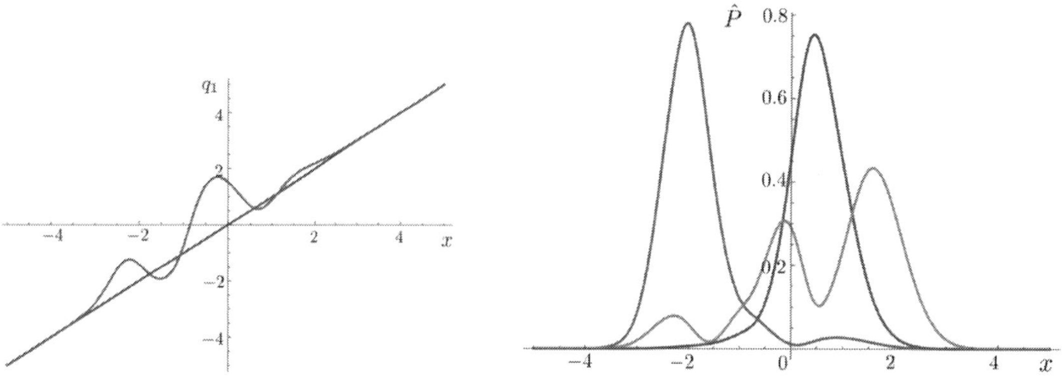

Figure 3. On the left, function q_1 (purple curve) and as reference q_0 (blue curve). On the right, probability densities $\hat{P}$ for $n = 0$ (blue curve), $n = 1$ (purple curve) and $n = 2$ (green curve). The parameters used are $\epsilon_k = \epsilon_1 = 4$, $C_3 = 0.04$ and $C_4 = 1$.

5. Concluding remarks

Two different exactly solvable families of systems were constructed departing from a Dirac-Moshinsky oscillator like system in one dimension using supersymmetric quantum mechanics. They were explicitly built to be isospectral. The first family obtained is characterized by a continuous parameter and, as the Dirac-Moshinsky Oscillator is related to the harmonic oscillator in non-relativistic quantum mechanics, the new systems are related to the Abraham-Moses-Mielnik oscillator. The second generated family is characterized by three parameters, two of them are continuous and the other one is discrete. This last family was obtained using the confluent supersymmetry algorithm and expressions for the solutions were given.

Acknowledgments

ACA acknowledges the Conacyt fellowship 207577.

References

[1] Abraham PB, Moses HE 1980 Changes in potentials due to changes in the point spectrum: Anharmonic oscillators with exact solutions *Phys. Rev.* A **22** 1333

[2] Abramowitz M, Stegun I 1964 *Handbook of mathematical functions with formulas, graphs, and mathematical tables* Dover Publications (New York)

[3] Bagrov VG, Samsonov BF 1996 Supersymmetry of a non stationary Schrödinger equation *Phys. Lett.* A **210** 60–64

[4] Bermudez D, Fernandez DJ 2014 Supersymmetric quantum mechanics and Painlev equations *AIP Conf. Proc.* **1575** 50

[5] Castanos O, Frank A, Lopez R, Urrutia LF 1991 Soluble extensions of the Dirac oscillator with exact and broken supersymmetry *Phys. Rev.* D **43** 544

[6] Contreras-Astorga A, Fernandez DJ 2008 Supersymmetric partners of the trigonometric Pöschl-Teller potentials *J. Phys.* A: *Math. Theor.* **41** 475303

[7] Contreras-Astorga A, Fernandez DJ and Negro J 2012 Solutions of the Dirac equation in a magnetic field and intertwining operators *SIGMA* **8** 082

[8] Contreras-Astorga A, Schulze-Halberg A 2014 The confluent supersymmetry algorithm for Dirac equations with pseudoscalar potentials *J. Math. Phys.* **55** 103506

[9] Cooper F, Khare A, Sukhatme U 1995 Supersymmetry and quantum mechanics *Phys. Rep.* **251** 267–388

[10] Correa F, Jakubsky V 2014 Twisted kinks, Dirac transparent systems, and Darboux transformations *Phys. Rev.* D **90** 125003

[11] Fernandez DJ 2010 Supersymmetric quantum mechanics *AIP Conference Proceedings* **1287** 3–36

[12] Junker G, Roy R 1998 Conditionally exactly solvable potentials: a supersymmetric construction method, *Ann. Phys.* **270** 155–177

[13] Midya B 2011 Nonsingular potentials from excited state factorization of a quantum system with position-dependent mass, *J. Phys.* A*: Math. Theor.* **44** 435306

[14] Midya B, Fernandez DJ 2014 Dirac electron in graphene under supersymmetry generated magnetic fields *J. Phys.* A: *Math. Theor.* **47** 285302

[15] Mielnik B 1984 Factorization method and new potentials with the oscillator spectrum *J. Math. Phys.* **25** 3387–3389

[16] Mielnik B, Rosas-Ortiz O 2004 Factorization: little or great algorithm? *J. Phys.* A: *Math. Gen.* **37** 10007–10035

[17] Moshinsky M, Szczepaniak A 1989 The Dirac oscillator *J. Phys.* A: *Math. Gen.* **22** L817

[18] Nieto LM, Pecheritsin AA, Samsonov BF 2003 Intertwining technique for the one-dimensional stationary Dirac equation *Ann. Phys.* **305** 151–189

[19] Peres NMR 2009 Scattering in one-dimensional heterostructures described by the Dirac equation *J. Phys.: Condens Matter* **21** 095501

[20] Quesne C 1991 Supersymmetry and the Dirac oscillator *Int. J. Mod. Phys.* A **06** 1567

[21] Rosas-Ortiz O 1998 Exactly solvable hydrogen-like potentials and the factorization method *J. Phys.* A: *Math. Gen.* **31** 10163–10179

[22] Rosu HC 1997 Supersymmetric Fokker-Planck strict isospectrality *Phys. Rev.* E **56** 2269–2271

[23] Sadurni E 2011 The Dirac-Moshinsky oscillator: theory and applications *AIP Conf. Proc.* **1334** 249

[24] Schulze-Halberg A, Morales Rivas J, Pea Gil JJ, Garca-Ravelo J, Roy P 2009 Exact solutions of the Fokker-Planck equation from an nth order supersymmetric quantum mechanics approach *Phys. Lett.* A **373** 1610–1615

Supercoherent states approach to the SUSY harmonic oscillator

Erik Díaz-Bautista and David J. Fernández C.
Physics Department, Cinvestav, A.P. 14-740, 07000 México D.F., México
E-mail: ediaz@fis.cinvestav.mx, david@fis.cinvestav.mx

Dedicated to Professor Bogdan Mielnik
for his 50 years of scientific career

Abstract. The nonlinear supercoherent states, associated with a nonlinear generalization of the Kornbluth-Zypman (KZ) supersymmetric annihilation operator (SAO) of the supersymmetric harmonic oscillator, will be studied. We discuss as well the Heisenberg uncertainty relation $\sigma_x^2\sigma_p^2$ for a special case which will allow us to compare our results with those obtained for the KZ linear supercoherent states.

1. Introduction

For the standard harmonic oscillator, the commutators between the Hamiltonian $\hat{H}$ with the annihilation and creation operators $\hat{a}, \hat{a}^\dagger$ generate the well known Heisenberg-Weyl algebra.

On the other hand, for the SUSY harmonic oscillator [1, 2] the analysis of the commutation relations between the Hamiltonian $\hat{H}_s$

$$\hat{H}_s = \omega \begin{pmatrix} \hat{a}^\dagger\hat{a} & 0 \\ 0 & \hat{a}\hat{a}^\dagger \end{pmatrix} \tag{1}$$

and the corresponding creation $\hat{A}^\dagger$ and annihilation operators $\hat{A}$ of the system requires the knowledge of the explicit form of these operators [3, 4].

Recently [5], a quite general expression for $\hat{A}$ was introduced and its eigenvectors $|Z\rangle$ with complex eigenvalues z were found (called supercoherent states), which became expressed in terms of the standard harmonic oscillator coherent states [6, 7].

Despite the expression proposed in [5] for $\hat{A}$ is very general, it is not unique. In this work, we will consider some specific deformations $\hat{Q}$, which maintain the structure given in [5]. We will analyze also a particular deformation of the SAO, and we will find the explicit form of its eigenvectors $|\mathcal{X}\rangle$ with complex eigenvalues X. Due to the deformation assumed, the eigenstates of $\hat{Q}$ turn out to be expressed in terms of nonlinear coherent states associated to deformed (nonlinear) Lie algebras [8–11], so they are called nonlinear supercoherent states.

Journal of Physics: Conference Series **624** (2015) 012014 doi:10.1088/1742-6596/624/1/012014

2. Standard coherent states (CS)

A standard coherent state $|\alpha\rangle$ can be defined as an eigenstate of the annihilation operator $\hat{a}$ with complex eigenvalue α.

In the Fock basis, the normalized coherent states read

$$|\alpha\rangle = e^{-\frac{1}{2}|\alpha|^2} \sum_{n=0}^{\infty} \frac{\alpha^n}{\sqrt{n!}} |n\rangle. \tag{2}$$

Other definitions can be used to build this type of quantum states, which are equivalent to each other for the harmonic oscillator; nevertheless, in this work we will use the previous definition.

3. Nonlinear coherent states (NLCS)

The nonlinear coherent states $|z\rangle_f$ can be defined as eigenstates of a deformed annihilation operator $\tilde{a} = f(\hat{N})\hat{a}$ with complex eigenvalue [12–15], where $f(\hat{N})$ is a well behaved real function of the number operator $\hat{N} = \hat{a}^\dagger\hat{a}$.

The structure of these states depends on which Fock states $|n\rangle$ are annihilated by $\tilde{a}$; this is determined by the explicit form of the function $f(\hat{N})$ since

$$\tilde{a}|n\rangle = \sqrt{n} f(n-1)|n-1\rangle, \quad n = 0, 1, 2, \ldots \tag{3}$$

We will consider the following linear form $f(\hat{N}) = \hat{N} - k\hat{1}$, where $k \in \mathbb{N} \cup \{-1, 0\}$. So, the nonlinear coherent states $|\alpha\rangle_{(k)}$ associated to the operator $\tilde{a}_{(k)} = (\hat{N} - k\hat{1})\hat{a}$ turn out to be

$$|\alpha\rangle_{(k)} = \left[\frac{(k+1)!}{{}_0F_2(1, k+2; r^2)}\right]^{1/2} \sum_{n=0}^{\infty} \frac{\alpha^n}{\Gamma(n+1)\sqrt{\Gamma(n+k+2)}} |n+k+1\rangle, \tag{4}$$

where $r = |\alpha|$ and ${}_pF_q$ is the generalized hypergeometric function

$${}_pF_q(a_1, \ldots, a_p, b_1, \ldots, b_q; x) = \frac{\Gamma(b_1)\ldots\Gamma(b_q)}{\Gamma(a_1)\ldots\Gamma(a_p)} \sum_{n=0}^{\infty} \frac{\Gamma(a_1+n)\ldots\Gamma(a_p+n)}{\Gamma(b_1+n)\ldots\Gamma(b_q+n)} \frac{x^n}{n!}. \tag{5}$$

Note that in Eq. (4) the contribution of the lowest energy Fock states to the nonlinear coherent state depends on the parameter k, i.e., somehow it is possible to isolate the lowest energy eigenstates according to the value of k.

For instance, for $f(\hat{N}) = \hat{N} + \hat{1}$ $(k = -1)$ it turns out that $f(n-1) = n \neq 0$, $\forall\ n = 1, 2, \ldots$ which implies that $\tilde{a}|0\rangle = 0$. Thus:

$$|\alpha\rangle_{nl} \equiv |\alpha\rangle_{(-1)} = [{}_0F_2(1, 1; r^2)]^{-1/2} \sum_{n=0}^{\infty} \frac{\alpha^n}{\Gamma(n+1)\sqrt{\Gamma(n+1)}} |n\rangle. \tag{6}$$

On the other hand, for $f(\hat{N}) = \hat{N}$ $(k = 0)$ we have that $f(0) = 0$. This implies that $\tilde{a}|0\rangle = \tilde{a}|1\rangle = 0$, and the corresponding NLCS become now

$$|\alpha\rangle_{NL} \equiv |\alpha\rangle_{(0)} = \left[{}_0F_2(1, 2; r^2)\right]^{-1/2} \sum_{n=0}^{\infty} \frac{\alpha^n}{\Gamma(n+1)\sqrt{\Gamma(n+2)}} |n+1\rangle. \tag{7}$$

This result indicates that the contribution to $|\alpha\rangle_{NL}$ of the ground state is removed, since this eigenstate is isolated from the remaining ones, due to $\tilde{a}|0\rangle = \tilde{a}^\dagger|0\rangle = 0$.

3.1. Heisenberg uncertainty relation

Knowing the explicit form of the standard coherent states $|\alpha\rangle$ and the nonlinear ones $|\alpha\rangle_{(k)}$, the corresponding Heisenberg uncertainty relations can be straightforwardly calculated.

For the standard coherent states, the mean values of the operators $\hat{x}$, $\hat{p}$, $\hat{x}^2$, $\hat{p}^2$ become (with $\hbar = m = \omega = 1$):

$$\begin{aligned} \langle\hat{x}\rangle &= \sqrt{2}\,\mathrm{Re}(\alpha), & \langle\hat{x}^2\rangle &= \frac{1}{2} + 2[\mathrm{Re}(\alpha)]^2, \\ \langle\hat{p}\rangle &= \sqrt{2}\mathrm{Im}(\alpha), & \langle\hat{p}^2\rangle &= \frac{1}{2} + 2[\mathrm{Im}(\alpha)]^2. \end{aligned} \tag{8}$$

Hence, the Heisenberg uncertainty relation turns out to be given by $(\sigma_x)^2_\alpha(\sigma_p)^2_\alpha = 1/4$.

For the nonlinear coherent states $|\alpha\rangle_{(k)}$, the mean values of the operators $\hat{x}$, $\hat{p}$, $\hat{x}^2$, $\hat{p}^2$ are:

$$\langle\hat{x}\rangle_{(k)} = \sqrt{2}\,\mathrm{Re}(\alpha)\frac{{}_0F_2(2,k+2;r^2)}{{}_0F_2(1,k+2;r^2)}, \tag{9a}$$

$$\langle\hat{p}\rangle_{(k)} = \sqrt{2}\mathrm{Im}(\alpha)\frac{{}_0F_2(2,k+2;r^2)}{{}_0F_2(1,k+2;r^2)}, \tag{9b}$$

$$\langle\hat{x}^2\rangle_{(k)} = \frac{2k+3}{2} + [\mathrm{Re}(\alpha)]^2\beta_{k,+}(r) + [\mathrm{Im}(\alpha)]^2\beta_{k,-}(r), \tag{9c}$$

$$\langle\hat{p}^2\rangle_{(k)} = \frac{2k+3}{2} + [\mathrm{Re}(\alpha)]^2\beta_{k,-}(r) + [\mathrm{Im}(\alpha)]^2\beta_{k,+}(r), \tag{9d}$$

where

$$\beta_{k,\pm}(r) = \frac{1}{k+2}\frac{{}_0F_2(2,k+3;r^2)}{{}_0F_2(1,k+2;r^2)} \pm \frac{1}{2}\frac{{}_0F_2(3,k+2;r^2)}{{}_0F_2(1,k+2;r^2)}. \tag{10}$$

The Heisenberg uncertainty relation becomes:

$$\begin{aligned} (\sigma_x)^2_{\alpha,k}(\sigma_p)^2_{\alpha,k} = &\left[\frac{2k+3}{2} - [\mathrm{Re}(\alpha)]^2\tau_k(r) + [\mathrm{Im}(\alpha)]^2\beta_{k,-}(r)\right] \\ &\times\left[\frac{2k+3}{2} - [\mathrm{Im}(\alpha)]^2\tau_k(r) + [\mathrm{Re}(\alpha)]^2\beta_{k,-}(r)\right], \end{aligned} \tag{11}$$

$$\tau_k(r) = 2\left[\frac{{}_0F_2(2,k+2;r^2)}{{}_0F_2(1,k+2;r^2)}\right]^2 - \beta_{k,+}(r).$$

When $k = -1$, the expression for the Heisenberg uncertainty relation is (see Figure 1):

$$\begin{aligned} (\sigma_x)^2_\alpha(\sigma_p)^2_\alpha = &\left[\frac{1}{2} - [\mathrm{Re}(\alpha)]^2\tau_{-1}(r) + [\mathrm{Im}(\alpha)]^2\beta_{-1,-}(r)\right] \\ &\times\left[\frac{1}{2} - [\mathrm{Im}(\alpha)]^2\tau_{-1}(r) + [\mathrm{Re}(\alpha)]^2\beta_{-1,-}(r)\right], \end{aligned} \tag{12}$$

where

$$\tau_{-1}(r) = 2\left[\frac{{}_0F_2(1,2;r^2)}{{}_0F_2(1,1;r^2)}\right]^2 - \frac{{}_0F_2(2,2;r^2)}{{}_0F_2(1,1;r^2)} - \frac{1}{2}\frac{{}_0F_2(1,3;r^2)}{{}_0F_2(1,1;r^2)}.$$

When $k = 0$, the expression for the Heisenberg uncertainty relation is now (see Figure 2 and [16]):

$$(\sigma_x)^2_\alpha(\sigma_p)^2_\alpha = \left[\frac{3}{2} - [\mathrm{Re}(\alpha)]^2\rho(r)\right]\left[\frac{3}{2} - [\mathrm{Im}(\alpha)]^2\rho(r)\right], \tag{13}$$

where

$$\rho(r) = 2\left[\frac{{}_0F_2(2,2;r^2)}{{}_0F_2(1,2;r^2)}\right]^2 - \frac{{}_0F_2(2,3;r^2)}{{}_0F_2(1,2;r^2)}.$$

For $k = 1$ and $k = 2$ we have shown the Heisenberg uncertainty relation in Figure 3 and Figure 4, respectively.

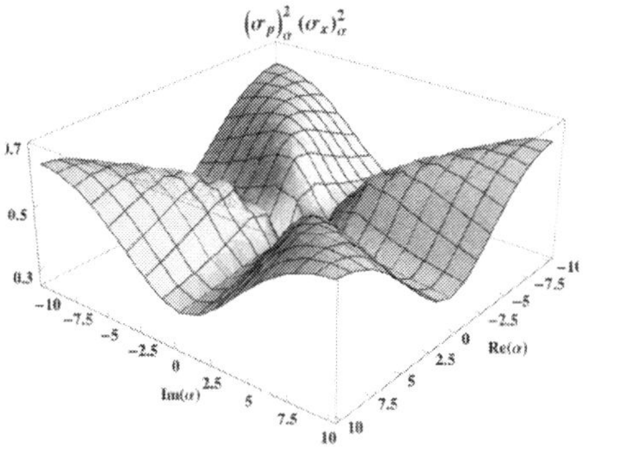

Figure 1. Heisenberg uncertainty relation $(\sigma_x)^2_\alpha(\sigma_p)^2_\alpha$ as function of α for the nonlinear coherent states of Eq. (6) ($k = -1$).

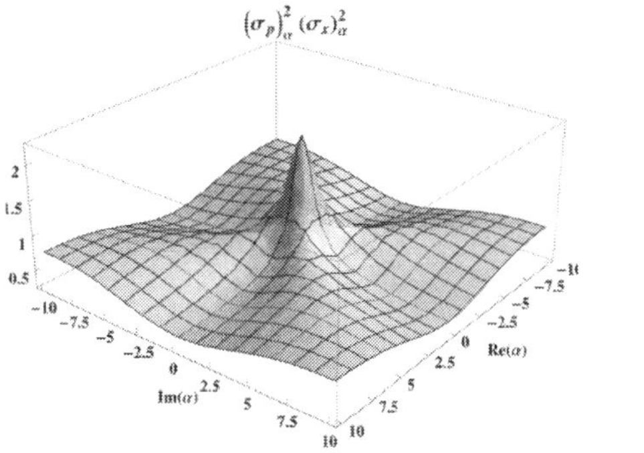

Figure 2. Heisenberg uncertainty relation $(\sigma_x)^2_\alpha(\sigma_p)^2_\alpha$ as function of α for the nonlinear coherent states of Eq. (7) ($k = 0$).

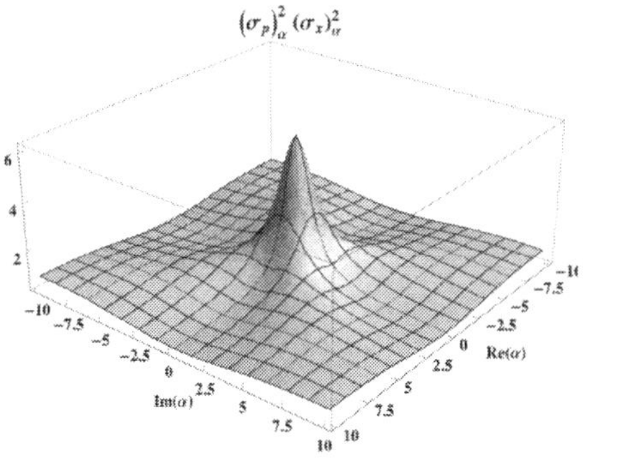

Figure 3. Heisenberg uncertainty relation $(\sigma_x)^2_\alpha(\sigma_p)^2_\alpha$ as function of α for the nonlinear coherent states of Eq. (4) with $k = 1$.

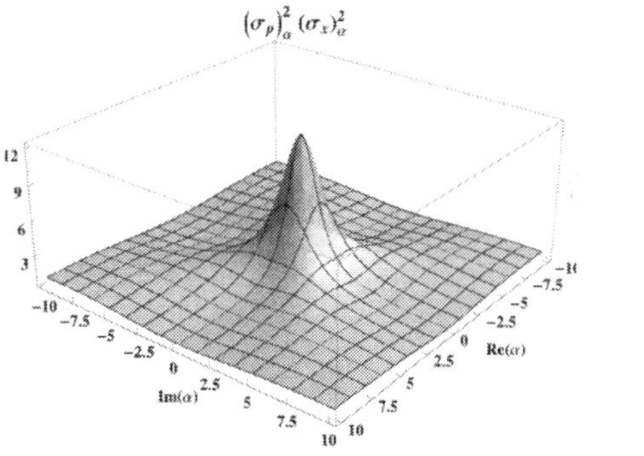

Figure 4. Heisenberg uncertainty relation $(\sigma_x)^2_\alpha(\sigma_p)^2_\alpha$ as function of α for the nonlinear coherent states of Eq. (4) with $k = 2$.

4. Nonlinear supercoherent states

A very general form for the SAO $\hat{A}$ [5] and its simplest deformation are respectively:

$$\hat{A} = \begin{pmatrix} k_1\hat{a} & k_2 \\ k_3\hat{a}^2 & k_4\hat{a} \end{pmatrix}, \quad \hat{Q} = \begin{pmatrix} k_1 f(\hat{N})\hat{a} & k_2 \\ k_3[f(\hat{N})\hat{a}]^2 & k_4 f(\hat{N})\hat{a} \end{pmatrix}, \tag{14}$$

where $k_i \in \mathbb{C}$ and we will take either $f(\hat{N}) = \hat{N} + \hat{1}$ or $f(\hat{N}) = \hat{N}$. We will label $\hat{Q} = \hat{A}'$ when $f(\hat{N}) = \hat{N} + \hat{1}$, and $\hat{Q} = \hat{A}''$ when $f(\hat{N}) = \hat{N}$. Note that $\hat{Q} = \hat{A}$ when $f(\hat{N}) = \hat{1}$.

Both linear and nonlinear supercoherent states are defined in the generic way as

$$\hat{Q}|\mathcal{X}\rangle = X|\mathcal{X}\rangle, \quad |\mathcal{X}\rangle = \begin{pmatrix} \sum_{n=0}^{\infty} a_n|n\rangle \\ \sum_{n=1}^{\infty} c_n|n-1\rangle \end{pmatrix}, \quad X \in \mathbb{C}. \tag{15}$$

From this equation an overall matrix relationship is found as

$$K\begin{pmatrix}\tilde{a}_{n+1}\\ \tilde{c}_{n+1}\end{pmatrix} = \begin{pmatrix}\tilde{a}_n\\ \tilde{c}_n\end{pmatrix}, \quad K = \begin{pmatrix}k_1 & k_2\\ k_3 & k_4\end{pmatrix}, \tag{16}$$

where the quantities $\tilde{a}_n$ and $\tilde{c}_n$ are defined as

$$\tilde{a}_n = \beta_n\sqrt{n!}X^{1-n}a_n, \quad \tilde{c}_n = \beta_{n-1}\sqrt{(n-1)!}X^{1-n}c_n, \tag{17}$$

with

$$X = \begin{cases} z & \text{for } \hat{Q} = \hat{A} \\ Y & \text{for } \hat{Q} = \hat{A}' \\ Z & \text{for } \hat{Q} = \hat{A}'' \end{cases}, \qquad \beta_n = \begin{cases} 1 & \text{for } \hat{Q} = \hat{A} \text{ and } n \geq 1 \\ n! & \text{for } \hat{Q} = \hat{A}' \text{ and } n \geq 1 \\ (n-1)! & \text{for } \hat{Q} = \hat{A}'' \text{ and } n \geq 2 \end{cases}. \tag{18}$$

As can be seen, both the linear and nonlinear supercoherent states depend on the eigenvalues $\psi_\pm$ of the matrix K. This induces a classification of the supercoherent states $|\mathcal{X}\rangle$ into three different families as follows: degenerate ($\psi_+ = \psi_- \equiv \psi \neq 0$); singular ($\psi_+\psi_- = 0$); generic (everything else).

Furthermore, by rewriting a_n and c_n in terms of $\psi_\pm$, the eigenstates of $\hat{A}$ turn out to be expressed in terms of the CS in Eq. (2) while those of $\hat{A}'$ and $\hat{A}''$ are determined by the ones of Eq. (6) or Eq. (7), when choosing $f(\hat{N}) = \hat{N} + \hat{1}$ or $f(\hat{N}) = \hat{N}$ respectively. Then we will identify the respective supercoherent states as follows:

$$|\mathcal{X}\rangle = \begin{cases} |Z\rangle & \text{for } \hat{Q} = \hat{A} \\ |\mathcal{Y}\rangle & \text{for } \hat{Q} = \hat{A}' \\ |\mathcal{Z}\rangle & \text{for } \hat{Q} = \hat{A}'' \end{cases}. \tag{19}$$

4.1. Supercoherent states classification

4.1.1. Generic The general solution of the matrix equation (16) leads to:

$$|\mathcal{X}\rangle = \mathcal{B}_1|\mathcal{X}_A\rangle + \mathcal{B}_2|\mathcal{X}_C\rangle, \tag{20}$$

where

$$\mathcal{B}_1 = \begin{cases} a_0/k_1 & \text{for } \hat{Q} = \hat{A}, \hat{A}' \\ a_1/k_1 & \text{for } \hat{Q} = \hat{A}'' \end{cases}, \quad \mathcal{B}_2 = \begin{cases} c_1/(Xk_1) & \text{for } \hat{Q} = \hat{A}, \hat{A}' \\ c_2/(Xk_1) & \text{for } \hat{Q} = \hat{A}'' \end{cases}, \tag{21a}$$

$$|\mathcal{X}_A\rangle = \frac{1}{\psi_+ - \psi_-}G_A\begin{pmatrix}|\chi_+\rangle\\ |\chi_-\rangle\end{pmatrix}, \quad |\mathcal{X}_C\rangle = \frac{1}{\psi_+ - \psi_-}G_C\begin{pmatrix}|\chi_+\rangle\\ |\chi_-\rangle\end{pmatrix}, \tag{21b}$$

with

$$\chi_\pm \equiv X\psi_\pm^{-1}, \quad |\chi_\pm\rangle = \begin{cases} |\beta_\pm\rangle & \text{for } \hat{Q} = \hat{A} \\ |\varphi_\pm\rangle_{nl} & \text{for } \hat{Q} = \hat{A}' \\ |\phi_\pm\rangle_{NL} & \text{for } \hat{Q} = \hat{A}'' \end{cases}, \tag{22a}$$

and

$$G_A = \begin{pmatrix} \psi_+(\psi_+ - k_4) & -\psi_-(\psi_- - k_4) \\ k_3 X & -k_3 X \end{pmatrix}, \tag{22b}$$

$$G_C = \begin{pmatrix} k_2\psi_+\psi_- & -k_2\psi_+\psi_- \\ X[k_1\psi_+ - (k_1^2 + k_2k_3)] & -X[k_1\psi_- - (k_1^2 + k_2k_3)] \end{pmatrix}. \tag{22c}$$

Besides the set $\{|\mathcal{X}_A\rangle, |\mathcal{X}_C\rangle\}$, we can choose another set of supercoherent states formed by the elements

$$|\mathcal{X}_\pm\rangle = \begin{pmatrix} k_2\psi_\pm|\chi_\pm\rangle \\ (\psi_\pm - k_1)X|\chi_\pm\rangle \end{pmatrix}, \tag{23}$$

whereby it is possible to pass from a parameter space $\{k_1, k_2, k_3, k_4\}$ to a new one, formed by $\{\psi_+, \psi_-, k_1, k_2\}$.

4.1.2. Degenerate The supercoherent states are explicitly

$$|\mathcal{X}\rangle = \mathcal{B}_1|\mathcal{X}_A^d\rangle + \mathcal{B}_2|\mathcal{X}_C^d\rangle, \tag{24}$$

where

$$|\mathcal{X}_A^d\rangle = G_A^{(d)}\begin{pmatrix} |\chi\rangle \\ |\chi'\rangle \end{pmatrix}, \quad |\mathcal{X}_C^d\rangle = G_C^{(d)}\begin{pmatrix} |\chi\rangle \\ |\chi'\rangle \end{pmatrix}, \tag{25}$$

with

$$|\chi'\rangle = \frac{d}{d\chi}|\chi\rangle, \tag{26a}$$

$$G_A^{(d)} = \begin{pmatrix} k_1 & -(\psi - k_4)\chi \\ 0 & -k_3\chi^2 \end{pmatrix}, \quad G_C^{(d)} = \psi\begin{pmatrix} 0 & -k_2\chi \\ k_1\chi & -(\frac{k_4-k_1}{2})\chi^2 \end{pmatrix}. \tag{26b}$$

4.1.3. Singular The corresponding supercoherent states become now

$$|\mathcal{X}_s\rangle = \begin{pmatrix} k_1|\chi\rangle \\ k_3\chi|\chi\rangle \end{pmatrix}, \quad \chi \equiv \frac{X}{(k_1 + k_4)}. \tag{27}$$

4.2. Superposition and uncertainties

Consider a superposition of states $|\mathcal{X}_\pm\rangle$ with parameters η and λ as follows:

$$|\mathcal{X}_m\rangle = \cos\eta|\mathcal{X}_+\rangle + e^{i\lambda}\sin\eta|\mathcal{X}_-\rangle = \begin{pmatrix} \gamma_{1+}|\chi_+\rangle + \gamma_{1-}|\chi_-\rangle \\ \gamma_{2+}|\chi_+\rangle + \gamma_{2-}|\chi_-\rangle \end{pmatrix}, \tag{28}$$

where $\gamma_{1\pm}$ and $\gamma_{2\pm}$ are given by

$$\begin{aligned} \gamma_{1+} &= k_2\psi_+\cos\eta, & \gamma_{1-} &= k_2\psi_- e^{i\lambda}\sin\eta, \\ \gamma_{2+} &= (\psi_+ - k_1)\cos\eta, & \gamma_{2-} &= (\psi_- - k_1)e^{i\lambda}\sin\eta. \end{aligned} \tag{29}$$

The mean value of an arbitrary observable $\hat{c}$ is then

$$\langle\hat{c}\rangle = \frac{\langle\mathcal{X}_m|\hat{c}|\mathcal{X}_m\rangle}{\langle\mathcal{X}_m|\mathcal{X}_m\rangle}, \tag{30a}$$

$$\langle\mathcal{X}_m|\hat{c}|\mathcal{X}_m\rangle = \Gamma^+\langle\chi_+|\hat{c}|\chi_+\rangle + \Gamma^-\langle\chi_-|\hat{c}|\chi_-\rangle + 2\,\mathrm{Re}(\Gamma^{+-}\langle\chi_+|\hat{c}|\chi_-\rangle), \tag{30b}$$

with

$$\Gamma^+ = |\gamma_{1+}|^2 + |\gamma_{2+}X|^2, \quad \Gamma^- = |\gamma_{1-}|^2 + |\gamma_{2-}X|^2, \quad \Gamma^{+-} = \gamma_{1+}^*\gamma_{1-} + \gamma_{2+}^*\gamma_{2-}|X|^2. \tag{31}$$

This allows us to find straightforwardly the expressions for the uncertainties of the operators $\hat{x}$ and $\hat{p}$ and their squares.

4.2.1. Uncertainties for $\hat{Q}=\hat{A}$. The expressions for the uncertainties are [5]:

$$\sigma_x^2 = \frac{\Gamma^+[1+(\beta_+ + \beta_+^*)^2]e^{|\beta_+|^2} + \Gamma^-[1+(\beta_- + \beta_-^*)^2]e^{|\beta_-|^2} + 2\,\mathrm{Re}[\Gamma^{+-}[1+(\beta_- + \beta_+^*)^2]e^{\beta_+^*\beta_-}]}{2[\Gamma^+ e^{|\beta_+|^2} + \Gamma^- e^{|\beta_-|^2} + 2\,\mathrm{Re}[\Gamma^{+-}e^{\beta_+^*\beta_-}]]}$$
$$- \left(\frac{\Gamma^+(\beta_+ + \beta_+^*)e^{|\beta_+|^2} + \Gamma^-(\beta_- + \beta_-^*)e^{|\beta_-|^2} + 2\,\mathrm{Re}[\Gamma^{+-}(\beta_- + \beta_+^*)e^{\beta_+^*\beta_-}]}{\sqrt{2}[\Gamma^+ e^{|\beta_+|^2} + \Gamma^- e^{|\beta_-|^2} + 2\,\mathrm{Re}[\Gamma^{+-}e^{\beta_+^*\beta_-}]]}\right)^2, \quad (32)$$

$$\sigma_p^2 = \frac{\Gamma^+[1-(\beta_+ - \beta_+^*)^2]e^{|\beta_+|^2} + \Gamma^-[1-(\beta_- - \beta_-^*)^2]e^{|\beta_-|^2} + 2\,\mathrm{Re}[\Gamma^{+-}[1-(\beta_- - \beta_+^*)^2]e^{\beta_+^*\beta_-}]}{2[\Gamma^+ e^{|\beta_+|^2} + \Gamma^- e^{|\beta_-|^2} + 2\,\mathrm{Re}[\Gamma^{+-}e^{\beta_+^*\beta_-}]]}$$
$$- \left(\frac{-i\Gamma^+(\beta_+ - \beta_+^*)e^{|\beta_+|^2} - i\Gamma^-(\beta_- + \beta_-^*)e^{|\beta_-|^2} + 2\,\mathrm{Re}[-i\Gamma^{+-}(\beta_- + \beta_+^*)e^{\beta_+^*\beta_-}]}{\sqrt{2}[\Gamma^+ e^{|\beta_+|^2} + \Gamma^- e^{|\beta_-|^2} + 2\,\mathrm{Re}[\Gamma^{+-}e^{\beta_+^*\beta_-}]]}\right)^2, \quad (33)$$

where the quantities in Eq. (31) are evaluated taking $X=z$.

4.2.2. Uncertainties for $\hat{Q}=\hat{A}'$. For this case, the expressions for the uncertainties are:

$$\begin{aligned}
\sigma_x^{2(nl)} = {} & \Delta_1^{-1}\Bigg\{\Gamma^+\left[[\mathrm{Re}(\varphi_+)]^2\beta_{-1,+}(|\varphi_+|) + [\mathrm{Im}(\varphi_+)]^2\beta_{-1,-}(|\varphi_+|) + \frac{1}{2}\right] \\
& + \Gamma^-\left[[\mathrm{Re}(\varphi_-)]^2\beta_{-1,+}(|\varphi_-|) + [\mathrm{Im}(\varphi_-)]^2\beta_{-1,-}(|\varphi_-|) + \frac{1}{2}\right] \\
& + 2\,\mathrm{Re}\Bigg(\frac{\Gamma^{+-}}{2\sqrt{{}_0F_2(1,1;|\varphi_+|^2)}\sqrt{{}_0F_2(1,1;|\varphi_-|^2)}} \\
& \times \left[\frac{(\varphi_-^2 + \varphi_+^{*2})}{2}\,{}_0F_2(1,3;\varphi_+^*\varphi_-) + 2\varphi_+^*\varphi_-\,{}_0F_2(2,2;\varphi_+^*\varphi_-) + {}_0F_2(1,1;\varphi_+^*\varphi_-)\right]\Bigg)\Bigg\} \\
& - \Delta_1^{-2}\Bigg\{\sqrt{2}\Gamma^+\,\mathrm{Re}(\varphi_+)\frac{{}_0F_2(1,2;|\varphi_+|^2)}{{}_0F_2(1,1;|\varphi_+|^2)} + \sqrt{2}\Gamma^-\,\mathrm{Re}(\varphi_-)\frac{{}_0F_2(1,2;|\varphi_-|^2)}{{}_0F_2(1,1;|\varphi_-|^2)} \\
& + \sqrt{2}\,\mathrm{Re}\left((\varphi_- + \varphi_+^*)\frac{\Gamma^{+-}\,{}_0F_2(1,2;\varphi_+^*\varphi_-)}{\sqrt{{}_0F_2(1,1;|\varphi_+|^2)}\sqrt{{}_0F_2(1,1;|\varphi_-|^2)}}\right)\Bigg\}^2, \qquad (34)
\end{aligned}$$

$$\begin{aligned}
\sigma_p^{2(nl)} = {} & \Delta_1^{-1}\Bigg\{\Gamma^+\left[[\mathrm{Re}(\varphi_+)]^2\beta_{-1,-}(|\varphi_+|) + [\mathrm{Im}(\varphi_+)]^2\beta_{-1,+}(|\varphi_+|) + \frac{1}{2}\right] \\
& + \Gamma^-\left[[\mathrm{Re}(\varphi_-)]^2\beta_{-1,-}(|\varphi_-|) + [\mathrm{Im}(\varphi_-)]^2\beta_{-1,+}(|\varphi_-|) + \frac{1}{2}\right] \\
& + 2\,\mathrm{Re}\Bigg(\frac{\Gamma^{+-}}{2\sqrt{{}_0F_2(1,1;|\varphi_+|^2)}\sqrt{{}_0F_2(1,1;|\varphi_-|^2)}} \\
& \times \left[-\frac{(\varphi_-^2 + \varphi_+^{*2})}{2}\,{}_0F_2(1,3;\varphi_+^*\varphi_-) + 2\varphi_+^*\varphi_-\,{}_0F_2(2,2;\varphi_+^*\varphi_-) + {}_0F_2(1,1;\varphi_+^*\varphi_-)\right]\Bigg)\Bigg\} \\
& - \Delta_1^{-2}\Bigg\{\sqrt{2}\Gamma^+\mathrm{Im}(\varphi_+)\frac{{}_0F_2(1,2;|\varphi_+|^2)}{{}_0F_2(1,1;|\varphi_+|^2)} + \sqrt{2}\Gamma^-\mathrm{Im}(\varphi_-)\frac{{}_0F_2(1,2;|\varphi_-|^2)}{{}_0F_2(1,1;|\varphi_-|^2)} \\
& + \sqrt{2}\,\mathrm{Re}\left(-i(\varphi_- - \varphi_+^*)\frac{\Gamma^{+-}\,{}_0F_2(1,2;\varphi_+^*\varphi_-)}{\sqrt{{}_0F_2(1,1;|\varphi_+|^2)}\sqrt{{}_0F_2(1,1;|\varphi_-|^2)}}\right)\Bigg\}^2 \qquad (35)
\end{aligned}$$

where $X=Y$ in Eq. (31), $\beta_{-1,\pm}(r)$ is given by Eq. (10) with $k=-1$ and

$$\Delta_1 = \Gamma^+ + \Gamma^- + 2\,\mathrm{Re}\left(\frac{\Gamma^{+-}\,{}_0F_2(1,1;\varphi_+^*\varphi_-)}{\sqrt{{}_0F_2(1,1;|\varphi_+|^2)}\sqrt{{}_0F_2(1,1;|\varphi_-|^2)}}\right). \quad (36)$$

Journal of Physics: Conference Series **624** (2015) 012014 doi:10.1088/1742-6596/624/1/012014

4.2.3. Uncertainties for $\hat{Q} = \hat{A}''$. Similarly, the expressions for the uncertainties are now:

$$
\begin{aligned}
\sigma_x^{2(NL)} &= \Delta_2^{-1}\left\{\Gamma^+\left[[\mathrm{Re}(\phi_+)]^2\frac{{}_0F_2(2,3;|\phi_+|^2)}{{}_0F_2(1,2;|\phi_+|^2)}+\frac{3}{2}\right]+\Gamma^-\left[[\mathrm{Re}(\phi_-)]^2\frac{{}_0F_2(2,3;|\phi_-|^2)}{{}_0F_2(1,2;|\phi_-|^2)}+\frac{3}{2}\right]\right.\\
&\quad+2\,\mathrm{Re}\left(\frac{\Gamma^{+-}}{2\sqrt{{}_0F_2(1,2;|\phi_+|^2)}\sqrt{{}_0F_2(1,2;|\phi_-|^2)}}\right.\\
&\quad\left.\left.\times\left[\frac{(\phi_-+\phi_+^*)^2}{2}\,{}_0F_2(2,3;\phi_+^*\phi_-)+3\,{}_0F_2(1,2;\phi_+^*\phi_-)\right]\right)\right\}\\
&\quad-\Delta_2^{-2}\left\{\sqrt{2}\Gamma^+\,\mathrm{Re}(\phi_+)\frac{{}_0F_2(2,2;|\phi_+|^2)}{{}_0F_2(1,2;|\phi_+|^2)}+\sqrt{2}\Gamma^-\,\mathrm{Re}(\phi_-)\frac{{}_0F_2(2,2;|\phi_-|^2)}{{}_0F_2(1,2;|\phi_-|^2)}\right.\\
&\quad\left.+\sqrt{2}\,\mathrm{Re}\left((\phi_-+\phi_+^*)\frac{\Gamma^{+-}\,{}_0F_2(2,2;\phi_+^*\phi_-)}{\sqrt{{}_0F_2(1,2;|\phi_+|^2)}\sqrt{{}_0F_2(1,2;|\phi_-|^2)}}\right)\right\}^2, \qquad (37)
\end{aligned}
$$

$$
\begin{aligned}
\sigma_p^{2(NL)} &= \Delta_2^{-1}\left\{\Gamma^+\left[[\mathrm{Im}(\phi_+)]^2\frac{{}_0F_2(2,3;|\phi_+|^2)}{{}_0F_2(1,2;|\phi_+|^2)}+\frac{3}{2}\right]+\Gamma^-\left[[\mathrm{Im}(\phi_-)]^2\frac{{}_0F_2(2,3;|\phi_-|^2)}{{}_0F_2(1,2;|\phi_-|^2)}+\frac{3}{2}\right]\right.\\
&\quad+2\,\mathrm{Re}\left(\frac{\Gamma^{+-}}{2\sqrt{{}_0F_2(1,2;|\phi_+|^2)}\sqrt{{}_0F_2(1,2;|\phi_-|^2)}}\right.\\
&\quad\left.\left.\times\left[-\frac{(\phi_--\phi_+^*)^2}{2}\,{}_0F_2(2,3;\phi_+^*\phi_-)+3\,{}_0F_2(1,2;\phi_+^*\phi_-)\right]\right)\right\}\\
&\quad-\Delta_2^{-2}\left\{\sqrt{2}\Gamma^+\mathrm{Im}(\phi_+)\frac{{}_0F_2(2,2;|\phi_+|^2)}{{}_0F_2(1,2;|\phi_+|^2)}+\sqrt{2}\Gamma^-\mathrm{Im}(\phi_-)\frac{{}_0F_2(2,2;|\phi_-|^2)}{{}_0F_2(1,2;|\phi_-|^2)}\right.\\
&\quad\left.+\sqrt{2}\,\mathrm{Re}\left(-i(\phi_--\phi_+^*)\frac{\Gamma^{+-}\,{}_0F_2(2,2;\phi_+^*\phi_-)}{\sqrt{{}_0F_2(1,2;|\phi_+|^2)}\sqrt{{}_0F_2(1,2;|\phi_-|^2)}}\right)\right\}^2 \qquad (38)
\end{aligned}
$$

with $X = Z$ in Eq. (31) and

$$\Delta_2 = \Gamma^+ + \Gamma^- + 2\,\mathrm{Re}\left(\frac{\Gamma^{+-}\,{}_0F_2(1,2;\phi_+^*\phi_-)}{\sqrt{{}_0F_2(1,2;|\phi_+|^2)}\sqrt{{}_0F_2(1,2;|\phi_-|^2)}}\right). \qquad (39)$$

5. A particular case

To analyze the behavior of the supercoherent states (linear and nonlinear) the following particular values for the parameters k_i are taken: $k_1 = k_4 = 1$, $k_2 = \cos\theta$ and $k_3 = \sin\theta$. Thus, the deformed SAO in Eq. (14) takes the generic form:

$$\hat{Q} = \begin{pmatrix} f(\hat{N})\hat{a} & \cos\theta \\ \sin\theta[f(\hat{N})\hat{a}]^2 & f(\hat{N})\hat{a} \end{pmatrix}, \qquad (40)$$

meanwhile the superposition (28) (with $\eta = \lambda = \pi/4$) becomes

$$|\mathcal{X}_\theta\rangle = \frac{1}{\sqrt{2}}(|\mathcal{X}_{+,\theta}\rangle + e^{i\pi/4}|\mathcal{X}_{-,\theta}\rangle). \qquad (41)$$

On the other hand, the eigenvalues $\psi_\pm$ of the matrix K of Eq. (16) become now:

$$\psi_{\pm,\theta} = 1 \pm \sqrt{\frac{1}{2}\sin(2\theta)}, \qquad (42)$$

so that for $0 < \theta < \pi/2$ both $\psi_\pm$ are real while for $\pi/2 < \theta < \pi$ they turn out to be complex. By substituting equations (41) and (42) in expressions (32)–(39), the Heisenberg uncertainty relation for each form of the operator $\hat{Q}$ is found.

5.1. Heisenberg uncertainty relation for $\hat{Q} = \hat{A}$.

The Heisenberg uncertainty relation associated with the linear supercoherentes states of [5] is shown in Figures 5 and 6. For z real (left plot, Figure 5) the uncertainty reaches a maximum value equal to 0.83 at $|z| \sim 0.5$ for real eigenvalues $\psi_\pm$ $(0 < \theta < \pi/2)$, while it shows a growing behavior as $|z|$ increases for complex eigenvalues $\psi_\pm$ $(\pi/2 < \theta < \pi)$. For complex z (right plot, Figure 6), the uncertainty relation behaves similarly as for real z.

5.2. Heisenberg uncertainty relation for $\hat{Q} = \hat{A}'$.

Figures 7 and 8 show the Heisenberg uncertainty relation $\sigma_x^{2(nl)}\sigma_p^{2(nl)}$ for the nonlinear supercoherent states $|\mathcal{Y}\rangle$ associated to the operator $\hat{A}'$. In case of Y real (left plot, Figure 8), for both real $(0 < \theta < \pi/2)$ as well as complex $(\pi/2 < \theta < \pi)$ eigenvalues $\psi_\pm$, the uncertainty starts from a minimum value and grows slowly. In the region of degenerate eigenvalues, for $\theta = \pi/2$, the uncertainty reaches a maximum value. For complex Y (right plot, Figure 8), the uncertainty relation behaves similarly as for real Y.

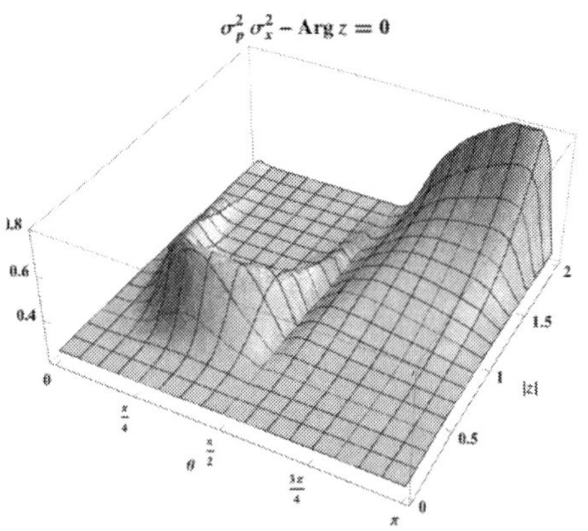

Figure 5. Heisenberg uncertainty relation $\sigma_x^2\sigma_p^2$ (Eqs. (32) and (33)) for the supercoherent states $|Z\rangle$ with z real.

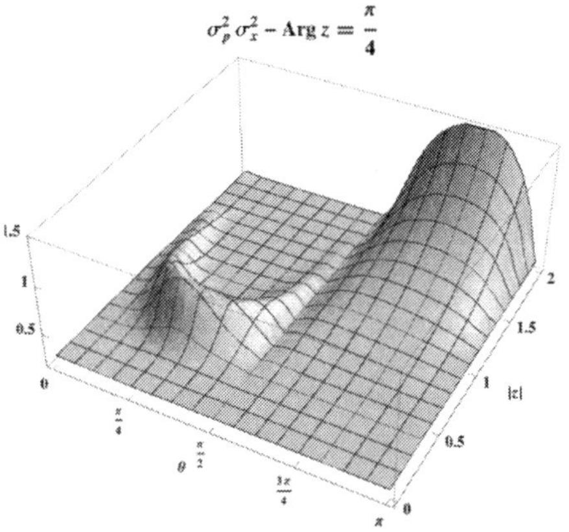

Figure 6. Heisenberg uncertainty relation $\sigma_x^2\sigma_p^2$ (Eqs. (32) and (33)) for the supercoherent states $|Z\rangle$ with z complex.

5.3. Heisenberg uncertainty relation for $\hat{Q} = \hat{A}''$.

Figures 9 and 10 show the uncertainty relations $\sigma_x^{2(NL)}\sigma_p^{2(NL)}$ for the nonlinear supercoherent states $|\mathcal{Z}\rangle$ associated to the operator $\hat{A}''$. In case of Z real (left plot, Figure 9), for both real $(0 < \theta < \pi/2)$ as well as complex $(\pi/2 < \theta < \pi)$ eigenvalues $\psi_\pm$, the uncertainty starts from a value equal to 2.25, then it decreases smoothly and grows again. In the region of degenerate eigenvalues $\psi_\pm$, for $\theta = \pi/2$, the uncertainty reaches a minimum value close to 0.25. For complex Z (right plot, Figure 10), the uncertainty relation behaves similarly as for real Z.

As shown in Figures 5–10, the behavior of the Heisenberg uncertainty relation for the linear supercoherent states $|Z\rangle$ becomes more involved than for the nonlinear ones, $|\mathcal{Y}\rangle$ and $|\mathcal{Z}\rangle$. Moreover, since the nonlinear supercoherent states $|\mathcal{Z}\rangle$ are expressed in terms of the nonlinear coherent states $|\alpha\rangle_{NL}$ which isolate the eigenstate $|0\rangle$, the corresponding uncertainty relation starts from a different value at $Z = 0$ than its counterparts associated to $\hat{A}$ and $\hat{A}'$.

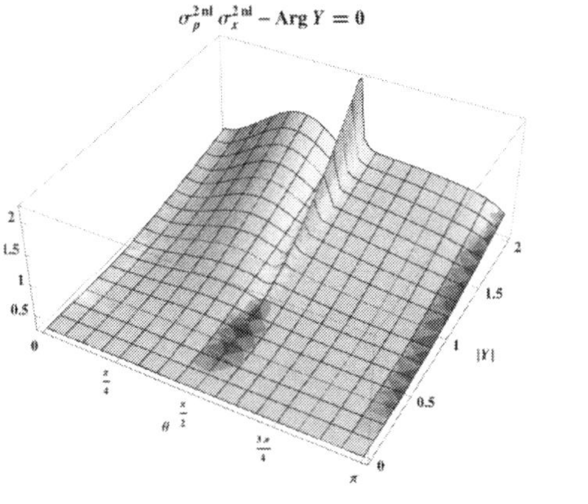

Figure 7. Heisenberg uncertainty relation $\sigma_x^{2(nl)}\sigma_p^{2(nl)}$ (Eqs. (34) and (35)) for the nonlinear supercoherent states $|\mathcal{Y}\rangle$ with Y real.

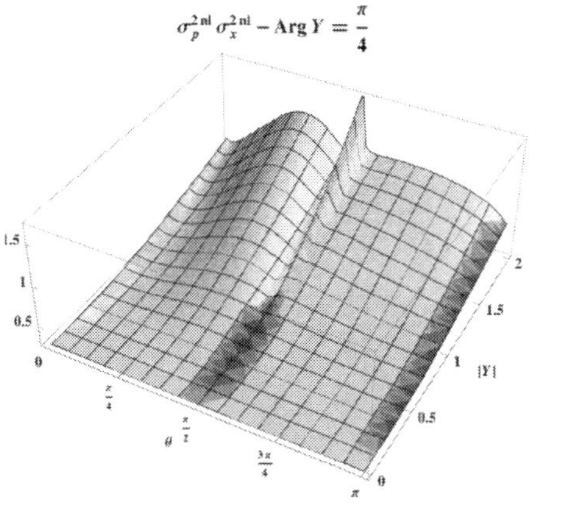

Figure 8. Heisenberg uncertainty relation $\sigma_x^{2(nl)}\sigma_p^{2(nl)}$ (Eqs. (34) and (35)) for the nonlinear supercoherent states $|\mathcal{Y}\rangle$ with Y complex.

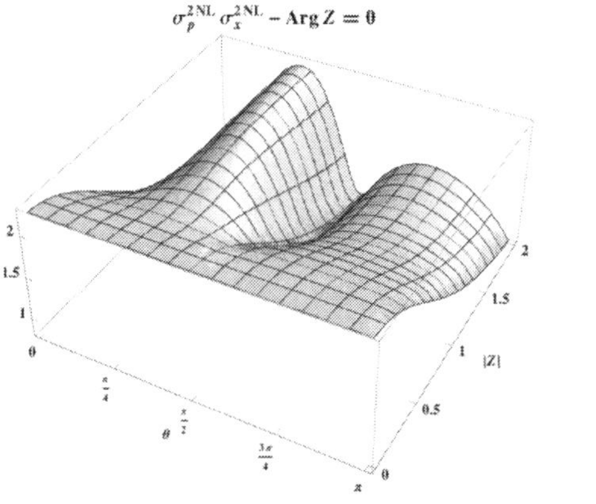

Figure 9. Heisenberg uncertainty relation $\sigma_x^{2(NL)}\sigma_p^{2(NL)}$ (Eqs. (37) and (38)) for nonlinear supercoherent states $|\mathcal{Z}\rangle$ with Z real.

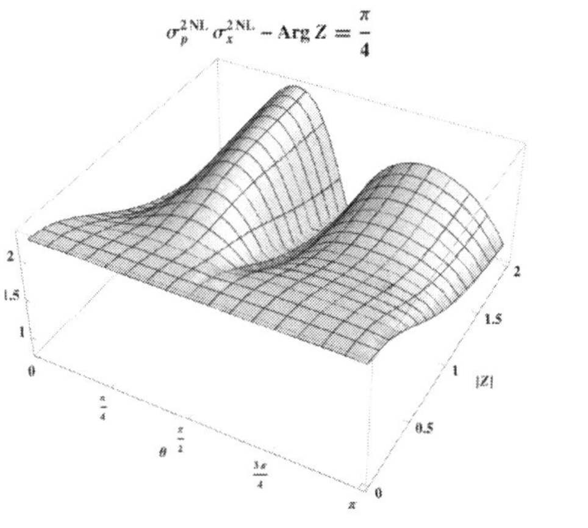

Figure 10. Heisenberg uncertainty relation $\sigma_x^{2(NL)}\sigma_p^{2(NL)}$ (Eqs. (37) and (38)) for nonlinear supercoherent states $|\mathcal{Z}\rangle$ with Z complex.

6. Conclusions

The deformation of the supersymmetric annihilation operator $\hat{A}$ given in [5] allowed us to find interesting results related with the nonlinear supercoherent states. As it was noted, the form of the operator $\tilde{a} = f(\hat{N})\hat{a}$ in the corresponding nonlinear coherent states, enabled us to isolate the ground state contribution, which has interesting implications for the properties of the nonlinear supercoherent states and their associated Heisenberg uncertainty relation.

Let us remark that all the results obtained in this paper obey the Heisenberg uncertainty principle ($\sigma_x^2\sigma_p^2 \geq 1/4$), which ensures that the states constructed here are consistent with the quantum theory.

Finally, the freedom we have for choosing the form of the function $f(\hat{N})$ will allow us to consider a more detailed algebraic study, focused on the commutation relations of the operators $\{\hat{H}_s, \hat{A}, \hat{A}^\dagger\}$ for the supersymmetric harmonic oscillator and the possible relation with the polynomial deformations of the Heisenberg algebra [17, 18], as it happens with the operators $\{\hat{H}, \hat{a}, \hat{a}^{\dagger}\}$ for the standard harmonic oscillator.

Acknowledgments
The authors acknowledge the support of Conacyt, Project 152574. EDB also acknowledges the Conacyt fellowship 290649.

References

[1] Wess J and Zumino B 1974 *Nucl. Phys. B* **70** 39
[2] Witten E 1981 *Nucl. Phys. B* **188** 513
[3] Aragone C and Zypman F 1986 *J. Phys. A: Math. Gen.* **19** 2267
[4] Bérubé-Lauzière Y and Hussin V 1993 *J. Phys. A: Math. Gen.* **26** 6271
[5] Kornbluth M and Zypman F 2013 *J. Math. Phys.* **54** 012101
[6] Glauber R J 1963 *Phys. Rev. Lett.* **10** 84; 1963 *Phys. Rev.* **130** 2529; ibid 2766.
[7] Glauber R J in *Quantum Optics and Electronics*, Les Houches, eds. C. DeWitt, A. Blandin and C. Cohen-Tannoudji (Gordon and Breach, New York 1965)
[8] Granovoskii Ya I, Lutzenko I M and Zhedanov A S 1992 *Ann. Phys.* **217** 1
[9] Vinet L and Letourneau P 1995 *Ann. Phys.* **243** 144
[10] Junker G and Roy P 1997 *Phys. Lett. A* **232** 155; ibid 1999 *A* **257** 113
[11] Karassiov V P 1994 *J. Phys. A: Math. Gen.* **27** 153
[12] Manko V I, Marmo G, Sudarshan E C G and Zaccaria F 1997 *Phys. Scrip.* **55**, 528
[13] Roy B and Roy P 2000 *J. Opt. B: Quantum Semiclass. Opt.* **2**, 65
[14] Dodonov V V 2002 *J. Opt. B: Quantum Semiclass. Opt.* **4**, R1
[15] Hong-Yi Fan and Hai-Ling Cheng 2002 *Commun. Theor. Phys.* **37** 655
[16] Fernández D J, Hussin V and Nieto L M 1994 *J. Phys. A: Math. Gen.* **27** 3547
[17] Fernández D J and Hussin V 1999 *J. Phys. A: Math. Gen.* **32**, 3603
[18] Carballo J M, Fernández D J, Negro J and Nieto L M 2004 *J. Phys. A: Math. Gen.* **37**, 10349

SUSY partners of the complex oscillator and Painlevé IV equation

David J. Fernández C. and J.C. González

Departamento de Física, Cinvestav, A.P. 14-740, 07000 México D.F., Mexico

E-mail: `david@fis.cinvestav.mx` , `jglopez@fis.cinvestav.mx`

Dedicated to Professor Bogdan Mielnik for his 50 years of scientific career.

Abstract. Supersymmetry transformations will be used to obtain new exactly solvable potentials from the complex oscillator. The corresponding Hamiltonians are ruled by polynomial Heisenberg algebras. By applying a mechanism to reduce to second the order of these algebras, the connection with the Painlevé IV equation is achieved, supplying us with an algorithm to generate solutions to such equation.

1. Introduction

Supersymmetric quantum mechanics (SUSY QM) is a simple method to generate exactly solvable potentials from a given initial one such that the spectra of the original and built Hamiltonians differ by a finite number of levels [1–7]. This procedure has been applied mainly to hermitian Hamiltonians, in order to generate either real or complex new potentials [8–13], with an associated spectrum which in the last case could contain a finite number of additional complex energy levels. The method has been used also for non-hermitian Hamiltonians with the so-called PT-symmetry [14–16]. Indeed, in the context of optics it is possible to find families of complex potentials with entirely real spectra with a missing single level, the one associated to an arbitrary initial mode [17]. However, up to our knowledge there are not many works employing Hamiltonians with purely complex eigenvalues.

In this work we will address this subject by taking as initial system the complex oscillator [18–21], whose potential has the harmonic oscillator form but with a complex frequency ω. We shall implement interesting first-order SUSY transformations which employ excited state eigenfunctions as transformation functions, leading to solvable potentials whose spectra are completely determined.

Let us note that, in general, the set of SUSY partners of the harmonic oscillator Hamiltonian realize the even-order polynomial Heisenberg algebras (PHA) [22–27]. In addition, some subsets are ruled simultaneously by second-order PHA, which are linked with the Painlevé IV (PIV) equation; hence, a simple method for generating solutions of such equation can be designed [28–31]. Since the complex oscillator is mapped in a simple way into the harmonic oscillator, thus complex solutions of the PIV equation will be found by applying supersymmetry transformations to the complex oscillator.

This paper is organized as follows. In Section 2 we will sketch the SUSY QM. Then, we will discuss the PHA in Section 3, paying special attention to those of second-order, which are linked with the PIV equation. In Section 4 we will generate the SUSY partners of the complex oscillator, while in Section 5 we will analyze the corresponding solutions to the PIV equation. Section 6 contains our conclusions and some discussion about the perspectives of our work.

2. Supersymmetric quantum mechanics

Let us consider a chain of intertwined Hamiltonians $\hat{H}_j$ such that [4, 5, 32–34]:

$$\hat{H}_j \hat{A}_j^+ = \hat{A}_j^+ \hat{H}_{j-1}, \quad \hat{H}_{j-1} \hat{A}_j^- = \hat{A}_j^- \hat{H}_j, \quad j = 1, \cdots, k, \tag{1a}$$

$$\hat{A}_j^{\pm} = \frac{1}{\sqrt{2}} \left[\mp \frac{\mathrm{d}}{\mathrm{d}x} + \beta_j (x, \varepsilon_j) \right], \tag{1b}$$

$$\hat{H}_j = -\frac{1}{2} \frac{\mathrm{d}^2}{\mathrm{d}x^2} + V_j (x), \tag{1c}$$

where $\beta_j (x, \varepsilon_j)$ and $V_j (x)$ are complex functions, i.e., $\hat{A}_j^+$ and $\hat{A}_j^-$ are not adjoint to each other. Thus, the following equations must be satisfied:

$$\beta_j' (x, \varepsilon_j) + \beta_j^2 (x, \varepsilon_j) = 2 [V_{j-1} (x) - \varepsilon_j], \tag{2}$$

$$V_j (x) = V_{j-1} (x) - \beta_j' (x, \varepsilon_j). \tag{3}$$

The solution $\beta_j (x, \varepsilon_j)$ of the j-th Riccati equation (2) is expressed in terms of two solutions of the $(j-1)$-th one:

$$\beta_j (x, \varepsilon_j) = -\beta_{j-1} (x, \varepsilon_{j-1}) - \frac{2 (\varepsilon_j - \varepsilon_{j-1})}{\beta_{j-1} (x, \varepsilon_j) - \beta_{j-1} (x, \varepsilon_{j-1})}, \quad j = 1, \cdots, k, \tag{4}$$

which implies that the method depends either of k solutions $\beta_1 (x, \varepsilon_j)$ of the initial Riccati equation, or of the associated Schrödinger solutions $u_j (x)$ such that $\beta_1 (x, \varepsilon_j) = u_j' (x) / u_j (x)$, i.e.,

$$-\frac{1}{2} u_j'' (x) + V_0 (x) u_j (x) = \varepsilon_j u_j (x). \tag{5}$$

The spectrum of the Hamiltonian $\hat{H}_k$ consists of the initial one plus k new energy levels [4,34–36]:

$$\mathrm{Sp}(\hat{H}_k) = \mathrm{Sp}(\hat{H}_0) \cup \{\varepsilon_1, \cdots, \varepsilon_k\}. \tag{6}$$

Moreover, the Hamiltonians $\hat{H}_0$ and $\hat{H}_k$ are intertwined as follows:

$$\hat{H}_k \hat{B}_k^+ = \hat{B}_k^+ \hat{H}_0, \quad \hat{H}_0 \hat{B}_k^- = \hat{B}_k^- \hat{H}_k, \tag{7}$$

with

$$\hat{B}_k^+ = \hat{A}_k^+ \cdots \hat{A}_1^+, \quad \hat{B}_k^- = \hat{A}_1^- \cdots \hat{A}_k^-, \tag{8}$$

being k-th order differential intertwining operators. The initial and final potentials, $V_0 (x)$ and $V_k (x)$ respectively, are related by

$$V_k (x) = V_0 (x) - [\ln W (u_1, \cdots, u_k)]'', \tag{9}$$

where $W (u_1, \cdots, u_k)$ is the Wronskian of the k seed solutions $\{u_1, \cdots, u_k\}$. Let us note that this technique has been employed successfully to generate new solvable potentials $V_k (x)$ departing from several interesting initial Hamiltonians (mainly Hermitian [5, 6, 34]).

3. Polynomial Heisenberg algebras

A $(m-1)$-th order PHA is a deformation of the Heisenberg-Weyl algebra with three generators $\{\hat{H}, \hat{\mathcal{L}}_m^+, \hat{\mathcal{L}}_m^-\}$ which obey:

$$[\hat{H}, \hat{\mathcal{L}}_m^{\pm}] = \pm \hat{\mathcal{L}}_m^{\pm}, \tag{10a}$$

$$[\hat{\mathcal{L}}_m^-, \hat{\mathcal{L}}_m^+] = \hat{N}_m(\hat{H} + \hat{1}) - \hat{N}_m(\hat{H}) = P_{m-1}(\hat{H}), \tag{10b}$$

$$\hat{N}_m(\hat{H}) = \prod_{j=1}^{m} (\hat{H} - \mathcal{E}_j \hat{1}), \tag{10c}$$

where $\hat{H}$ has the form given in Eq. (1c), $\hat{\mathcal{L}}_m^{\pm}$ are m-th order differential ladder operators, $\hat{N}_m(\hat{H}) = \hat{\mathcal{L}}_m^{+}\hat{\mathcal{L}}_m^{-}$ is a generalization of the number operator, $P_{m-1}(\hat{H})$ is a polynomial in $\hat{H}$ of degree $m-1$. Note that the zeros $\mathcal{E}_j$ of $\hat{N}_m(\hat{H})$ correspond to the associated extremal state energies.

The PHA generated by $\{\hat{H}, \hat{\mathcal{L}}_m^{+}, \hat{\mathcal{L}}_m^{-}\}$ gives information about $\mathrm{Sp}(\hat{H})$ as follows. First the extremal states of the system, such that

$$\hat{\mathcal{L}}_m^{-}\psi = 0, \tag{11}$$

have to be chosen. From Eqs. (10c) and (11) we obtain

$$\hat{\mathcal{L}}_m^{+}\hat{\mathcal{L}}_m^{-}\psi = \prod_{j=1}^{m}(\hat{H} - \mathcal{E}_j\hat{1})\psi = 0, \tag{12}$$

which suggests that, in the diagonalizable case, we can choose the $\psi_{\mathcal{E}_j}$ satisfying the stationary Schrödinger equation with complex eigenvalues $\mathcal{E}_j$:

$$\hat{H}\psi_{\mathcal{E}_j} = \mathcal{E}_j\psi_{\mathcal{E}_j}. \tag{13}$$

Departing now from these extremal states, by acting iteratively with $\hat{\mathcal{L}}_m^{+}$ onto them, we can build m mathematical energy ladders with the same spacing $\Delta E = 1$ starting from each $\mathcal{E}_j$.

A particularly interesting case appears for $m = 3$, when the ladder operators are of third order. Next, it is analized the corresponding second-order PHA.

3.1. Second order PHA

The connection between PIV equation and PHA appears for $m = 3$ [27]. Hence, let $\hat{\mathcal{L}}_3^{+}$ be a third-order differential ladder operator factorized as the product of a first and a second-order differential one $\hat{\mathcal{L}}_3^{+} = \hat{\mathcal{L}}_1^{+}\hat{\mathcal{L}}_2^{+}$, where

$$\hat{\mathcal{L}}_1^{+} = \frac{1}{\sqrt{2}}\left[-\frac{\mathrm{d}}{\mathrm{d}x} + f(x)\right], \tag{14a}$$

$$\hat{\mathcal{L}}_2^{+} = \frac{1}{2}\left[\frac{\mathrm{d}^2}{\mathrm{d}x^2} + g(x)\frac{\mathrm{d}}{\mathrm{d}x} + h(x)\right]. \tag{14b}$$

The complex functions f, g, h and the potential V are unknown. For simplicity, we suppose the existence of an auxiliary Hamiltonian $\hat{H}_{\mathrm{a}}$ which is intertwined with $\hat{H}$ in the way

$$\hat{H}\hat{\mathcal{L}}_1^{+} = \hat{\mathcal{L}}_1^{+}(\hat{H}_{\mathrm{a}} + \hat{1}), \quad \hat{H}_{\mathrm{a}}\hat{\mathcal{L}}_2^{+} = \hat{\mathcal{L}}_2^{+}\hat{H}. \tag{15}$$

These relationships lead to a coupled system of equations for f, g, h, V and its derivatives. Its solution supplies an expression for the potential:

$$V = \frac{x^2}{2} - \frac{g'}{2} + \frac{g^2}{2} + xg + \mathcal{E}_3 - \frac{1}{2}, \tag{16}$$

where g satisfies the PIV equation with parameters $a, b \in \mathbb{C}$ [37], namely

$$g'' - \frac{g'^2}{2g} + \frac{3}{2}g^3 + 4xg^2 + 2\left(x^2 - a\right)g + \frac{b}{g}, \tag{17a}$$

$$a := \mathcal{E}_1 + \mathcal{E}_2 - 2\mathcal{E}_3 - 1, \quad b := -2\left(\mathcal{E}_1 - \mathcal{E}_2\right)^2. \tag{17b}$$

We need to know the function g for a pair of parameters a, b; then, the extremal energies $\mathcal{E}_j$ and the potential V in (16) can be found. Moreover, the three extremal states are expressed in terms of g as:

$$\psi_{\mathcal{E}_1} \propto \left(\frac{g'}{2g} - \frac{g}{2} - \frac{1}{g}\sqrt{-\frac{b}{2}} - x\right) e^{\int \left(\frac{g'}{2g} + \frac{g}{2} - \frac{1}{g}\sqrt{-\frac{b}{2}}\right) \mathrm{d}x}, \tag{18a}$$

$$\psi_{\mathcal{E}_2} \propto \left(\frac{g'}{2g} - \frac{g}{2} + \frac{1}{g}\sqrt{-\frac{b}{2}} - x\right) e^{\int \left(\frac{g'}{2g} + \frac{g}{2} + \frac{1}{g}\sqrt{-\frac{b}{2}}\right) \mathrm{d}x}, \tag{18b}$$

$$\psi_{\mathcal{E}_3} \propto e^{-\frac{1}{2}x^2 - \int g \, \mathrm{d}x}. \tag{18c}$$

Conversely, if we are able to identify Hamiltonians having third-order differential ladder operators, it is possible to design a simple mechanism to obtain solutions of the PIV equation (see [27] and Section 5 in this paper).

4. SUSY partners of the complex oscillator

In order to implement the SUSY technique, we need to find the general solution $u(x)$ of the stationary Schrödinger equation for an arbitrary complex factorization energy ε. The potential associated to the complex oscillator reads:

$$V_0(x) = \frac{1}{2}\omega^2 x^2, \quad \omega = e^{i\theta}, \quad -\frac{\pi}{2} \le \theta < \frac{3\pi}{2}, \tag{19}$$

ω being a dimensionless complex frequency and θ its phase. A direct calculation leads to [38]

$$u(x) = e^{-\frac{1}{2}\omega x^2}\left[{}_1F_1\left(\frac{1}{4} - \frac{\varepsilon}{2\omega}, \frac{1}{2}; \omega x^2\right) + 2\nu x \frac{\Gamma\left(\frac{3}{4} - \frac{\varepsilon}{2\omega}\right)}{\Gamma\left(\frac{1}{4} - \frac{\varepsilon}{2\omega}\right)}\, {}_1F_1\left(\frac{3}{4} - \frac{\varepsilon}{2\omega}, \frac{3}{2}; \omega x^2\right)\right], \tag{20}$$

where we will take $|\nu| < 1$ and ${}_1F_1$ is the confluent hypergeometric function (Kummer). In general $\lim_{|x|\to\infty} |u(x)| = 0$ is not satisfied. In order to ensure the square-integrability of u, any of the two series ${}_1F_1$ in Eq. (20) must reduce to a polynomial. Thus, three different cases arise, depending on the value of θ: *(i)* for $-\frac{\pi}{2} < \theta < \frac{\pi}{2}$ the energy levels of the complex oscillator are given by

$$E_n(\theta) = \left(n + \frac{1}{2}\right) e^{i\theta}, \quad n = 0, 1, 2, \cdots, \tag{21a}$$

and the corresponding eigenfunctions read

$$\phi_n(x) = C_n \mathrm{H}_n(\sqrt{\omega}x) e^{-\frac{1}{2}\omega x^2}, \quad \sqrt{\omega} = e^{i\frac{\theta}{2}}, \tag{21b}$$

where C_n are normalization factors and $\mathrm{H}_n(\sqrt{\omega}x)$ are the Hermite polynomials. *(ii)* For $\frac{\pi}{2} < \theta < \frac{3\pi}{2}$ the eigenfunctions and eigenvalues are

$$\phi_n(x) = D_n \mathrm{H}_n(\sqrt{-\omega}x) e^{\frac{1}{2}\omega x^2}, \quad \sqrt{-\omega} = e^{i\frac{\theta-\pi}{2}}, \tag{22a}$$

$$E_n(\theta) = \left(n + \frac{1}{2}\right) e^{i(\theta-\pi)}, \quad n = 0, 1, 2, \ldots \tag{22b}$$

(iii) For $\theta = \pm\frac{\pi}{2}$ there do not exist square-integrable eigenfunctions of $\hat{H}$, since $V_0(x)$ corresponds to the repulsive oscillator [39].

The eigenvalues of the complex oscillator lie either in the first or in the fourth quadrant of the complex energy plane, since $V_0(x)$ is invariant under the change $\omega \to -\omega$, thus the eigenfunctions and eigenvalues for both ω and $-\omega$ are the same. Moreover, since

$$E_n(-\theta) = [E_n(\theta)]^\star, \tag{23}$$

we will work from now in with the domain $\theta \in \left[0, \frac{\pi}{2}\right)$.

Let us define now the annihilation and creation operators $\hat{a}^{\pm}_{\omega}$ in analogy to the harmonic oscillator:

$$\hat{a}^{\pm}_{\omega} := \frac{1}{\sqrt{2}}\left(\mp\frac{\mathrm{d}}{\mathrm{d}x} + \omega x\right). \tag{24}$$

Note that they define a fixed direction on the complex energy plane (see Figure 1). In addition, $\hat{a}^{+}_{\omega} \neq (\hat{a}^{-}_{\omega})^{\dagger}$, since $\omega \in \mathbb{C}$. The algebra of these operators is similar to the harmonic oscillator one [40]:

$$[\hat{a}^{-}_{\omega}, \hat{a}^{+}_{\omega}] = \omega\hat{1}, \tag{25a}$$

$$\{\hat{a}^{-}_{\omega}, \hat{a}^{+}_{\omega}\} = 2\hat{H}_0, \tag{25b}$$

$$[\hat{H}_0, \hat{a}^{\pm}_{\omega}] = \pm\omega\hat{a}^{\pm}_{\omega}. \tag{25c}$$

By using this algebra, it is possible to find the square-integrable eigenfunctions of $\hat{H}_0$ and their corresponding eigenvalues in a very elegant way. They turn out to be given by:

$$\phi_n(x) = \frac{1}{\sqrt{n!}}\left(\hat{a}^{+}_{\omega}\right)^n \phi_0(x) = C_n \mathrm{H}_n(\omega^{\frac{1}{2}}x)e^{-\frac{1}{2}\omega x^2}, \tag{26}$$

where $\phi_0 \propto e^{-\frac{1}{2}\omega x^2}$ is the ground state. Note that when $\varepsilon = E_n$, the general solution in Eq. (20) can be reduced to the expression in Eq. (26).

A diagram of the complex energy plane, the eigenvalues E_n of the complex oscillator for a fixed θ and the action of the annihilation and creation operators $\hat{a}^{\pm}_{\omega}$ can be seen in Figure 1.

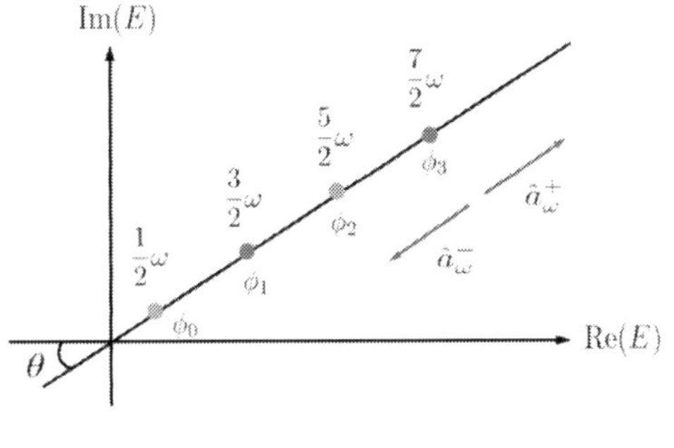

Figure 1. Diagram of the complex energy plane. Some eigenvalues for a fixed θ are shown at their right positions.

Let us apply now a non-singular k-th order supersymmetry transformation which creates k new levels, additional to E_n, for the new Hamiltonian $\hat{\mathsf{H}}_k = -\frac{1}{2}\frac{\mathrm{d}^2}{\mathrm{d}x^2} + V_k$, i.e.,

$$\mathrm{Sp}(\hat{\mathsf{H}}_k) = \mathrm{Sp}(\hat{\mathsf{H}}_0) \cup \{\varepsilon_1, \cdots, \varepsilon_k\}.$$

In order to do that, we take k solutions $\{u_j;\ j = 1, \cdots, k\}$ of the initial Schrödinger equation. Thus, the expression for the new potential V_k is given by Eq. (9) with $V_0(x) = \frac{1}{2}\omega^2 x^2$.

The algebra for systems described by $\hat{\mathsf{H}}_k$ is generated by its natural ladder operators

$$\hat{\mathcal{L}}^{\pm}_{2k+1} = \hat{B}^{+}_{k}\hat{a}^{\pm}_{\omega}\hat{B}^{-}_{k}, \tag{27}$$

where $\hat{B}_k^{\pm}$ are the k-th order differential intertwining operators of Eq. (8). The standard commutation relations of the polynomial Heisenberg algebra are satisfied:

$$[\mathsf{H}_k, \hat{\mathcal{L}}_{2k+1}^{\pm}] = \pm\omega\hat{\mathcal{L}}_{2k+1}^{\pm}, \tag{28}$$

which, up to the factor ω, are the same as in Eq. (10a). Thus, it is natural to redefine the Hamiltonian as $\hat{H}_k = \mathsf{H}_k/\omega$, so that Eq. (28) is transformed into

$$[\hat{H}_k, \hat{\mathcal{L}}_{2k+1}^{\pm}] = \pm\hat{\mathcal{L}}_{2k+1}^{\pm}. \tag{29}$$

The Hamiltonians $\hat{H}_k$ associated with this algebra read:

$$\hat{H}_k = -\frac{1}{2\omega}\frac{\mathrm{d}^2}{\mathrm{d}x^2} + \frac{1}{\omega}V_k(x) = -\frac{1}{2\omega}\frac{\mathrm{d}^2}{\mathrm{d}x^2} + \frac{1}{2}\omega x^2 - \frac{1}{\omega}\frac{\mathrm{d}^2}{\mathrm{d}x^2}\left[\ln W(u_1, \cdots, u_k)\right], \tag{30}$$

which take the standard Schrödinger form through the substitution $y = \omega^{\frac{1}{2}}x$.

In conclusion, the $(2k+1)$-th order operators $\hat{\mathcal{L}}_{2k+1}^{\pm}$ and the Hamiltonian $\hat{H}_k$ (in fact a family) generate a $2k$-th order PHA. As we saw in Section 3, systems which are connected with the PIV equation must have third-order ladder operators and satisfy a second-order PHA. Thus, we need to identify the subfamily of $\hat{H}_k$ (if any) having as well third-order ladder operators. In fact, there is a theorem which enables us to reduce to second the initial order $(2k)$ of the PHA, producing then solutions of the PIV equation [28]. Some results of this formalism, for first and second-order SUSY, are next illustrated.

4.1. First-order transformation

For a transformation function as given in Eq. (20) with $\varepsilon \in \mathbb{C}$, the first-order SUSY partner potential of the complex oscillator turns out to be:

$$V_1(x) = \frac{1}{2}\omega^2 x^2 - \frac{u''}{u} + \left(\frac{u'}{u}\right)^2. \tag{31}$$

Figure 2 shows some of these potentials for several values of the factorization energy ε. The probability densities associated to some eigenfunctions for fixed values of ε and ν, are plotted in Figure 3.

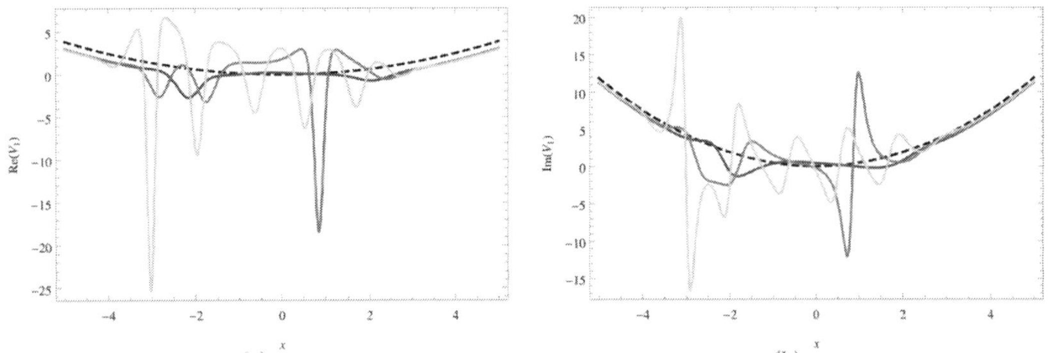

(a) (b)

Figure 2. Real (a) and imaginary parts (b) of the potential V_1 of Eq. (31) for $\varepsilon = 0.01 + 0.2i$ (blue), $\varepsilon = 0.9 + 0.2i$ (red) and $\varepsilon = 3.5 + 0.2i$ (green). The discontinuous line represents the complex oscillator potential $V_0 = \frac{1}{2}\omega^2 x^2$. We have taken $\nu = 0.7 + 0.3i$ and $\theta = \frac{\pi}{5}$.

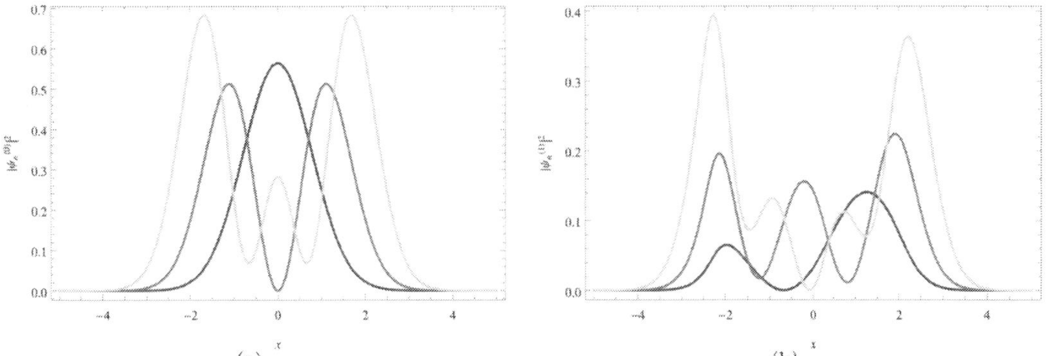

(a) (b)

Figure 3. Normalized probability densities for: (a) the eigenstates of $\hat{H}_0$ with $n = 0$ (blue), $n = 1$ (red) and $n = 2$ (green); (b) their corresponding densities for $\hat{H}_1$. We have taken $\varepsilon = 0.01 + 0.2i$, $\nu = 0.7 + 0.3i$ and $\theta = \frac{\pi}{5}$.

4.2. Higher-order transformation

As it was said previously, a k-th order SUSY transformation requires, in general, k independent seed solutions of the Schrödinger equation. In order to induce the reduction theorem, however, we have to consider just connected seed solutions satisfying:

$$u_j = (\hat{a}_\omega^-)^{j-1} u_1, \tag{32a}$$

$$\varepsilon_j = \varepsilon_1 - (j-1)\,\omega, \quad j = 1, \cdots, k, \tag{32b}$$

i.e., the only free function is a nodeless solution u_1 of the stationary Schrödinger equation for the complex factorization energy ε_1, with form given in Eq. (20). For instance, the second-order SUSY partner potential of the complex oscillator (for $k = 2$) is given by:

$$V_2(x) = \frac{1}{2}\omega^2 x^2 - \frac{W''}{W} + \left(\frac{W'}{W}\right)^2, \tag{33}$$

where

$$W =: W(u_1, \hat{a}_\omega^- u_1) \propto \left(\omega^2 x^2 + \omega - 2\varepsilon_1\right) u_1^2 - (u_1')^2. \tag{34}$$

In Figure 4 we have plotted the real and imaginary parts of $V_0(x)$ and $V_2(x)$ for different values of ε_1, with the other parameters ν_1 and θ remaining fixed.

5. Solutions of the Painlevé IV equation

Systems having third-order ladder operators, ruled then by second-order PHA, are connected with the PIV equation (see Section 3). Thus, the first-order SUSY partners of the complex oscillator are linked directly with the PIV equation. Moreover, from Eq. (18c) we can obtain an expression to generate solutions to such equation in terms of the extremal states $\psi_{\mathcal{E}_j}$ by making carefully the change from the harmonic to the complex oscillator:

$$g_1^{(j)}(x) = -\omega^{\frac{1}{2}} x - \omega^{-\frac{1}{2}} \left[\ln \psi_{\mathcal{E}_j}(x)\right]', \quad j = 1, 2, 3, \tag{35}$$

where the subscripting indicates the order of the SUSY transformation. This expression already includes the fact that, by making cyclic permutations of the indices of the extremal states, we can generate three different solutions to the PIV equation. For a fixed value of ε_1 the solutions (35) of the PIV equation constitute a 3-parametric family.

In Figure 5 we show some complex solutions to the PIV equation for several values of the factorization energy ε_1 and a fixed ν_1, with $|\nu_1| < 1$.

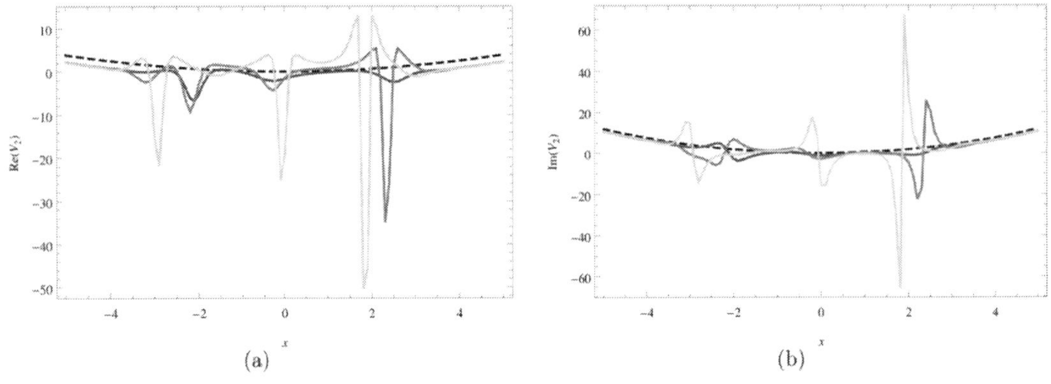

Figure 4. Real (a) and imaginary parts (b) of the potential V_2 of Eq. (33) for $\varepsilon_1 = 0.01 + 0.1i$ (blue), $\varepsilon_1 = 0.4 + 0.1i$ (red) and $\varepsilon_1 = 0.95 + 0.1i$ (green). The discontinuous line represents the complex oscillator potential $V_0 = \frac{1}{2}\omega^2 x^2$. We have taken $\nu_1 = 0.7 - 0.3i$ and $\theta = \frac{\pi}{5}$.

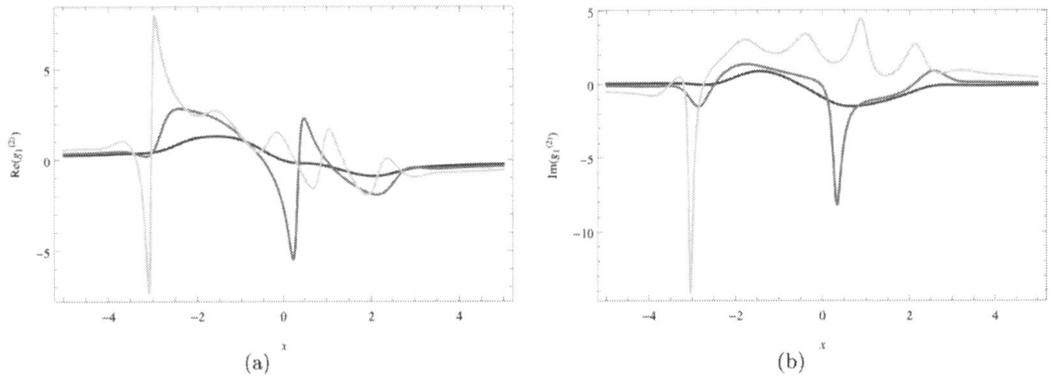

Figure 5. Real (a) and imaginary (b) parts of $g_1^{(2)}$ for $\varepsilon_1 = 0.04 + 0.8i$ (blue), $\varepsilon_1 = 0.9 + 0.8i$ (red) and $\varepsilon_1 = 3 + 0.8i$ (green). We have taken $\theta = \frac{\pi}{5}$ and $\nu_1 = 0.9 - 0.1i$.

Note that the solution $g_1^{(j)}$ is non-singular if the extremal state $\psi_{\mathcal{E}_j}$ does not have zeros. In addition, these solutions have a null asymptotic behavior when $|x| \to \infty$.

In a more general context, for k-th order SUSY an explicit expression for a solution of the PIV equation appears after the reduction theorem is applied, which is given by:

$$g_k(x) = -\omega^{\frac{1}{2}} x - \omega^{-\frac{1}{2}} \left[\ln \frac{W(u_1, \cdots, u_{k-1})}{W(u_1, \cdots, u_k)} \right]', \quad k \geq 2, \tag{36}$$

where the u_j, $j = 1, \cdots, k$ satisfy Eqs. (32). In Figure 6 we illustrate the typical behavior of g_2 for several values of ε_1 and $\nu_1 = 0.8 + 0.5i$.

6. Conclusions

In this paper we have introduced a complex generalization of the standard factorization method to analyze the SUSY partners of the non-Hermitian complex oscillator Hamiltonian. The corresponding potentials were generated through SUSY transformations for complex factorization energies. An algorithm to generate complex solutions of the Painlevé IV equation, from the extremal state eigenfunctions of these Hamiltonians, was as well explored.

Let us note that the physical interpretation of complex energies remains open, although possible applications related with absortive (dissipative) systems have been noticed recently [13].

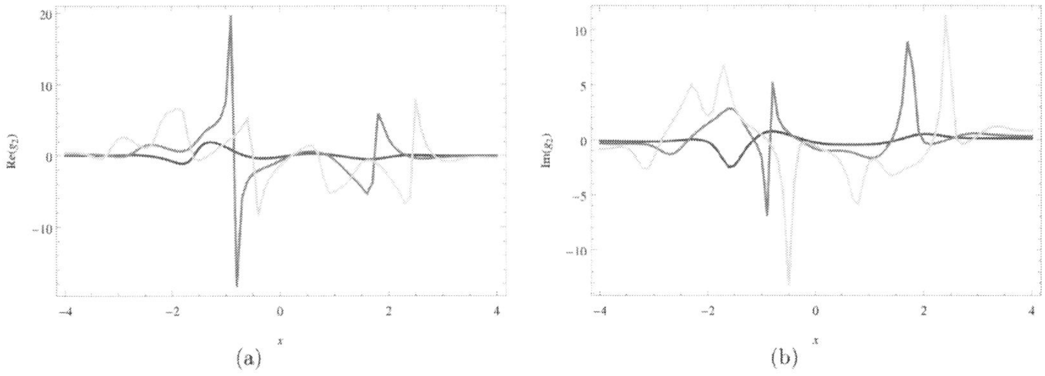

Figure 6. Real (a) and imaginary (b) parts of g_2 for $\varepsilon_1 = 0.03 - 0.4i$ (blue), $\varepsilon_1 = 1 - 0.4i$ (red) and $\varepsilon_1 = 3 - 0.4i$ (green). We have taken $\theta = \frac{\pi}{5}$ and $\nu_1 = 0.4 - 0.3i$.

Acknowledgements

The authors acknowledge the support of Conacyt (México), project 152574. JCG also acknowledges the Conacyt PhD scholarship 262669.

References

[1] Andrianov A.A., Cannata F. Nonlinear supersymmetry for spectral design in quantum mechanics. *J. Phys. A: Math. Gen.* **37** 10297-10321 (2004).

[2] Bagchi B.K.. Supersymmetry in quantum and classical mechanics. *Monographs and Surveys in Pure and Applied Mathematics* Vol. 116, Chapman & Hall/CRC, Boca Raton, FL (2001).

[3] Dong S.H. Factorization method in quantum mechanics. *Fundamental Theories of Physics* Vol. 150, Springer, Dordrecht (2007).

[4] Fernández D.J., Fernández-García N. Higher-order supersymmetric quantum mechanics. *AIP Conf. Proc.* **744** 236-273 (2005).

[5] Fernández D.J. Supersymmetric quantum mechanics. *AIP Conf. Proc.* **1287** 3-36 (2010).

[6] Mielnik B., Rosas-Ortiz O. Factorization: little or great algorithm?. *J. Phys. A: Math. Gen.* **37** 10007-10035 (2004).

[7] Sukumar C.V. Supersymmetric quantum mechanics and its applications. *AIP Conf. Proc.* **744** 166-235 (2005).

[8] Andrianov A.A., Ioffe M.V., Cannata F., Dedonder J.P. SUSY quantum mechanics with complex superpotentials and real energy spectra. *Int. J. Mod. Phys. A* **14** 2675-2688 (1999).

[9] Fernández D.J., Muñoz R., Ramos A. Second Order SUSY transformations with 'complex energies'. *Phys. Lett. A* **308** 11-16 (2003).

[10] Rosas-Ortiz O., Muñoz R. Non-Hermitian SUSY hydrogen-like Hamiltonians with real spectra. *J. Phys. A: Math. Gen.* **36** 8497-8506 (2003).

[11] Fernández-García N. Darboux-deformed barriers and resonances in quantum mechanics. *Rev. Mex. Fis.* **53** S4 42-45 (2007).

[12] Fernández-García N., Rosas-Ortiz O. Gamow-Siegert functions and Darboux-deformed short range potentials. *Ann. Phys.* **323** 1397-1414 (2008).

[13] Fernández-García N., Rosas-Ortiz O. Optical potentials using resonance states in supersymmetric quantum mechanics. *J. Phys.: Conf. Ser.* **128** 012044 (14 pages) (2008).

[14] Bender C.M., Boettcher S. Real spectra in non-Hermitian Hamiltonians having PT symmetry. *Phys. Rev. Lett.* **80** 5243-5246 (1998).

[15] Znojil M. PT symmetric harmonic oscillators. *Phys. Lett. A* **259** 220-223 (1999).

[16] Mostafazadeh A. Pseudo-Hermiticity versus PT symmetry: The necessary condition for the reality of the spectrum of a non-Hermitian Hamiltonian. *J. Math. Phys.* **43** 205-214 (2002).

[17] Mohammad-Ali M., Heinrich M., Christodoulides D. Supersymmetry-generated complex optical potentials with real spectra. *Phys. Rev. A* **87** 043819-5 (2013).

[18] Barton G. Quantum mechanics of the inverted oscillator potential. *Ann. Phys.* **166** 322-363 (1986).

[19] Jannussis A., Skuras E. Harmonic oscillator with complex frequency. *Nuovo Cimento B* **94** 29-36 (1986).

[20] Baskoutas S., Jannussis A. Quantum Mechanics of the inverted Caldirola-Kanai oscillator. *Nuovo Cimento B* **107** 255-267 (1992).

[21] Baskoutas S., Jannussis A., Mignani R., Papatheou V. Tunnelling process for non-Hermitian systems: the complex frequency inverted oscillator. *J. Phys. A: Math. Gen.* **26** L819-L824 (1993).

[22] Fernández D.J., Hussin V., Nieto L.M. Coherent states for isospectral oscillator Hamiltonians *J. Phys. A: Math. Gen.* **27** 3547-3564 (1994).

[23] Fernández D.J., Nieto L.M., Rosas-Ortiz O. Distorted Heisenberg algebra and coherent states for isospectral oscillator Hamiltonian *J. Phys. A: Math. Gen.* **28** 2693-2708 (1995).

[24] Fernández D.J., Negro J., del Olmo M.A. Group Approach to the Factorization of the Radial Oscillator Equation *Ann. Phys.* **252** 386-412 (1996)

[25] Fernández D.J., Hussin V, Mielnik B. A simple generation of exactly solvable anharmonic oscillators. *Phys. Lett. A* **244** 309-316 (1998).

[26] Fernández D.J., Hussin V. Higher-order SUSY, linearized nonlinear Heisenberg algebras and coherent states. *J. Phys. A: Math. Gen.* **32** 3603-3619 (1999).

[27] Carballo J.M., Fernández D.J., Negro J., Nieto L.M. Polinomial Heisenberg algebras. *J. Phys. A: Math. Gen.* **37** 10349-10362 (2004).

[28] Bermúdez D., Fernández D.J. Supersymmetric quantum mechanics and Painlevé IV equation. *SIGMA* **7** 025 (14 pages) (2011).

[29] Bermúdez D., Fernández D.J. Non-hermitian Hamiltonians and the Painlevé IV equation with real parameters. *Phys. Lett. A* **375** 2974-2978 (2011).

[30] Bermúdez D. Complex SUSY transformations and the Painlevé IV equation. *SIGMA* **8** 069 (10 pages) (2012).

[31] Bermúdez D., Fernández D.J. Supersymmetric quantum mechanics and Painlevé equations. *AIP Conf. Proc.* **1575** 50-88 (2014).

[32] Mielnik B. Factorization method and new potentials with the oscillator spectrum. *J. Math. Phys.* **25** 3387-3389 (1984).

[33] Sukumar C.V. Supersymmetric quantum mechanics of one-dimensional systems. *J. Phys. A: Math. Gen.* **18** 2917-2936 (1985).

[34] Fernández D.J., Glasser M.L., Nieto L.M. New isoespectral oscillator potentials. *Phys. Lett. A* **240** 15-20 (1998).

[35] Andrianov A.A., Ioffe M.V., Spiridonov V.P. Higher-derivative supersymetry and the Witten index. *Phys. Lett. A* **174** 273-279 (1993).

[36] Mielnik B., Nieto L.M., Rosas-Ortiz O. The finite difference algorithm for higher-order supersymmetric. *Phys. Lett. A* **269** 70-78 (2000).

[37] Iwasaki K., Kimura H., Shimomura S., Yoshida M. From Gauss to Painlevé. Braunschwig Vieweg (1991).

[38] Junker G., Roy P. Conditionally exactly solvable potentials: a supersymmetric construction method. *Ann. Phys.* **270** 155-177 (1998).

[39] Bermúdez D., Fernández D.J. Factorization method and new potentials from the inverted oscillator. *Ann. Phys.* **333** 290-306 (2013).

[40] Ballesteros A., Herranz F.J., Parashar P. Quantum Heisenberg-Weyl algebras. *J. Phys. A: Math. Gen.* **30** L149-L154 (1997).

Supersymmetric infinite wells and coherent states

M-A Fiset[1] **and V Hussin**[2,3]

[1]Department of Physics, McGill University, Montréal, QC H3A 2T8, Canada.
[2]Département de mathématiques et de statistique, Université de Montréal, C. P. 6128, Succ. Centre-ville, Montréal (Québec) H3C 3J7, Canada.
[3] Centre de recherches mathématiques, Université de Montréal, C. P. 6128, Succ. Centre-ville, Montréal (Québec) H3C 3J7, Canada.

E-mail: marc-antoine.fiset@mail.mcgill.ca, veronique.hussin@umontreal.ca

Abstract. Gaussian Klauder coherent states are discussed in the context of the infinite well quantum model, otherwise known as the particle in a box. A supersymmetric partner system is also presented, as well as a construction of coherent states in this new system. We show that these states can be chosen, in both systems to have many properties usually expected for coherent states. In particular, they yield highly localised wave packets for a short period of time, which evolve in a quasi-classical manner and which saturate approximately Heisenberg uncertainty relation. These studies are elaborated in one- and two-dimensional contexts. Finally, some relations are established between the Gaussian states being mostly used here and the generalised coherent states, which are more standardly found in the literature.

1. Introduction

The study of coherent states in quantum mechanics is a subject very well documented [1–6]. These states have been approached from different points of view and we can cite, for example, their definition as eigenstates of an annihilation operator and as minimium uncertainty states. They have also been defined for a large amount of solvable quantum systems and their properties have been analysed [6–8].

One approach, that has not deserved much attention until now, is the one using the so-called Gaussian Klauder coherent states [9]. They are given as a superposition of energy eigenstates that leads to a good localisation in the phase space of the system under consideration. For the harmonic oscillator, these last states are a good approximation of the usual coherent states. It has been shown that it is also the case for the infinite well and the Morse systems [8–11]. It means that they are good states for investigating physical properties of the systems under consideration (behaviour of the position and momentum observables and uncertainty relation, for example).

Even if most of the works were focusing on one dimensional (1D) systems, some results have been given in the two dimensional (2D) case [10]. In particular, the 2D infinite well, having a quadratic degenerate energy spectrum, has attracted some attention and the construction of coherent states had to be adjusted [10, 12] .

All these considerations have been extended using a supersymmetric (SUSY) approach [13]. Indeed, SUSY partner Hamiltonians have been constructed in 1D and 2D using intertwining relations and factorisation methods [13–15]. These new Hamiltonians are shown to be deeply related to the original ones. The coherent states systems are as well closely related to the original ones.

In this work, we consider the infinite well quantum system both in 1D and 2D and their SUSY partners. We then construct the Gaussian Klauder coherent states (GCS) and investigate their properties. In Section 2, we show how the SUSY partners of the 1D infinite well are obtained. New potentials are thus produced that depend on two parameters (one is an integer and the other one is a real number). The complete energy spectra of the new Hamiltonians are obtained. The GCS are then constructed and their properties are exhibited. In Section 3, we generalise the preceding approach to 2D where the SUSY partners exhibit more freedom. Complete energy spectra are also produced. We generalise the definition of GCS to this context. We discuss as well another type of coherent states, the so-called generalised coherent states (GeCS). Because of the degeneracy of the energy spectrum, the usual definition of those states as the eigenstates of an annihilation operator has to be adjusted. We finish the paper with some conclusions and questions for future work.

2. The 1D infinite well and SUSY partners

2.1. Description of the models

Let us first set our notational convention concerning the infinite square well [9] to be used throughout this work. A particle of mass M is subject to a potential taken to be

$$V(x) = \begin{cases} 0, & 0 < x < \pi \\ \infty, & \text{otherwise.} \end{cases} \tag{1}$$

The stationary eigenstates and the discrete energies of this system are

$$\psi_n(x) = \sqrt{\frac{2}{\pi}} \sin nx, \qquad E_n = \frac{\hbar^2}{2M} n^2, \quad n = 1, 2, \ldots \tag{2}$$

In the following, we will use dimensionless units, setting $\hbar = 1$, $M = 1/2$, such that the Hamiltonian is $H_x = -\frac{d^2}{dx^2} + V(x)$.

SUSY partners of the infinite well have been constructed [13] starting from usual intertwining relations involving the supercharges Q_x, $Q_x^\dagger$. These can be defined, in particular, as differential operators of second order. Let us summarize the results.

Starting from the Hamiltonian H_x, a SUSY partner Hamiltonian $\tilde{H}_x = -\frac{d^2}{dx^2} + \tilde{V}(x)$ is obtained from the relations

$$\tilde{H}_x Q_x = Q_x H_x, \quad Q_x^\dagger \tilde{H}_x = H_x Q_x^\dagger. \tag{3}$$

For the infinite well, with $V(x) = 0$ in the domain $x \in]0, \pi[$, internal consistency constrains the supercharges to take the form

$$Q_x = \frac{d^2}{dx^2} + \eta(x)\frac{d}{dx} + \epsilon + \frac{1}{2}(\eta^2(x) - \eta'(x)) \tag{4}$$

and

$$Q_x^\dagger = \frac{d^2}{dx^2} - \eta(x)\frac{d}{dx} + \epsilon + \frac{1}{2}(\eta^2(x) - 3\eta'(x)), \tag{5}$$

where ϵ is an arbitrary constant. The function $\eta(x)$ satisfies (in the so-called confluent case [14]):

$$2\eta(x)\eta''(x) - (\eta'(x))^2 - 4\eta^2(x)\eta'(x) + \eta^4(x) + 4\epsilon\eta^2(x) = 0 \tag{6}$$

and the new potential is given as

$$\tilde{V}(x) = 2\eta'(x). \tag{7}$$

Moreover, the products $Q_x^\dagger Q_x$ and $Q_x Q_x^\dagger$ are respectively polynomials in H_x and $\tilde{H}_x$:

$$Q_x^\dagger Q_x = (H_x - \epsilon)^2, \quad Q_x Q_x^\dagger = (\tilde{H}_x - \epsilon)^2. \tag{8}$$

The resolution of (6) leads to admissible solutions for $\epsilon = k^2$ with $k = 1, 2, \ldots$ A particular energy $\epsilon = E_k$ within the original spectrum thus needs to be chosen for the supersymmetry. We get [14]

$$\eta(x; k, \omega) = \frac{4k\sin^2(kx)}{\sin(2kx) + 2k(\pi\omega - x)}, \tag{9}$$

where ω is an arbitrary constant. The corresponding potentials are given as

$$\tilde{V}(x; k, \omega) = \begin{cases} \dfrac{32k^2\sin(kx)[\sin(kx) + k(\pi\omega - x)\cos(kx)]}{[\sin(2kx) + 2k(\pi\omega - x)]^2}, & 0 < x < \pi \\ \infty, & \text{otherwise.} \end{cases} \tag{10}$$

These potentials can be shown to be non singular if $\omega \in]-\infty, 0[\cup]1, \infty[$. Two instances are illustrated on Figure 1, one for each of the disconnected subsets of the parameter space of ω, i.e. $\omega = 2$ and $\omega = -1$. We thus observe that changing the sign of $(\omega - 1/2)$ reverses this potential about its vertical axis. Decreasing the magnitude of $|\omega - 1/2|$ accentuates the departure from a sinusoidal potential. The parameter k controls the amount of oscillations.

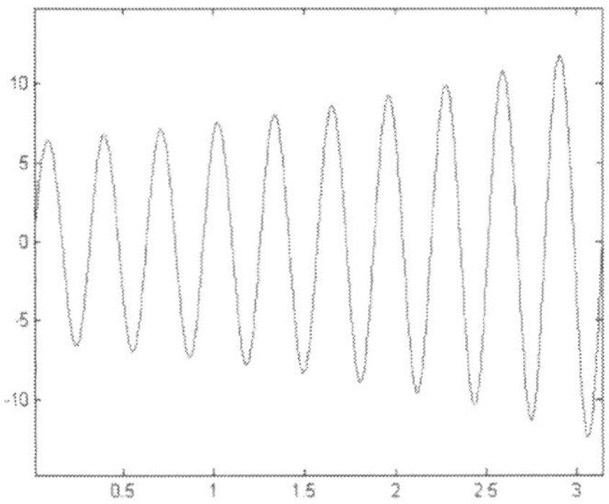

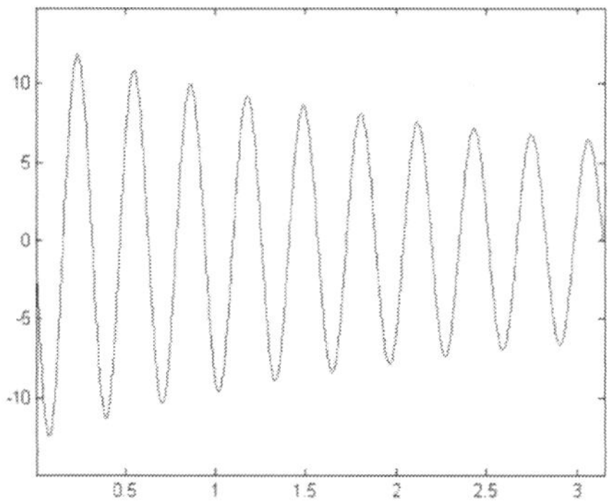

Figure 1 - SUSY potential $\tilde{V}(x; k = 10, \omega)$ as a function of x, for (left) $\omega = 2$ and (right) $\omega = -1$.

Notice the striking symmetry under the $\mathbb{Z}_2$ action $\omega \to 1 - \omega$, $x \to \pi - x$. This symmetry is accidental in the sense that it is not a consequence of the SUSY algebra, nor the imposed intertwining relations (3). Still, it will have interesting effects on the coherent states built out of SUSY potential functions, as we shall discuss in Section 2.3.

The normalised SUSY eigenstates $\tilde{\psi}_n(x)$ are obtained from the intertwining relations (3) and the expressions (8):

$$\tilde{\psi}_n(x; k, \omega) = (\epsilon - E_n)^{-1} Q_x \psi_n(x), \quad (n \neq k). \tag{11}$$

More explicitly, we get

$$\tilde{\psi}_n(x; k, \omega) = \sqrt{\frac{2}{\pi}} \frac{\sin(nx)[\sin(2kx)(n^2 + k^2) + 2k(\pi\omega - x)(n^2 - k^2)] - 4nk\cos(nx)\sin^2(kx)}{(n^2 - k^2)[\sin(2kx) + 2k(\pi\omega - x)]} \tag{12}$$

For $n \neq k$, they are physical states, i.e. they are normalisable and such that $\tilde{\psi}_n(0;k,\omega) = \tilde{\psi}_n(\pi;k,\omega) = 0$, since $\eta(0;k,\omega) = \eta(\pi;k,\omega) = 0$. The corresponding energies are $E_n = n^2$ as in the original case.

For $n = k$, since $\epsilon = E_k = k^2$, this procedure does not yield $\tilde{\psi}_k(x)$ because $Q_x\psi_k(x) = 0$. The completeness of the spectrum of $\tilde{H}_x$ has been investigated [14] and a single additional state was found. For the sake of completeness, we now explain how this missing state $\tilde{\psi}_k(x)$ could be obtained. It is, in fact, the simultaneous solution to the system of equations

$$Q_x^\dagger\tilde{\psi}_k(x) = 0, \quad H_x\tilde{\psi}_k(x) = \epsilon\tilde{\psi}_k(x), \tag{13}$$

that reduces to a first order differential equation on $\tilde{\psi}_k(x)$:

$$-\eta(x;k,\omega)\frac{d\tilde{\psi}_k(x)}{dx} + (\tilde{V}(x;k,\omega) + \frac{1}{2}(\eta^2(x;k,\omega) - 3\frac{d\eta(x;k,\omega)}{dx}))\tilde{\psi}_k(x) = 0. \tag{14}$$

From [14], we know that $\eta'(x) = \eta^2(x) + 2\beta(x)\eta(x)$, with $\beta(x)$ solving the Riccati equation $\beta'(x) + \beta^2(x) = -\epsilon$. Equation (14) thus becomes

$$\frac{d\tilde{\psi}_k(x)}{dx} = (\eta(x;k,\omega) + \beta(x;k))\tilde{\psi}_k(x) \tag{15}$$

and the normalised solution of energy $\epsilon = E_k = k^2$ is given as

$$\tilde{\psi}_k(x;k,\omega) = \sqrt{\frac{2}{\pi}}\sin(kx)\frac{2\pi k\sqrt{\omega(\omega-1)}}{\sin(2kx) + 2k(\pi\omega - x)}. \tag{16}$$

It is normalisable and such that $\tilde{\psi}_k(0,k,\omega) = \tilde{\psi}_k(\pi,k,\omega) = 0$. With this additional state the spectrum of $\tilde{H}_x$ is thus complete.

2.2. Gaussian Klauder coherent states for the infinite well

As mentioned in the introduction, the Gaussian Klauder coherent states (GCS) can be built for many different systems as a special superposition of energy eigenstates in order to get a reasonably well localised probability density distribution for a short period of time [9]. They have proven to be relevant for the study of the harmonic oscillator and the infinite well. For this last system, we recently [11] formalised the relation between those states and the generalised coherent states (GeCS), constructed as eigenstates of an annihilation operator of the system under consideration. We have shown in particular that the GCS can be chosen to reproduce approximately the GeCS in some specific area of their parameter space. In this section, we thus deal only with those GCS and summarise their properties.

For real constants ϕ_0, $n_0 \geq 0$ and $\sigma_0 > 0$, they are defined as the Gaussian combination

$$\Psi_G(x,t;n_0,\sigma_0,\phi_0) = \sum_{n=1}^{\infty} C_n^G(n_0,\sigma_0,\phi_0)e^{-iE_nt}\psi_n(x), \quad C_n^G(n_0,\sigma_0,\phi_0) = \frac{e^{-\frac{(n-n_0)^2}{4\sigma_0^2} - in\phi_0}}{\sqrt{N_G(n_0,\sigma_0)}}, \tag{17}$$

where the normalisation factor is

$$N_G(n_0,\sigma_0) = \sum_{n=0}^{\infty} e^{-\frac{(n-n_0)^2}{2\sigma_0^2}}. \tag{18}$$

The resolution of the identity is satisfied for theses states, as well as time stability and continuity in n_0 and σ_0 [9].

Interestingly, not only (17) has a Gaussian distribution in n, the sum can be carried out approximately to yield a Gaussian *wavefunction*. Indeed, as demonstrated in [11],

$$\Psi_{\mathrm{G}}(n_0, \sigma_0, \phi_0; x, t) \simeq \frac{1}{(\sqrt{2\pi}s)^{1/2}} \exp\left[-\frac{(x-X)^2}{4s^2} + iPx\right], \tag{19}$$

(up to a x-independent phase factor), for $x \in [0, \pi]$, $t > 0$ and with the following shorthands:

$$X = \phi_0 + \frac{Pt}{1/2}, \quad P = n_0, \quad s = \frac{1}{2\sigma}, \quad \sigma^2 = \frac{\tau}{4(\tau^2 + t^2)}, \quad \tau = (4\sigma_0^2)^{-1}.$$

Here "$\simeq$" stresses an approximate validity under the conditions $n_0 \gg \sigma_0 \gg 1$, $X \gg s$, $\pi - X \gg s$, and $t \ll \tau$.

This result implies, as alluded to earlier, that the state is a well localised wave packet bouncing back and forth on the walls of the well, in a quasi-classical fashion. Its initial width is set by the parameter σ_0, which also determines the time scale τ of the Lorentzian decay of the packet. On the other hand, the velocity of the motion scales as the parameter n_0. Finally, the phase ϕ_0 establishes the initial position of the wave packet. Here, of course, we naturally identify $X \simeq \langle x \rangle$ and $P \simeq \langle p \rangle$.

Another remarkable consequence of (19) is that it explains why the GCS almost saturate Heisenberg uncertainty relation, when the conditions of validity are satisfied [11]. This is readily understood by recalling the standard fact that the most general wave function minimizing $\Delta x \Delta p$ is [5]

$$\Psi(x,t) = A(t)\mathrm{e}^{-\frac{(x-\langle x \rangle)^2}{4s^2} + i\langle p \rangle x}. \tag{20}$$

2.3. Gaussian Klauder coherent states for the SUSY partners

The construction of the GCS can easily be attempted for the SUSY partner in just the same way it was performed in the original system:

$$\tilde{\Psi}_{\mathrm{G}}(x, t; n_0, \sigma_0, \phi_0) = \sum_{n=1}^{\infty} C_n^{\mathrm{G}}(n_0, \sigma_0, \phi_0)\mathrm{e}^{-iE_n t}\tilde{\psi}_n(x). \tag{21}$$

However, the algebraic construction reviewed in section 2.1 a priori does not enforce (21) to retain the well-understood quasi-classical behaviour summarised in (19).

We notice on the other hand that the SUSY stationary states (11) with $n \neq k$, can be regarded as perturbed versions of the initial eigenmodes (2), provided the overall sign is chosen as in (12). This is especially obvious at $k \gg 1$, i.e. by using a high energy mode for the SUSY mapping.

The case $n = k$ deserves here, like in section 2.1, some special care. As clear from (16), the perturbative equivalence $\psi_n(x) \simeq \tilde{\psi}_n(x; k \gg 1, \omega)$ no longer holds for this state. Thus, the mode $\tilde{\psi}_k(x; k, \omega)$ corresponding to the energy used for the SUSY mapping is always fundamentally different from its original homologue $\psi_k(x)$.

The consequences of these remarks for the GCS now appear more distinctly. Whenever n_0 is chosen far (i.e. out of a few standard deviations) from k, the GCS of the SUSY partner agree with the GCS of the original system. The analysis of section 2.2 then carries on, and a localised Gaussian quasi-classical wave packet minimizing approximately Heisenberg relation is generated.

If n_0 and k are of the same order, on the other hand, some discrepant features are expected for the SUSY GCS. As a concrete example, Figure 2 exhibits the time-evolution of the probability density distribution calculated from (21), in the *coincident* case $n_0 = k$. Clearly, an effect of the supersymmetry is to generate superimposed wavelets on the principal Gaussian wave packet. As time evolves, moreover, some side modes appear next to the dominant packet. Numerical calculations show that these "harmonics" coherently follow the principal peak in its displacements inside the well as time evolves. The wave packet slowly spreads and bounces on the walls as in the original construction.

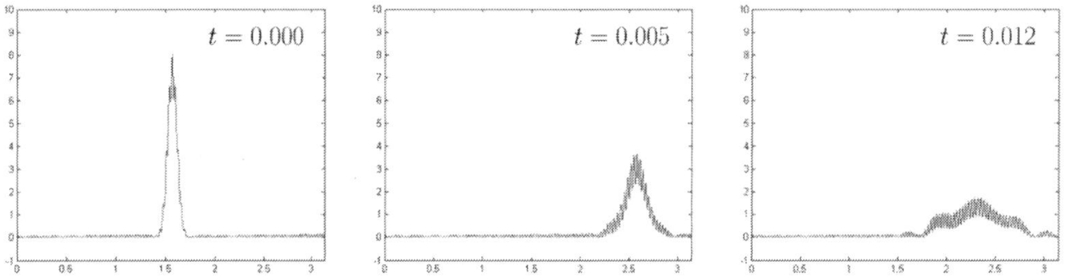

(a) Initial distribution (b) Before 1st bounce (c) After 1st bounce

Figure 2 - Time-evolution of the probability density distribution inside the well for the SUSY coherent state $\tilde{\Psi}_G(x, t; n_0 = 100, \sigma_0 = 10, \phi_0 = \pi/2)$. The SUSY parameters are: $k = 100$, $\omega = 2$.

Let us now turn to the interesting question of the effect of the SUSY parameter ω on the GCS. Our numerical calculations seem to suggest that the magnitude of $|\omega - 1/2|$ weakly affect the behaviour of the states. Thus, the above discussion continues to apply. What is more subtle is the effect of changing the sign of $(\omega - 1/2)$. In spite of its simple mirror effect on the potential, as discussed in section 2.1, it tremendously affects the GCS. As shown on Figure 3, the $\mathbb{Z}_2$ action suppresses the wavelets and side-modes, even for the coincident case $n_0 = k$. This is surprising since the above discussion is still applicable, so we might still have expected some disagreement due to the exceptional state $\tilde{\psi}_k(x; k, \omega)$. At the level of the wave functions (12) and (16), the $\mathbb{Z}_2$ action acts as an alternating phase factor (except at energy E_k, where the alternating pattern is broken): $\tilde{\psi}_n \to -\cos(n\pi)\tilde{\psi}_n$, $\tilde{\psi}_k \to +\cos(k\pi)\tilde{\psi}_k$ (provided we also map $x \to \pi - x$).

Note that although the GCS of the SUSY Hamiltonian can be chosen to be almost identical to the GCS of the original system, the potential functions remain very different. Indeed, the limit $k \gg 1$ emphatically does not lead to $\tilde{V}(x; k, \omega) \simeq 0$ in $[0, \pi]$. Thus, the construction of well behaved coherent states for this nontrivial system is a remarkable consequence of the SUSY approach.

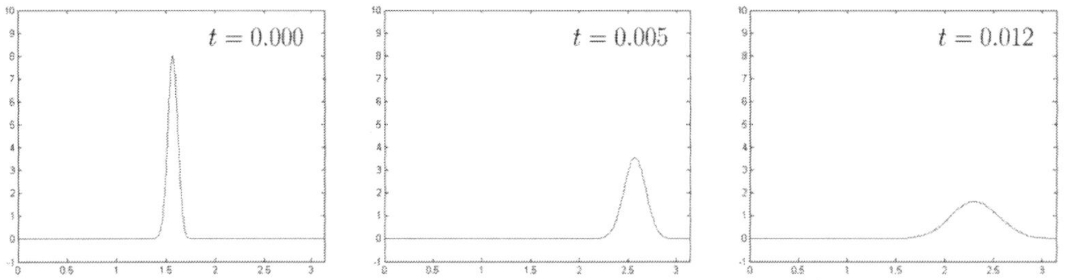

(a) Initial distribution (b) Before 1st bounce (c) After 1st bounce

Figure 3 - Time-evolution of the probability density distribution inside the well for the SUSY coherent state $\tilde{\Psi}_G(x, t; n_0 = 100, \sigma_0 = 10, \phi_0 = \pi/2)$. The SUSY parameters are: $k = 100$, $\omega = -1$.

As a concluding side comment, let us note that the classical dynamics in the SUSY system (10), which is generally expected for coherent state constructions, is not immediately transparent. A minimal criteria for classicality might thus be chosen to be a relatively slowly evolving and highly localized wave envelope. We have adhered to this point of view in the current work.

3. The 2D infinite well and SUSY partners

3.1. Description of the original model

We consider again a particle of mass 1/2, now assumed to move in a square 2D box of size π. The Hamiltonian is given as

$$H = H_x + H_y, \tag{22}$$

where H_x and H_y are the 1D Hamiltonians of the infinite well in the directions x and y respectively. The corresponding normalised eigenstates and discrete energies are

$$\Psi_{n,m}(x,y) = \psi_n(x)\psi_m(y) = \frac{2}{\pi}\sin nx \sin my \tag{23}$$

and

$$E_{n,m} = n^2 + m^2, \tag{24}$$

where $n, m = 1, 2, \ldots$

We see that this quadratic energy spectrum presents two types of degeneracies. The first type is a permutation degeneracy since the eigenstates $\Psi_{n,m}(x,y)$ and $\Psi_{m,n}(x,y)$, $n \neq m$, are distinct with the same energy $E_{n,m}$. The second type is called accidental [10,12] or arithmetic [16]. For example, we have $E_{5,5} = E_{1,7} = 50$ with $\Psi_{5,5}(x,y)$ and $\Psi_{1,7}(x,y)$ as distinct eigenfunctions and $E_{1,8} = E_{4,7} = 65$ with $\Psi_{1,8}(x,y)$ and $\Psi_{4,7}(x,y)$ as distinct eigenfunctions. The problem of identifying the number and type of these degeneracies for the infinite well has already been considered [10,12,16].

We present here some notations that will be useful in the discussion of coherent states of such a system:

- We rank the energies (24) in increasing order and write them as E_ν, with index $\nu = 0, 1, \ldots$ We also introduce the shifted energies $\mathcal{E}_\nu = E_\nu - E_0$.
- As alluded to above, more than one state may correspond to E_ν. We thus define μ_ν as the index for the summation of all the eigenstates associated to the degenerate energy E_ν and d_ν as the associated number of degeneracies. The index μ_ν goes from 0 to $d_\nu - 1$. Although they are irrelevant for our purposes, let us point out that explicit formula for the existence and number of accidental degeneracies exist in the number theoretic literature. They involve the number of divisors of E_ν of the form $4k+1$ and $4k+3$, $k \in \mathbb{N}$.
- The indices (n, m) corresponding to a state alternatively referred to with (ν, μ_ν) are written $(n_{\nu,\mu_\nu}, m_{\nu,\mu_\nu})$.
- We rename the eigenstate $\Psi_{n_{\nu,\mu_\nu}, m_{\nu,\mu_\nu}}(x,y)$, the state corresponding to the energy E_ν and degeneracy index μ_ν, as $\Psi^{\nu,\mu_\nu}(x,y)$.
- To each energy E_ν, we can thus associate a single "cumulative" eigenstate constructed as a superposition of the states $\Psi^{\nu,\mu_\nu}(x,y)$. It takes the form

$$\Phi^\nu(x,y) = \sum_{\mu_\nu=0}^{d_\nu-1} \gamma_{\nu,\mu_\nu} \Psi^{\nu,\mu_\nu}(x,y), \tag{25}$$

  where the coefficients γ_{ν,μ_ν} are arbitrary non zero complex numbers. Note that we have taken into account the permutation degeneracy in the superposition formula.
- Finally, a polar parametrization ($\rho_\nu = \sqrt{E_\nu}$, $\tan\theta_{\nu,\mu_\nu} = \frac{n_{\nu,\mu_\nu}}{m_{\nu,\mu_\nu}}$) is introduced. It provides an intuitive picture of the indices ν and μ_ν. As exemplified in Figure 4, the energy index ν is associated with the *radius* ρ_ν in (m,n)-plane, while the degeneracy index μ_ν is associated with the *angle* θ_{ν,μ_ν}. For definiteness, let us assume that μ_ν increases as n_ν increases, while m_ν decreases (for fixed ν). Put differently, μ_ν increases as the angle θ_{ν,μ_ν} increases.

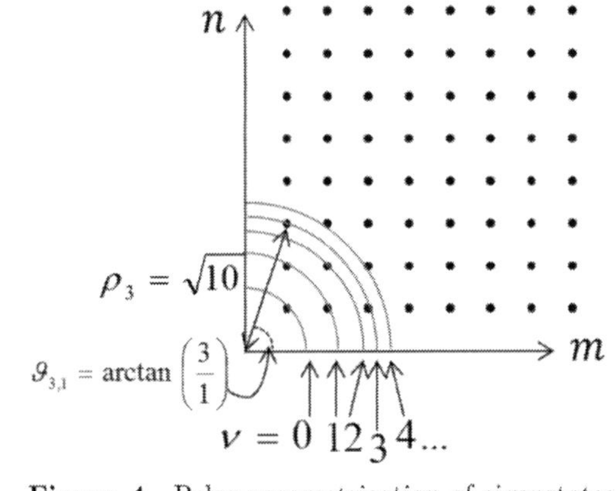

Figure 4 - Polar parametrisation of eigenstates

3.2. Construction of SUSY partners
2D generalisations of SUSY quantum mechanics have been proposed some time ago and, in particular, the approach of Ioffe and collaborators [15] has given some relevant results. Unfortunately, using supercharges that are essentially a linear combination of the ones used in 1D does not lead to the new potentials we have already obtained in 1D. An alternative approach is considered here.

A SUSY partner Hamiltonian which is separable in x and y could be obtained from SUSY 1D Hamiltonians as constructed in Section 2. Indeed, let us take $Q = Q_x Q_y$ and $Q^\dagger = Q_x^\dagger Q_y^\dagger$ where Q_x, $Q_x^\dagger$ and Q_y, $Q_y^\dagger$ satisfy the interwining 1D relations (3) in the corresponding variables. We thus easily get

$$\tilde{H} Q^\dagger = Q^\dagger H, \quad Q\tilde{H} = HQ, \tag{26}$$

where $\tilde{H} = \tilde{H}_x + \tilde{H}_y$ with $\tilde{H}_x = -\frac{d^2}{dx^2} + V(x, k_1, \omega_1)$ and $\tilde{H}_y = -\frac{d^2}{dy^2} + V(y, k_2, \omega_2)$. The corresponding potentials are explicitly given in (10).

Eigenstates of the SUSY partner Hamiltonian $\tilde{H}$ are thus easily constructed from

$$\tilde{\Psi}_{n,m}(x, y; k_1, k_2, \omega_1, \omega_2) = ((k_1^2 - n^2)(k_2^2 - m^2))^{-1} Q^+ \Psi_{n,m}(x, y) = \tilde{\psi}_n(x; k_1, \omega_1)\tilde{\psi}_m(y; k_2, \omega_2), \tag{27}$$

for $n \neq k_1, m \neq k_2$. The corresponding energies are $E_{n,m} =$ (24). In 1D, we had to add one state (16) in order to get a complete spectrum. In 2D, we have to add more states. Indeed, we also get the following states:

$$\tilde{\Psi}_{k_1,m}(x, y; k_1, k_2, \omega_1, \omega_2) = \tilde{\psi}_{k_1}(x; k_1, \omega_1)\tilde{\psi}_m(y; k_2, \omega_2), \tag{28}$$

with energy $E_{k_1,m}$,

$$\tilde{\Psi}_{n,k_2}(x, y; k_1, k_2, \omega_1, \omega_2) = \tilde{\psi}_n(x; k_1, \omega_1)\tilde{\psi}_{k_2}(y; k_2, \omega_2), \tag{29}$$

with energy E_{n,k_2}, and finally the eigenstate with energy E_{k_1,k_2} is given as

$$\tilde{\Psi}_{k_1,k_2}(x, y; k_1, k_2, \omega_1, \omega_2) = \tilde{\psi}_{k_1}(x; k_1, \omega_1)\tilde{\psi}_{k_2}(y; k_2, \omega_2). \tag{30}$$

3.3. Coherent states
Generalised and Gaussian coherent states have been constructed for the usual 2D infinite well [10, 12]. The generalised coherent states (GeCS) are as usual defined as eigenstates of an

annihilation operator of the system under consideration. But in the 2D case, due to the existence of a quadratic degenerate energy spectrum, we had to express the energies in increasing order and also to make a superposition of the different eigenstates with the same energy as given in (25). The construction of an annihilation operator and numerical calculations giving the behaviour of those states can be found in [12]. They have been shown to be closely related to GCS for some values of the coherent states parameters.

We will slightly adjust these constructions in this work.

3.3.1. Gaussian coherent states The definition of GCS is proposed as a direct generalisation of the 1D case. Indeed, we take

$$\Psi_{\mathrm{G}}^{2D}(x, y, t; n_0, m_0, \sigma_{n_0}, \sigma_{m_0}, \phi_{n_0}, \phi_{m_0}) = \Psi_{\mathrm{G}}(x, t; n_0, \sigma_{n_0}, \phi_{n_0})\Psi_{\mathrm{G}}(y, t; m_0, \sigma_{m_0}, \phi_{m_0}), \quad (31)$$

with $\Psi_{\mathrm{G}} = (17)$. Explicitly, we get

$$\Psi_{\mathrm{G}}^{2D}(x, y, t; n_0, m_0, \sigma_{n_0}, \sigma_{m_0}, \phi_{n_0}, \phi_{m_0}) = \sum_{n=1}^{\infty}\sum_{m=1}^{\infty} C_{n,m}^{\mathrm{G}} \mathrm{e}^{-iE_{n,m}t}\Psi_{m,n}(x, y), \quad (32)$$

where

$$C_{n,m}^{\mathrm{G}} = C_{n,m}^{\mathrm{G}}(n_0, m_0, \sigma_{n_0}, \sigma_{m_0}, \phi_{n_0}, \phi_{m_0}) = C_n^{\mathrm{G}}(n_0, \sigma_{n_0}, \phi_{n_0})C_m^{\mathrm{G}}(m_0, \sigma_{m_0}, \phi_{m_0}). \quad (33)$$

This definition pays off since the analysis of [11] reviewed in section 2.2 continues to apply here. As in the 1D case, we get a well localised state and quasi-classical behaviour. Approximate saturation of the Heisenberg uncertainty bound is also achieved, as in the 1D setting.

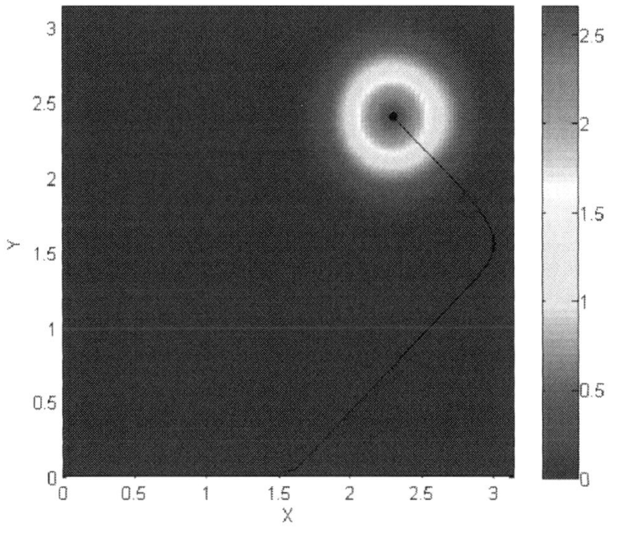

Figure 5 - Probability density distribution and time evolution of the position expectation value of the SUSY GCS in 2D until $t = 0.012$. State parameters are chosen as: $n_0 = m_0 = 100$, $\sigma_{n_0} = \sigma_{m_0} = 10$, $\phi_{n_0} = \pi/2$, and $\phi_{n_0} = 0$. SUSY parameters are: $k_1 = 100$, $k_2 = 50$, $\omega_1 = -1$, and $\omega_2 = 2$.

Such a definition of GCS is easily extended to the SUSY case where the eigenstates $\Psi_{m,n}(x, y)$ in (32) are replaced by the eigenstates $\tilde{\Psi}_{m,n}(x, y)$. Again, the factorization of the state yields a straightforward generalisation of the 1D results. As an example, Figure 5 shows a trace of the time evolution of the position expectation value for the SUSY GCS in 2D. A neat localized packet bouncing on the walls is obtained once again.

3.3.2. Generalised coherent states The GeCs are eigenstates of an annihilation operator of the 2D infinite well and they have been defined as [12]

$$\Psi_{\rm Ge}(x,y,t;z) = \frac{1}{\sqrt{N_{\rm Ge}(z)}} \sum_{\nu=0}^{\infty} \frac{z^{\nu}}{\sqrt{\rho(\nu)}} {\rm e}^{-i\mathcal{E}_{\nu}t}\Phi^{\nu}(x,y), \tag{34}$$

with the normalisation factor

$$N_{\rm Ge}(z) \equiv \sum_{\nu=0}^{\infty} \frac{|z|^{2\nu}}{\rho(\nu)}$$

and

$$\rho(\nu) = \begin{cases} 1, & \text{if } \nu = 0 \\ \Pi_{i=1}^{\nu}\mathcal{E}_i, & \text{if } \nu \neq 0. \end{cases} \tag{35}$$

These states depend on a continuous complex parameter z as in the 1D case. With respect to an alternative approach [12], more freedom is given to the states $\Phi^{\nu}(x,y)$ being a superposition of states with same energy $\mathcal{E}_{\nu}$.

4. Conclusion and future work

SUSY partners of the infinite well have been constructed and have shown to satisfy relevant properties compared with the ones of the original quantum system. In particular, for the 1D system, we have constructed a set of coherent states (GCS) which depends on real and discrete parameters. The relation between those states and the generalised ones (GeCS) has been formally given [11] and we have shown the behaviour of the GCS case.

In the 2D case, we have extended the construction of SUSY partners to get similar potentials as in 1D. More parameters are involved in this context and the existence of degeneracies in the energy spectrum has lead us to adjust the definition of coherent sates with respect with the ones used in the 1D case. The GCS are constructed in order to have a good behaviour with respect to localisation in the usual case as well as in the SUSY case. We have also constructed GeCS in the 2D setting.

It remains to make a link between both types of states. The use of the polar parametrisation is clearly a way to solve the problem. Indeed, the GCS may be written as

$$\Psi_{\rm G}(x,y,t) \equiv \frac{1}{\sqrt{N_{\rm G}}} \sum_{\nu=0}^{\infty} {\rm e}^{\frac{(\rho_{\nu}-\bar{\rho}_0)^2}{4\sigma_0^2}} {\rm e}^{-i\mathcal{E}_{\nu}t}\Psi^{{\rm G},\nu}(x,y). \tag{36}$$

The Gaussian terms depending on the radial variables could thus be related to the factor $z^{\nu}(\rho(\nu))^{-\frac{1}{2}}$ in (34) as it was the case in 1D [11]. The Gaussian terms involving the angular variables are related to the coefficients of the superposition of eigenstates with same energy. This way, the states $\Psi^{G,\nu}(x,y)$ will be related to $\Phi^{\nu}(x,y)$ in (34). As mentioned earlier, a particular approach of this question in 2D [10, 12] has shown that GeCS are in fact a good approximation of GCS. In a future work, we hope to formally solve this problem.

Acknowledgements

This work has been supported in part by research grants from Natural sciences and engineering research council of Canada (NSERC). Marc-Antoine Fiset acknowledges a NSERC fellowship.

References

[1] Schrödinger E 1926 *Naturwiss*, **14**, 664.
[2] Glauber R J 1963, *Phys. Rev.*, **130**, 2529; **131**, 2766.

[3] Klauder J R and Skagerstam B S, *Coherent States- Applications in Physics and Mathematical Physics*, World Scientific, Singapore, 1985.
[4] Perelomov A M, *Generalized Coherent states and their applications*, Springer, Berlin,1986.
[5] Cohen-Tannoudji C., Diu B. and Laloe F., *Mécanique quantique I*, Hermann, Paris, 1997.
[6] Gazeau J.P., *Coherent States in Quantum Physics*, Wiley-VCH, Weinheim, 2009.
[7] Dong S-H *Factorization Method in Quantum Mechanics, Fundamental theories in physics 150*, Springer, Dortrecht, 2008.
[8] Angelova M and Hussin V 2008 *J Phys A* **41** 30416.
[9] Fox R F and Choi M H 2000, *Phys. Rev. A*, **61**, 032107.
[10] Fox R F and Choi M H 2001, *Phys. Rev. A*, **64**, 042104.
[11] Fiset M A and Hussin V 2014, arXiv:1410.0305 (math-ph).
[12] Dello Sbarba L and Hussin V 2007, *J. Math. Phys.* **48**,012110.
[13] Fernandez D J, Hussin V and Rosas-Ortiz O 2007, *J. Phys. A: Math. Theor.* **40** 6491.
[14] Fernandez D J and Salinas-Hernandez E 2003, *J. Phys. A: Math. Gen.* **36** 2537.
[15] Cannata, F. M.V. Ioffe and D.N. Nishnianidze, *J. Phys. A: Math. Gen.* **35**, 1389, 2002.
[16] Itzykson C and Luck J M, *J.Phys. A: Math. Gen.* **19** (1986) 211.

Journal of Physics: Conference Series **624** (2015) 012017 doi:10.1088/1742-6596/624/1/012017

Truncated harmonic oscillator and Painlevé IV and V equations

David J Fernández C[1] and VS Morales-Salgado[2]

Departamento de Física, Cinvestav, A.P. 14-740, *07000 México D.F. Mexico*

E-mail: [1]`david@fis.cinvestav.mx`, [2]`vmorales@fis.cinvestav.mx`

Dedicated to Professor Bogdan Mielnik for his 50 years of scientific career.

Abstract. Quantum systems described by second and third order polynomial Heisenberg algebras are obtained applying supersymmetric quantum mechanics to the harmonic oscillator with an infinite potential barrier. These systems are linked with the Painlevé IV and V equations, respectively, thus several solutions of these non-linear second-order differential equations will be found, along with a chain of Bäcklund transformations connecting such solutions.

1. Introduction

The methods and techniques of supersymmetric quantum mechanics [1–22] can be applied in a straightforward manner to the harmonic oscillator [9, 23–25]. As a result, new systems are obtained, called supersymmetric (SUSY) partners of the harmonic oscillator. A well known fact about them is that they realize the polynomial Heisenberg algebras (PHA). These are deformations of the harmonic oscillator algebra, characterized by a commutator between the ladder operators which is a polynomial in the Hamiltonian.

Now, it has been shown that systems which realize a second-order PHA are connected to the Painlevé IV (PIV) equation while those ruled by a third-order PHA are linked with the Painlevé V (PV) equation [23, 26–30]. Thus, once a family of potentials characterized by the required PHA is obtained, for example, through the techniques of supersymmetric quantum mechanics, said connections can be exploited to produce non-singular solutions to the PIV and PV equations [13, 31, 32].

In previous works we implemented this analysis for the harmonic oscillator with an infinite barrier at the origin, or truncated harmonic oscillator for short [6, 33, 34]. Solutions to the PIV and PV equations have been found, along with the required supersymmetric partners of the initial system. In this paper we shall study several connections among such solutions, in the form of Bäcklund transformations.

The organization of this work is the following: in Section 2 we review the results of applying supersymmetric quantum mechanics to the truncated harmonic oscillator. In Section 3 we discuss the connection between polynomial Heisenberg algebras and the Painlevé IV and V differential equations. In Section 4 we obtain several solutions to the PIV equation using realizations of second order PHA's and relate them using Bäcklund transformations. In Section 5,

Journal of Physics: Conference Series **624** (2015) 012017 doi:10.1088/1742-6596/624/1/012017

using realizations of third order PHA's, we obtain solutions to the PV equation and relate them using Bäcklund transformations. Section 6 contains our conclusions and some final remarks.

2. Truncated harmonic oscillator

As was stated in the introduction, we will obtain solutions of the Painlevé IV and V equations using the supersymmetric partners of the truncated harmonic oscillator, the last being described by the Hamiltonian

$$H = -\frac{1}{2}\frac{d^2}{dx^2} + V, \qquad V = \begin{cases} \frac{x^2}{2} & \text{if } x > 0, \\ \infty & \text{if } x \leq 0. \end{cases}$$

The eigenfunctions of H, corresponding to the eigenvalues $\mathcal{E}_n = 2n + \frac{3}{2}$ are the odd eigenfunctions of the harmonic oscillator, i.e.,

$$\psi_n(x) \propto x\, e^{-x^2/2}\, {}_1F_1\Big(-n, \frac{3}{2}, x^2\Big),$$

since they have to satisfy the boundary conditions, particularly at the origin. The even eigenfunctions of the harmonic oscillator fail to do so, and this is the reason why the quantities $\mathcal{E}_n = 2n + \frac{1}{2}$ do not belong to the spectrum of H.

Anyway, they will be useful for the SUSY technique, thus we present them now:

$$\chi_n(x) \propto e^{-x^2/2}\, {}_1F_1\Big(-n, \frac{1}{2}, x^2\Big), \quad \mathcal{E}_n = 2n + \frac{1}{2}.$$

In order to obtain the supersymmetric partners of this system, we need to "intertwine" H with another Hamiltonian $\tilde{H} = -\frac{1}{2}\frac{d^2}{dx^2} + \tilde{V}(x)$ through the operator relation

$$\tilde{H} A^+ = A^+ H, \tag{1}$$

where A^+ is a differential operator of order k.

If $\psi(x)$ are eigenfunctions of H with eigenvalues E_n, such a relation implies that $\tilde{H} A^+ \psi_n(x) = E_n A^+ \psi_n(x)$, i.e.,

$$\phi_n \propto A^+ \psi_n \tag{2}$$

are eigenfunctions of $\tilde{H}$ with eigenvalues E_n.

The first-order supersymmetric partners of the truncated harmonic oscillator, or 1-SUSY partners for short, are obtained through a first-order operator A^+. Let us suppose now that we take a solution $u(x)$ of the stationary Schrödinger equation for H, i.e., $Hu = \epsilon u$, where $u(x)$ does not necessarily satisfy the associated boundary conditions. Then, the expressions

$$\begin{aligned} A^+ &= \frac{1}{\sqrt{2}}\left[-\frac{d}{dx} + [\ln(u)]'\right], \\ \tilde{V} &= V - [\ln(u)]'', \end{aligned} \tag{3}$$

fulfill equation (1) for the first-order case and provides $\tilde{V}$ once V is known. It is important to prevent that new singularities appear in $\tilde{V}$. Such a prescription imposes restrictions on $u(x)$, in particular, $u(x)$ must not have zeros in the domain of interest.

In analogy to the previous case, the second-order supersymmetric partners of the truncated harmonic oscillator, or 2-SUSY partners, are obtained using a pair of Schrödinger solutions

$u_1(x)$, $u_2(x)$, such that $Hu_j = \epsilon_j u_j$, with $j = 1, 2$. Remember that each $u_j(x)$ does not need to satisfy the boundary conditions. Then

$$\begin{aligned} A^+ &= \frac{1}{2}\left\{\frac{d^2}{dx^2} - [\ln W(u_1,u_2)]'\frac{d}{dx} + \frac{1}{2}\left([\ln W(u_1,u_2)]'' + [\ln W(u_1,u_2)]'^2\right) - 2V + \epsilon_1 + \epsilon_2\right\}, \\ \tilde{V} &= V - [\ln W(u_1,u_2)]'', \end{aligned} \tag{4}$$

where $W(u_1, u_2)$ is the Wronskian of u_1 and u_2. Once again, we need to consider those cases where the term $[\ln W(u_1, u_2)]''$ does not introduce new singularities to the supersymmetric partner, which translates in requiring that $W(u_1, u_2)$ does not have zeros in the domain of interest.

The k-th order case can be obtained by iteration of the procedure which led to the previous two cases, and requires k Schrödinger solutions such that $Hu_j = \epsilon_j u_j$, $j = 1, 2, ..., k$. The general form of $u_j(x)$ is given by

$$u_j(x) = e^{-x^2/2}\left[{}_1F_1\left(\frac{1-2\epsilon_j}{4}, \frac{1}{2}, x^2\right) + 2\nu_j \frac{\Gamma(\frac{3-2\epsilon_j}{4})}{\Gamma(\frac{1-2\epsilon_j}{4})}\, x\, {}_1F_1\left(\frac{3-2\epsilon_j}{4}, \frac{3}{2}, x^2\right)\right],$$

which is a linear combination of the odd and even solutions for the harmonic oscillator.

Since singularities are allowed at $x = 0$, but nowhere else, then we can proceed by cases with definite parity for $u(x)$. The results, which appear in more detail in [6, 33, 34], are summarized next.

2.1. *First-order supersymmetric partners*

When a first order intertwining operator A^+ is used with an odd transformation function u, the SUSY partner potential is given by

$$\tilde{V} = V + 1 + \frac{1}{x^2} - \left\{\ln\left[{}_1F_1\left(\frac{3-2\epsilon}{4}, \frac{3}{2}, x^2\right)\right]\right\}'', \tag{5}$$

the eigenenergies of $\tilde{H}$ being $\mathcal{E}_n = 2n + \frac{3}{2}$ and $\phi_n \propto A^+\psi_n$ the corresponding eigenfunctions. The allowed values for ϵ are those which satisfy $\epsilon < \frac{3}{2}$, i.e., below the ground state of H. However, in the limit $\epsilon \to \frac{3}{2}$ the level $E_0 = \frac{3}{2}$ is expunged from the spectrum of $\tilde{H}$.

On the other hand, the use of an even transformation function u produces new Hamiltonians $\tilde{H}$ with potentials of the form

$$\tilde{V} = V + 1 - \left\{\ln\left[{}_1F_1\left(\frac{1-2\epsilon}{4}, \frac{1}{2}, x^2\right)\right]\right\}''. \tag{6}$$

In this case, the energy levels of $\tilde{H}$ are given by $\mathcal{E}_n = 2n + \frac{1}{2}$, with $\phi_n \propto A^+\chi_n$ being the corresponding eigenfunctions. For $\epsilon < \frac{1}{2}$ no new singularities appear, i.e., these are the allowed values of ϵ. In the limit $\epsilon = \frac{1}{2}$ the level $\mathcal{E}_0 = \frac{1}{2}$ is expunged from the energy spectrum of $\tilde{H}$.

Notice that, for both cases, $\tilde{H}$ is isospectral to H, up to a shift in the ground state energy.

2.2. *Second-order supersymmetric partners*

For a second-order intertwining operator A^+ let us suppose, without lost of generality, that $\epsilon_2 < \epsilon_1$. There are four distinct parity combinations for $u_1(x)$ and $u_2(x)$.

Taking both u_1 and u_2 odd, the SUSY partner potential is

$$\tilde{V} = V + \frac{3}{x^2} + 2 - [\ln \omega_1(x)]''.$$

The absence of zeros of the function $\omega_1(x)$ sets the requirements on ϵ_1, ϵ_2. Then, no new singularities appear whenever $\epsilon_1, \epsilon_2 < \frac{3}{2}$ or $\frac{3+4n}{2} < \epsilon_1, \epsilon_2 < \frac{3+4(n+1)}{2}$. The energy spectrum of $\tilde{H}$ is given by $\{\mathcal{E}_n = 2n + \frac{3}{2}, \quad n = 0, 1, ...\}$ with $\phi_n(x) \propto A^+\psi_n(x)$ being the corresponding eigenfunctions. As a limiting case, when either ϵ_1 or ϵ_2 takes the value $\mathcal{E}_n$, then this level is expunged from the set of eigenvalues of $\tilde{H}$.

When u_1 and u_2 are both even we get

$$\tilde{V} = V + \frac{1}{x^2} + 2 - [\ln \omega_2(x)]'',$$

where $\omega_2(x)$ has no zeros in $(0, \infty)$ as long as $\epsilon_1, \epsilon_2 < \frac{1}{2}$ or $\frac{1+4n}{2} < \epsilon_1, \epsilon_2 < \frac{1+4(n+1)}{2}$. The eigenvalues of $\tilde{H}$ are $\mathcal{E}_n = 2n + \frac{1}{2}$, $n = 0, 1, ...$, and $\phi_n(x) \propto A^+\chi_n(x)$ are the corresponding eigenfunctions. One can erase one level $\mathcal{E}_n$ from the energy spectrum of $\tilde{H}$ by choosing ϵ_1 or ϵ_2 equal to said eigenvalue.

Taking now u_1 odd and u_2 even leads to the new potential

$$\tilde{V} = V + 2 - [\ln \omega_3(x)]''.$$

The energy spectrum is composed by the levels $\mathcal{E}_n = 2n + \frac{3}{2}$, the corresponding eigenfunctions are $\phi_n(x) \propto A^+\psi_n(x)$, and the condition on ϵ_1, ϵ_2 is $\frac{1+4n}{2} < \epsilon_2 < \epsilon_1 < \frac{3+4n}{2}$, with a limiting case $\epsilon_1 = \mathcal{E}_n$ for expunging the level $\mathcal{E}_n$. We can also add a level at ϵ_2 by requesting that $\epsilon_2 \neq \mathcal{E}_n$, the corresponding eigenfunction being $\phi_{\epsilon_2} \propto \frac{u_1}{W(u_1, u_2)}$.

As a final case, if u_1 is even and u_2 is odd, it turns out that

$$\tilde{V} = V + 2 - [\ln \omega_4(x)]'',$$

where ϵ_1 and ϵ_2 are required now to satisfy $\epsilon_2 < \epsilon_1 < \frac{1}{2}$ or $\frac{3+4n}{2} < \epsilon_2 < \epsilon_1 < \frac{5+4n}{2}$. The eigenfunctions of $\tilde{H}$ are $\phi_n(x) \propto A^+\psi_n$ with eigenvalues $\mathcal{E}_n = 2n + \frac{3}{2}$. The choice $\epsilon_2 = \mathcal{E}_n$ expunges such a level, while for $\epsilon_1 \neq \mathcal{E}_n$ one adds a level at ϵ_1 with eigenfunction $\phi_{\epsilon_1} \propto \frac{u_2}{W(u_1, u_2)}$.

3. Polynomial Heisenberg algebras

The polynomial Heisenberg algebras (PHA) are generated by three operators $\{\mathbb{H}, L^+, L^-\}$ satisfying the following commutation relations

$$[\mathbb{H}, L^{\pm}] = \pm L^{\pm},$$
$$[L^-, L^+] = P_m(\mathbb{H}),$$

where $\mathbb{H}$ is a Hamiltonian with Schrödinger form, $P_m(\mathbb{H})$ is a polynomial of degree m on the Hamiltonian and $L^{\pm}$ are $(m+1)$-th order differential ladder operators.

In particular, systems with third order ladder operators $L^{\pm}$ realize a second order PHA. By factorizing them as $L^+ = L_1^+ L_2^+$ and $L^- = L_2^- L_1^-$ where $L_{1,2}^- = (L_{1,2}^+)^\dagger$ and $L_1^+ = \frac{1}{\sqrt{2}}\left[-\frac{d}{dx} + f(x)\right]$, $L_2^+ = \frac{1}{2}\left[\frac{d^2}{dx^2} + g(x)\frac{d}{dx} + h(x)\right]$, we find at the end the following set of equations [9]:

$$f(x) = x + g(x),$$
$$h = \frac{g'}{2} - \frac{g^2}{2} - 2xg - x^2 + \varepsilon_2 + \varepsilon_3 - 2\varepsilon_1 - 1,$$
$$\mathbb{V} = \frac{x^2}{2} - \frac{g'}{2} + \frac{g^2}{2} + xg + \varepsilon_1 - \frac{1}{2},$$
$$\frac{d^2g}{dx^2} = \frac{1}{2g}\left(\frac{dg}{dx}\right)^2 + \frac{3}{2}g^3 + 4xg^2 + 2(x^2 - a)g + \frac{b}{g}.$$

The last one is the Painlevé IV (PIV) equation with parameters $a = \varepsilon_2 + \varepsilon_3 - 2\varepsilon_1 - 1$ and $b = -2(\varepsilon_2 - \varepsilon_3)^2$. Thus, we have gotten a connection between second order PHA's and PIV equation. We can either solve such a non-linear differential equation for realizing a second order PHA, or take a known realization of a second order PHA and obtain then solutions to the PIV equation.

The latter is done by using the factorization

$$N \equiv L^+L^- = (\mathbb{H} - \varepsilon_1)(\mathbb{H} - \varepsilon_2)(\mathbb{H} - \varepsilon_3), \tag{7}$$

to see that there are three extremal states ϕ_λ such that $L^-\phi_\lambda = L_2^-L_1^-\phi_\lambda = 0$ satisfying also $\mathbb{H}\phi_\lambda = \lambda\phi_\lambda$, with $\lambda = \varepsilon_1, \varepsilon_2, \varepsilon_3$. In particular, a ϕ_λ such that

$$L_1^-\phi_\lambda = \frac{1}{\sqrt{2}}\left[\frac{d}{dx} + f(x)\right]\phi_\lambda = 0$$

satisfies the extremal state condition. Here we have fixed λ, but keep in mind that there are three possible identifications for such value. Let us recall also that $f(x) = x + g(x)$ to obtain

$$g_\lambda(x) = -x - [\ln\phi_\lambda]', \tag{8}$$

which is a solution of the PIV equation with parameters a, b.

On the other hand, systems with fourth order ladder operators $L^+ = L_1^+L_2^+$ and $L^- = L_2^-L_1^-$ realize a third order PHA [9, 30]. Choosing $L_1^+ = \frac{1}{2}\left[\frac{d^2}{dx^2} + g_1(x)\frac{d}{dx} + h_1(x)\right]$, $L_2^+ = \frac{1}{2}\left[\frac{d^2}{dx^2} + g(x)\frac{d}{dx} + h(x)\right]$ and $L_{1,2}^- = (L_{1,2}^+)^\dagger$ we find another set of equations:

$$\begin{aligned} g_1(x) &= -g(x) - x, \\ g(x) &= \frac{x}{w-1}, \\ \frac{d^2w}{dz^2} &= \left(\frac{1}{2w} + \frac{1}{w-1}\right)\left(\frac{dw}{dz}\right)^2 - \frac{1}{z}\frac{dw}{dz} + \frac{(w-1)^2}{z^2}(aw + \frac{b}{w}) + c\frac{w}{z} + d\frac{w(w+1)}{w-1}, \end{aligned} \tag{9}$$

where $z = x^2$. Expression (9) is the Painlevé V (PV) equation, with $a = \frac{(\varepsilon_1-\varepsilon_2)^2}{2}$, $b = -\frac{(\varepsilon_3-\varepsilon_4)^2}{2}$, $c = \frac{\varepsilon_1+\varepsilon_2-\varepsilon_3-\varepsilon_4-1}{2}$, $d = -\frac{1}{8}$ being fixed parameters.

Now we are concerned with the factorization

$$N \equiv L^+L^- = (\mathbb{H} - \varepsilon_1)(\mathbb{H} - \varepsilon_2)(\mathbb{H} - \varepsilon_3)(\mathbb{H} - \varepsilon_4) \tag{10}$$

which indicates that there are four extremal states such that $L^-\phi_\lambda = L_2^-L_1^-\phi_\lambda = 0$ and $\mathbb{H}\phi_\lambda = \lambda\phi_\lambda$, $\lambda = \varepsilon_1, \varepsilon_2, \varepsilon_3, \varepsilon_4$.

In a similar way as in the previous case, but with lengthier calculations, we can write $g_1 = (\ln[W(\phi_\lambda, \phi_\sigma)])'$ for a fixed choice of a pair of distinct extremal states $\phi_\lambda, \phi_\sigma$, thus obtaining $g(x) = -x - (\ln[W(\phi_\lambda, \phi_\sigma)])'$. The use of two extremal states in this construction allows us to choose among six different identifications for ϕ_λ and ϕ_σ leading to six different functions g.

Finally, after the change of variable $x = \sqrt{z}$, it turns out that

$$w(z) = 1 + \frac{\sqrt{z}}{g(\sqrt{z})}$$

is the solution to the PV equation we are looking for, with fixed parameters a, b, c and d [27,35].

Journal of Physics: Conference Series **624** (2015) 012017 doi:10.1088/1742-6596/624/1/012017

4. Solutions to the Painlevé IV equation

Now we want to connect the SUSY partners of H obtained in Section 2 with the PIV equation. This is done by first noting that, if $a^{\pm}$ are the standard raising and lowering operators for the harmonic oscillator, then $[H, a^{\pm}] = \pm a^{\pm}$.

On the other hand, if we define a pair of new operators for $\tilde{H}$ as $L^{\pm} \equiv A^{+}a^{\pm}A$, where $A = (A^{+})^{\dagger}$, then $[\tilde{H}, L^{\pm}] = \pm L^{\pm}$, which is a commutator characteristic for ladder operators. Moreover, the commutator between L^{+} and L^{-} is a polynomial of degree $2k$ in $\tilde{H}$, i.e.,

$$[L^{-}, L^{+}] = P_{2k}(\tilde{H}), \tag{11}$$

which defines a PHA of $2k$-th order.

For a first order A^{+}, the ladder operators $L^{\pm}$ are of third order. In addition, for the 1-SUSY partner of the truncated harmonic oscillator we recognize that the extremal state energies for $\tilde{H}$ are given by

$$L^{+}L^{-} = (\tilde{H} - \epsilon)(\tilde{H} - \epsilon - 1)\left(\tilde{H} - \frac{1}{2}\right),$$

where the identification $\varepsilon_1 = \epsilon$, $\varepsilon_2 = \epsilon + 1$, $\varepsilon_3 = \frac{1}{2}$ has been done.

As a result, when one uses an odd transformation function $u(x)$, the solution of the PIV equation obtained from the first extremal state is

$$g_{\varepsilon_1} = \frac{1}{x} - 2x + \left(1 - \frac{2}{3}\epsilon\right) x \, \frac{{}_1F_1(\frac{7-2\epsilon}{4}; \frac{5}{2}; x^2)}{{}_1F_1(\frac{3-2\epsilon}{4}; \frac{3}{2}; x^2)}.$$

Meanwhile, for an even function $u(x)$ the solution of the PIV equation obtained from the first extremal state is

$$g_{\varepsilon_1} = -2x + (1 - 2\epsilon)\, x \, \frac{{}_1F_1(\frac{5-2\epsilon}{4}; \frac{3}{2}; x^2)}{{}_1F_1(\frac{1-2\epsilon}{4}; \frac{1}{2}; x^2)}.$$

In both cases the other two PIV solutions can be calculated from the first one through the following expressions

$$g_{\varepsilon_2} = -g_{\epsilon} - 2x - 2\left[\frac{x + (2\epsilon - x^2)(g_{\epsilon} + x) + (g_{\epsilon} + x)^3}{x^2 - 2\epsilon - 1 - (g_{\epsilon} + x)^2}\right],$$

$$g_{\varepsilon_3} = -\frac{g_{\epsilon}' + 2}{g_{\epsilon} + 2x}.$$

For a second order A^{+}, the ladder operators $L^{\pm}$ are of fifth order. So, in order to repeat the procedure carried out for the 1-SUSY partners of H we need to reduce the order of $L^{\pm}$. For this purpose we use a reduction theorem [13] which states that under the condition that $u_2(x) = a^{-}u_1(x)$, $\epsilon_2 = \epsilon_1 - 1$, there exists a pair of third order ladder operators $l^{\pm}$ for $\tilde{H}$ which product can be factorized into

$$l^{+}l^{-} = (\tilde{H} - \epsilon_1 + 1)(\tilde{H} - \epsilon_1 - 1)(\tilde{H} - 1/2). \tag{12}$$

Then, the three extremal states are those associated with $\varepsilon_1 = \epsilon - 1$, $\varepsilon_2 = \epsilon + 1$, $\varepsilon_3 = \frac{1}{2}$.

Recall that the reduction theorem required that ϵ_2 depended on ϵ_1, and $u_2(x)$ on $u_1(x)$. Thus, we can write $u_1(x) = u(x)$ and $\epsilon_1 = \epsilon$ for simplicity. So, after repeating the method used for the 1-SUSY partners of H, in the reduced case for the 2-SUSY partners the three different solutions of the PIV equation are given by

$$G_{\varepsilon_1} = -x - \alpha + 2\left[\frac{x+\alpha}{x^2+1-2\epsilon-\alpha^2}\right],$$

$$G_{\varepsilon_2} = G_{\varepsilon_1} + \frac{2\alpha^2 - 2x^2 + 2(2\epsilon+1)}{\alpha - G_{\varepsilon_1} - x},$$

$$G_{\varepsilon_3} = \frac{(x+\alpha)G_{\varepsilon_1}^2 + \left[2\epsilon - 1 + (x+\alpha)^2\right]G_{\varepsilon_1} + (2\epsilon-3)(x+\alpha)}{(x+\alpha)^2 + (x+\alpha)G_{\varepsilon_1} + 2\epsilon - 1},$$

where $\alpha = [\ln(u)]'$ and the solutions G_{ε_2} and G_{ε_3} are given in terms of G_{ε_1}. We must remark that in this case we have used a capital g for the PIV solutions in order to distinguish them from those obtained through the first-order SUSY partners.

4.1. Bäcklund transformations

Now we would like to connect the six solutions of PIV equation we have just found by transformations converting one into another. With this purpose let us recall that heuristically a Bäcklund transformation relates one solution of a given equation either to another solution of the same equation, possibly with different values of the parameters, or to a solution of another equation [36]; for a more formal definition you can see [37].

For the PIV equation, [36] presents a list of Bäcklund transformations relating PIV solutions among them. In particular, using the notation of this reference, we found that the solutions of the PIV equation g_σ, G_σ with $\sigma = \varepsilon_1, \varepsilon_2, \varepsilon_3$ are related by Bäcklund transformations according to the following diagram:

$$\overbrace{g_{\varepsilon_1} \underset{W^{\ddagger+}W^{\dagger+}}{\longrightarrow} g_{\varepsilon_2} \underset{W^{\ddagger-}}{\longrightarrow} g_{\varepsilon_3}}^{\text{1-SUSY}} \underset{\tilde{W}^{+}}{\longrightarrow} \overbrace{G_{\varepsilon_1} \underset{W^{\ddagger-}}{\longrightarrow} G_{\varepsilon_3} \underset{W^{\ddagger+}}{\longrightarrow} G_{\varepsilon_2}}^{\text{2-SUSY}}$$

where g_{ε_2} is obtained by applying two transformations $W^{\ddagger+}$ and $W^{\dagger+}$ in this order to g_{ε_1}, i.e., $g_{\varepsilon_2} = W^{\ddagger+}W^{\dagger+}g_{\varepsilon_1}$, and similarly for the rest of the diagram.

Notice that we could relate not only the set of solutions obtained through 1-SUSY, and those obtained by 2-SUSY, but we could also connect both sets of solutions through the expression $G_{\varepsilon_1} = \tilde{W}^{+}g_{\varepsilon_3}$

5. Solutions to the Painlevé V equation

To establish the connection between the PV equation and $\tilde{H}$ we need to take into account that, although $L^{\pm} = A^{+}a^{\pm}A$ (again $A = (A^{+})^{\dagger}$) have the right commutators with $\tilde{H}$ to be called ladder operators, it is $\mathcal{L}^{\pm} = A^{+}(a^{\pm})^2A$ which defines genuine ladder operators, which move a physical state of $\tilde{H}$ to another physical state of $\tilde{H}$. This makes $\mathcal{L}^{\pm}$ fourth-order ladder operators and thus the set $\{\tilde{H}, \mathcal{L}^{+}, \mathcal{L}^{-}\}$ supplies a realization of a third-order PHA, whenever A^{+} is of first-order.

This time the factorization

$$\mathcal{L}^{+}\mathcal{L}^{-} = \left(\tilde{H} - \epsilon\right)\left(\tilde{H} - \epsilon - 2\right)\left(\tilde{H} - \frac{1}{2}\right)\left(\tilde{H} - \frac{3}{2}\right)$$

indicates that there are four extremal states satisfying $\mathcal{L}^{-}\phi_\lambda = 0$ with $\lambda = \epsilon,\ \epsilon+2,\ \frac{1}{2},\ \frac{3}{2}$, then we have six inequivalent choices of two of them for a given $\tilde{H}$. The six solutions of the PV equation are:

$$w_{1a} = 1 + \frac{2\sqrt{2z}(1 + 2\epsilon - z + \sqrt{2z}\alpha)}{\sqrt{2z}(2+z) - 4(1+z)\alpha + 2\sqrt{2z}\alpha^2},$$

for $\lambda = \epsilon, \sigma = \epsilon + 2$.

$$w_{1b} = \frac{\left(2\alpha + \sqrt{2z}\right)\left(4\alpha - 4\epsilon\sqrt{2z} + \sqrt{2}z^{3/2} - 2\sqrt{2}\alpha^2\sqrt{z} - 4\sqrt{2z}\right)}{\left(2\alpha - \sqrt{2z}-\right)\left(4\alpha - 4\epsilon\sqrt{2z} + \sqrt{2}z^{3/2} - 2\sqrt{2}\alpha^2\sqrt{z} + 4\sqrt{2z}\right)},$$

for $\lambda = \epsilon + 2, \sigma = \frac{1}{2}$.

$$w_{1c} = 1 + \frac{\sqrt{2z}\left(8\epsilon z - 2z^2 + 4\alpha^2 z + 4\sqrt{2z}\alpha - 16\right)}{8\alpha - 2\sqrt{2}z^{3/2}\left(\alpha^2 + 2\epsilon - 3\right) + 4\alpha z\left(\alpha^2 + 2\epsilon - 1\right) - 8\sqrt{2z}\left(\alpha^2 + \epsilon - 1\right) + \sqrt{2}z^{5/2} - 2\alpha z^2},$$

for $\lambda = \epsilon + 2, \sigma = \frac{3}{2}$.

$$w_{1d} = \frac{2\sqrt{2} - 2\alpha\sqrt{z} - \sqrt{2}z}{2\sqrt{2} - 2\alpha\sqrt{z} + \sqrt{2}z},$$

for $\lambda = \epsilon, \sigma = \frac{3}{2}$.

$$w_{1e} = \frac{2\alpha + \sqrt{2z}}{2\alpha - \sqrt{2z}},$$

for $\lambda = \epsilon, \sigma = \frac{1}{2}$.

$$w_{1f} = -\frac{-4\alpha + \sqrt{2}z^{3/2} + 2\sqrt{2}\left(\alpha^2 - 1\right)\sqrt{z} + 4\alpha z}{4\alpha - 2\sqrt{2}\sqrt{z}\left(\alpha^2 + 2\epsilon - 2\right) + \sqrt{2}z^{3/2}}$$

for $\lambda = \frac{1}{2}, \sigma = \frac{3}{2}$. We just need to remember that $\alpha = [\ln(u)]'$.

A straightforward generalization of the reduction theorem used in the case of the PIV equation requires that $u_2 = (a^-)^2 u_1$, such that $\epsilon_2 = \epsilon_1 - 2$, thus supplying us with ladder operators $l^\pm$ such that

$$l^+ l^- = \left(H_2 - \frac{1}{2}\right)\left(H_2 - \frac{3}{2}\right)(H_2 - \epsilon + 2)(H_2 - \epsilon - 2).$$

Notice that the reduction theorem leaves only one free parameter out of the two original ϵ_1, ϵ_2. Thus we have chosen to set $\epsilon_1 = \epsilon$ and $\epsilon_2 = \epsilon - 2$.

The notation we use for these solutions of the PV equation is the following: w_{2a} if $\lambda = \epsilon - 2, \sigma = \epsilon + 2$, w_{2b} if $\lambda = \epsilon + 2, \sigma = \frac{1}{2}$, w_{2c} if $\lambda = \epsilon + 2, \sigma = \frac{3}{2}$, w_{2d} if $\lambda = \epsilon - 2, \sigma = \frac{3}{2}$, w_{2e} if $\lambda = \epsilon - 2, \sigma = \frac{1}{2}$, and w_{2f} if $\lambda = \frac{1}{2}, \sigma = \frac{3}{2}$.

Explicit expressions for the solutions obtained through 2-SUSY applied to the truncated oscillator are too long to be put in here, but let us give some simple examples.

For the choice $\mathcal{E}_1 = \epsilon - 2$, $\mathcal{E}_2 = \epsilon + 2$, $\mathcal{E}_3 = \frac{1}{2}$, $\mathcal{E}_4 = \frac{3}{2}$, where $\epsilon = \frac{5}{2}$ and $u(x)$ has even parity:

$$w_{2f} = \frac{4(z-1)}{z^2 + 2z - 1}.$$

For the permutation $\mathcal{E}_1 = \epsilon - 2$, $\mathcal{E}_2 = \epsilon + 2$, $\mathcal{E}_3 = \frac{1}{2}$, $\mathcal{E}_4 = \frac{3}{2}$, where $\epsilon = \frac{1}{2}$ and $u(x)$ has odd parity:

$$w_{2f} = \frac{3\sqrt{2}\sqrt{z} - 2(z+3)F\left(\frac{\sqrt{z}}{\sqrt{2}}\right)}{\sqrt{2}\sqrt{z}(z+3) - 2\left(z^2 + 2z + 3\right)F\left(\frac{\sqrt{z}}{\sqrt{2}}\right)}.$$

In the last equation $F(x)$ is the Dawson function, also called Dawson integral.

5.1. *Bäcklund transformations*

In [38] the authors studied the Bäcklund transformations for the fifth Painlevé equation. As presented in these paper, they considered a solution of the PV equation $w = w(z; \alpha, \beta, \gamma, \delta)$ with parameters α, β, γ, $\delta \neq 0$, such that

$$F_1(w) = zw' - k_1 cw^2 + (k_1 c - k_2 a + k_3 hz)w + k_2 a \neq 0, \tag{13}$$

where $c^2 = 2\alpha$, $a^2 = -2\beta$, $h^2 = -2\delta$. Then they found that the transformation

$$T_{k_1,k_2,k_3} : w(z; \alpha, \beta, \gamma, \delta) \to w_1(z; \alpha_1, \beta_1, \gamma_1, \delta_1) = 1 - \frac{2k_3 hzw}{F_1} \tag{14}$$

defines another solution $w_1(z; \alpha_1, \beta_1, \gamma_1, \delta_1)$ of the PV equation with parameters

$$\alpha_1 = -\frac{1}{16\delta}\left[\gamma + k_3 h(1 - k_2 a - k_1 c)\right]^2, \qquad \beta_1 = \frac{1}{16\delta}\left[\gamma - k_3 h(1 - k_2 a - k_1 c)\right]^2,$$
$$\gamma_1 = k_3 h(k_2 a - k_1 c), \qquad \delta_1 = \delta, \tag{15}$$

where $k_i^2 = 1$, $i \in \{1, 2, 3\}$.

Now let us go back to the solutions obtained in the previous subsection. Then for fixed value of ϵ one can show that T_{k_1,k_2,k_3} is a BT for the solutions previously found in the following cases:

- $T_{k_1,k_2,k_3} : w_{1b} \to w_{2a}$ whenever $k_1 = -1$, $k_2 = 1$, $k_3 = 1$ if $\epsilon < -\frac{3}{2}$; or $k_1 = -1$, $k_2 = -1$, $k_3 = 1$ if $-\frac{3}{2} < \epsilon < \frac{3}{2}$.
- $T_{k_1,k_2,k_3} : w_{1b} \to w_{2e}$ whenever $k_1 = -1$, $k_2 = -1$, $k_3 = 1$ if $\epsilon < -\frac{3}{2}$; or $k_1 = -1$, $k_2 = 1$, $k_3 = 1$ if $-\frac{3}{2} < \epsilon < \frac{3}{2}$.
- $T_{k_1,k_2,k_3} : w_{1c} \to w_{2a}$ whenever $k_1 = 1$, $k_2 = -1$, $k_3 = 1$ and $\frac{1}{2} < \epsilon$.
- $T_{k_1,k_2,k_3} : w_{1c} \to w_{2d}$ whenever $k_1 = 1$, $k_2 = 1$, $k_3 = 1$ if $\frac{1}{2} < \epsilon$; $k_1 = -1$, $k_2 = 1$, $k_3 = 1$ if either $-\frac{1}{2} < \epsilon < \frac{1}{2}$ or $\epsilon = \frac{1}{2}$; or $k_1 = -1$, $k_2 = -1$, $k_3 = 1$ if $\epsilon < -\frac{1}{2}$.
- $T_{k_1,k_2,k_3} : w_{1f} \to w_{2d}$ whenever $k_1 = -1$, $k_2 = -1$, $k_3 = 1$ and for all values of ϵ.
- $T_{k_1,k_2,k_3} : w_{1f} \to w_{2e}$ whenever $k_1 = -1$, $k_2 = 1$, $k_3 = 1$ and for all values of ϵ.
- $T_{k_1,k_2,k_3} : w_{2d} \to w_{1c}$ whenever $k_1 = 1$, $k_2 = 1$, $k_3 = -1$ if $\epsilon < -\frac{3}{2}$; or $k_1 = -1$, $k_2 = -1$, $k_3 = -1$ if $\frac{7}{2} < \epsilon$.
- $T_{k_1,k_2,k_3} : w_{2d} \to w_{1f}$ whenever $k_1 = -1$, $k_2 = 1$, $k_3 = -1$ if $\epsilon < -\frac{3}{2}$; $k_1 = 1$, $k_2 = 1$, $k_3 = -1$ if $-\frac{3}{2} < \epsilon$; or $k_1 = 1$, $k_2 = -1$, $k_3 = -1$ if $\frac{7}{2} < \epsilon$.
- $T_{k_1,k_2,k_3} : w_{2e} \to w_{1b}$ whenever $k_1 = 1$, $k_2 = 1$, $k_3 = -1$ if $\epsilon \leq -\frac{1}{2}$; $k_1 = -1$, $k_2 = 1$, $k_3 = -1$ if $-\frac{1}{2} \leq \epsilon < \frac{5}{2}$; or $k_1 = -1$, $k_2 = -1$, $k_3 = -1$ if $\frac{5}{2} < \epsilon$.
- $T_{k_1,k_2,k_3} : w_{2e} \to w_{1f}$ whenever $k_1 = -1$, $k_2 = 1$, $k_3 = -1$ if $\epsilon \leq -\frac{1}{2}$; or $k_1 = 1$, $k_2 = 1$, $k_3 = -1$ if $-\frac{1}{2} \leq \epsilon < \frac{5}{2}$.

6. Conclusions

In this work we implemented a simple and direct procedure to obtain several explicit solutions of the Painlevé IV and V equations. We used the SUSY partners of the harmonic oscillator with an infinite potential barrier at the origin to construct realizations of the polynomial Heisenberg algebras of second and third order, which are connected to the Painlevé IV and V equations, respectively. Such connections established a simple dependence of the particular solutions of both differential equations on the extremal states of the SUSY partners of the truncated oscillator.

Finally, we found some Bäcklund transformations relating the obtained solutions of the Painlevé IV and V equations. This was done in order to connect the results of distinct instances, such as different order of the SUSY transformation involved, or inequivalent permutations of the extremal states used.

Acknowledgments
The authors acknowledge the support of Conacyt (México), Project 152574. VSMS also acknowledges the Conacyt fellowship 243374.

References
[1] Witten E 1981 Nucl. Phys. B 185 513
[2] Witten E 1982 Nucl. Phys. B 202 253
[3] Mielnik B 1984 *J. Math. Phys.* **25** 3387
[4] Cooper F, Khare A and Sukhatme U 1995 *Phys. Rep.* **251** 267
[5] Junker G and Roy P 1998 *Ann. Phys.* **270** 155
[6] Márquez I F, Negro J and Nieto L M 1998 *J. Phys. A: Math. Gen.* **31** 4115
[7] Quesne C and Vansteenkiste N 1999 *Helv. Phys. Acta* **72** 71
[8] Mielnik B and Rosas-Ortiz O 2004 *J. Phys. A: Math. Gen.* **37** 10007
[9] Carballo J M, Fernández D J, Negro J and Nieto L M 2004 *J. Phys. A: Math. Gen.* **37** 10349
[10] Contreras-Astorga A and Fernández D J 2008 *J. Phys. A: Math. Theor.* **41** 475303
[11] Marquette I 2009 *J. Math. Phys.* **50** 095202
[12] Fernández D J 2010 *AIP Conf. Proc.* **1287** 3
[13] Bermúdez D and Fernández D J 2011 *SIGMA* **7** 025
[14] Andrianov A A and Ioffe M V 2012 *J. Phys. A: Math. Theor.* **45** 503001
[15] Gómez-Ullate D, Grandati Y and Milson R 2014 *J. Phys. A: Math. Theor.* **47** 015203
[16] Aoyama H, Sato M and Tanaka T 2001 *Nucl. Phys. B* **619** 105
[17] Fernández D J and Fernández-García N 2005 *AIP Conference Proceedings* **744** 236
[18] Andrianov A A, Ioffe M V and Spiridonov V P 1993 *Phys. Lett. A* **174** 273
[19] Andrianov A A, Ioffe M V, Cannata F and Dedonder J P 1995 *Int. J. Mod. Phys. A* **10** 2683
[20] Bagrov V G and Samsonov B F 1997 *Phys. Part. Nucl.* **28** 374
[21] Fernández D J, Hussin V and Mielnik B 1998 *Phys. Lett. A* **244** 309
[22] Cariñena J F, Ramos A and Fernández D J 2001 Ann. Phys. **292** 42
[23] Mateo J and Negro J 2008 *J. Phys. A: Math. Theor.* **41** 045204
[24] Fernández D J and Hussin V 1999 *J. Phys. A, Math. Gen.* **32** 3603
[25] Fernández A J, Negro J and Nieto L M 2004 *Phys. Lett. A* **324** 139
[26] Veselov A P and Shabat A B 1993 *Funct. Anal. Appl.* **27** 81
[27] Adler V E 1994 *Physica D* **73** 335
[28] Dubov S Y, Eleonskii V M and Kulagin N E 1994 *Chaos* **4** 47
[29] Andrianov A A, Cannata F, Ioffe M and Nishnianidze D 2000 *Phys. Lett. A* **266** 341
[30] Bermúdez D, Fernández D J and Negro J 2014 preprint Cinvestav
[31] Bermúdez D, Fernández D J 2011 *Phys. Lett. A* **375** 2974
[32] Bermúdez D 2012 *SIGMA* **8** 069
[33] Fernández D J and Morales-Salgado V S 2014 J. Phys. A: Math. Theor. **47** 035304
[34] Fernández D J and Morales-Salgado V S 2014 J. Phys: Conf. Ser. **512** 012024
[35] Iwasaki K, Kimura H, Shimomura S and Yoshida M 1991 *From Gauss to Painlevé* (Braunschweig: Friederich Vieweg and Sohn)
[36] Bassom A P, Clarkson P A and Hicks A C 1995 *Stud. Appl. Math.* **95** 1
[37] Ablowitz MJ and Segur H 1981, *Solitons and the Inverse Scattering Transform* SIAM p 154.
[38] Gromak V, Filipuk G 2001 *Math. Model. Anal.* 6 221

Wave propagation in the presence of a dielectric slab: The paraxial approximation

S. Cruz y Cruz, R. Razo

UPIITA, Instituto Politécnico Nacional, Av. Instituto Politécnico Nacional 2580 CP 07340 México D.F., Mexico

E-mail: sgcruzc@ipn.mx

Abstract. Light propagation processes in a homogeneous dielectric slab are considered from the point of view of the optical analog of the Schrödinger equation. The reflection and transmission coefficients and longitudinal shifts of the wavefront are determined by considering the analytical properties of the external field amplitudes as functions of the wave number. An analytic approach to obtain the resonances in the "long lifetime" limit is also presented.

1. Introduction

The analogies between light propagation processes and mechanical systems have been a matter of intense study over time. Many problems of geometrical optics in graded-index media can be easily solved [1] and a Hamiltonian (phase space) description of geometrical optics can be, as well, developed [2] by taking full advantage of the similarities between ray description of light and the Newtonian dynamics. The resemblance extends also to the regime of wave optics and quantum mechanical descriptions due to the formal equivalence between the scalar Helmholtz equation describing the propagation, in space, of light beams in the paraxial regime and the Schrödinger equation governing the time-evolution of a point particle in an interaction potential [3–6]. This equivalence has been used as a tool to simulate and test multiple processes involving quantum mechanical particles, such as Rabi oscillations, population transfer and atom optics [6–8], by means of optical-wave systems. Particular interest has been paid to the theoretical and experimental study of the transmission times in barrier and well potentials [9–11]. In this context, the similarities between the Helmholtz and Schrödinger equations have been convenient to design experiments that allow to measure the negative tunneling times by using optical systems in the microwave regime [12–15]. In the same way, many quantum formalisms have been also adapted for the description of light propagation in free space as well as in the presence of optical devices [3, 5–7, 16–19]. In particular, quantum-optical analogies have been useful to determine the bound modal fields propagating in waveguides and optical fibers [6, 7, 20] by modeling these devices as short range interaction potentials. Yet, it is well known that, under some circumstances, this kind of potentials may host bound as well as scattering and resonant states. Very recently the spectrum properties of particles subject to one dimensional short range potentials have been considered [21–29]. In those works the bound and resonant states, as well as the corresponding reflection and transmission coefficients and delay times can be obtained by inspecting the analytical properties of the reflection and transmission amplitudes as functions of the moment k. In this work we present a description of wave propagation through a dielectric homogeneous slab in the same scheme. Thus, the spectral properties of the system will be stated and some light guiding and scattering processes will be described. In Section 2 the paraxial approximation for linear inhomogeneous media is derived starting from the Maxwell's equations. In Section 3 the scattering of a light beam by a dielectric homogeneous slab is considered. The reflection and transmission coefficients are determined as well as the longitudinal shifts by means of the stationary phase approximation. Section 4 is devoted to determine the bound and resonant modes as the solutions to the paraxial Helmholtz equation, corresponding to complex eigenvalues, fulfilling the purely outgoing wave condition. Some concluding remarks are presented in Section 5.

2. The paraxial wave equation

We depart from the Maxwell equations without sources for wave propagation in the presence of an inhomogeneous dielectric material

$$\nabla\cdot\mathbf{D}(\mathbf{q},t)=0,\quad \nabla\cdot\mathbf{B}(\mathbf{q},t)=0, \tag{1}$$

$$\frac{\partial}{\partial t}\mathbf{D}(\mathbf{q},t)=\nabla\times\mathbf{H}(\mathbf{q},t),\quad -\frac{\partial}{\partial t}\mathbf{B}(\mathbf{q},t)=\nabla\times\mathbf{E}(\mathbf{q},t). \tag{2}$$

For linear media the constitutive relations read

$$\mathbf{D}(\mathbf{q},t)=\epsilon_0\epsilon(\mathbf{q})\mathbf{E}(\mathbf{q},t),\quad \mathbf{B}(\mathbf{q},t)=\mu_0\mathbf{H}(\mathbf{q},t), \tag{3}$$

where $\epsilon(\mathbf{q})$ is the relative permittivity connected to the refractive index through $n^2(\mathbf{q})=\epsilon(\mathbf{q})$. We are assuming that the magnetic polarizability and the dispersive properties of the medium are negligible. For one monochromatic component of the electromagnetic field the corresponding wave equations now contain terms bearing information about the inhomogeneity of the optical medium

$$\nabla^2\mathbf{E}(\mathbf{q})+k_0^2n^2\mathbf{E}(\mathbf{q})-\nabla\ln\epsilon(\mathbf{q})\cdot\mathbf{E}(\mathbf{q})=0, \tag{4}$$

$$\nabla^2\mathbf{H}(\mathbf{q})+k_0^2n^2\mathbf{H}(\mathbf{q})-\nabla\ln\epsilon(\mathbf{q})\times(\nabla\times\mathbf{H}(\mathbf{q}))=0, \tag{5}$$

with $k_0=\omega/c$ the wave number in vacuum. The solution of this problem is, in general, non trivial since the presence of these terms implies a coupling between the components of each field. Though, a simple inspection of equation (4) allows us to observe that it is possible to reduce it to a scalar wave equation for a representative component of the electric field whenever

$$\nabla\epsilon(\mathbf{q})\cdot\mathbf{E}=0. \tag{6}$$

In particular, we are interested in wave packets consisting of superpositions of transversal electric waves traveling in a guiding stratified medium, for which the refractive index depends only on one transversal coordinate, say x. If we choose the $Z-$axis as the optical axis, then the incident plane corresponds to the $XZ-$plane, meaning that the transversal electric field must have the form $\mathbf{E}(\mathbf{q})=E(\mathbf{q})\mathbf{e}_y$, with $\mathbf{e}_y$ a unit vector in the $Y-$direction. In this case, the condition (6) is fulfilled automatically. The Maxwell equations (1-2) lead to

$$E=E(x,z),\quad H_x(x,z)=\frac{i}{\mu_0\omega}\frac{\partial}{\partial z}E(x,z),\quad H_y(x,z)=0,\quad H_z(x,z)=-\frac{i}{\mu_0\omega}\frac{\partial}{\partial x}E(x,z) \tag{7}$$

with $E(x,z)$ the electric field amplitude that must be determined by solving the scalar Helmholtz equation

$$\frac{\partial^2}{\partial x^2}E(x,z)+\frac{\partial^2}{\partial z^2}E(x,z)+k_0^2n^2(x)E(x,z)=0. \tag{8}$$

In the paraxial regime, we expect a nearly harmonic behavior in the longitudinal coordinate. Then, we may write

$$E(x,z) = F(x,z)\,\mathrm{e}^{ik_0 n_0 z} \tag{9}$$

where $F(x,z)$ must be a function almost constant in z and n_0 is a reference refractive index. Then, it is natural to make the approximation [4–6, 30]

$$\left|\frac{\partial^2 F}{\partial z^2}\right| \ll k_0 \left|\frac{\partial F}{\partial z}\right|, \tag{10}$$

in which case equation (8) is reduced to the time-dependent Schrödinger-type equation

$$-\frac{1}{2k_0^2 n_0}\frac{\partial^2 F}{\partial x^2} - \frac{n^2 - n_0^2}{2n_0}F = \frac{i}{k_0}\frac{\partial F}{\partial z}. \tag{11}$$

It is well known that the paraxial approximation (10) is valid in the weakly guiding regime, where the refractive index is nearly constant [6], *i.e.*, $n = n_0 + \Delta n, \quad \Delta n \ll 1, \quad (\Delta n)^2 \approx 0$. Thus, to first order in Δn

$$\frac{n^2 - n_0^2}{2n_0} \approx \Delta n = n - n_0. \tag{12}$$

Furthermore, since n does not depend on the longitudinal coordinate, the amplitude F can be written in the form

$$F(x,z) = \mathrm{e}^{-i\beta k_0 z}\,\psi(x), \tag{13}$$

where $\psi(x)$ fulfills the stationary Schrödirnger-type equation

$$-\frac{1}{2k_0^2 n_0}\frac{d^2\psi}{dx^2} - n\psi = \varepsilon\psi, \quad \varepsilon = \beta - n_0, \tag{14}$$

known as the paraxial Helmholtz equation, for a point particle of mass n_0 exposed to an interaction potential $-n(x)$. Observe that the inverse of the wave number takes the place of the Planck's constant, implying that the "semiclassical limit", which in the electromagnetic case correspond to the ray description of light, appears for very small wavelengths.

Introducing the operators [3]

$$\hat{x} = x\cdot, \quad \hat{p}_x = -\frac{i}{k_0}\frac{\partial}{\partial x}, \quad \hat{p}_z = -\frac{i}{k_0}\frac{\partial}{\partial z}, \quad \hat{H} = -\frac{1}{2n_0}\hat{p}_x^2 - n(\hat{x}), \tag{15}$$

we are led to the eigenvalue problem

$$\hat{H}\mathbf{E} = -\hat{p}_z\mathbf{E} = \varepsilon\mathbf{E}. \tag{16}$$

The parameter ε is then related to the $z-$component of the generalized momentum $\mathbf{p}$ (see, *e.g.* [2, 3]) defining the slope of the light beam

$$\varepsilon = -p_z = -n(x)\cos\theta(x) \tag{17}$$

with θ the angle between the propagation direction and the optical axis. The eigenvalue equation (14) then defines the modal fields propagating inside and through the optical medium as the eigenstates of $\hat{H}$ (electromagnetic modes with well defined values of θ). It is worthwhile to note that, depending on the profile $n(x)$, the spectrum may include a set of isolated points (bound modes) and a continuous interval (scattering modes) with, possibly, an internal discrete structure (resonant modes) [21], in a complete analogy to the Schrödinger problem.

3. Wave propagation through a dielectric homogeneous slab: the scattering modes

In order to present some of the consequences of the analogies between the Schrödinger and the paraxial Helmoltz equations we will consider the simple case of electromagnetic propagation through a finite, dielectric homogeneous material in this framework. This problem is equivalent to the square well potential in Quantum Mechanics which has been studied in diverse contexts and has been used as a basis of models of a number of interactions (see *e.g.* [31–33]). Let us rewrite (14) in the form

$$-\frac{1}{2k_0^2 n_0}\frac{d^2\psi}{dx^2} + (1-n)\psi = (\varepsilon+1)\psi. \tag{18}$$

We are interested on wave propagation through a finite medium characterized by a real, constant refractive index n extending on the x-axis from $x = -a$ to $x = a$ tagged as region II (see Figure 1). Since n_0 is the reference refractive index characteristic of the optical

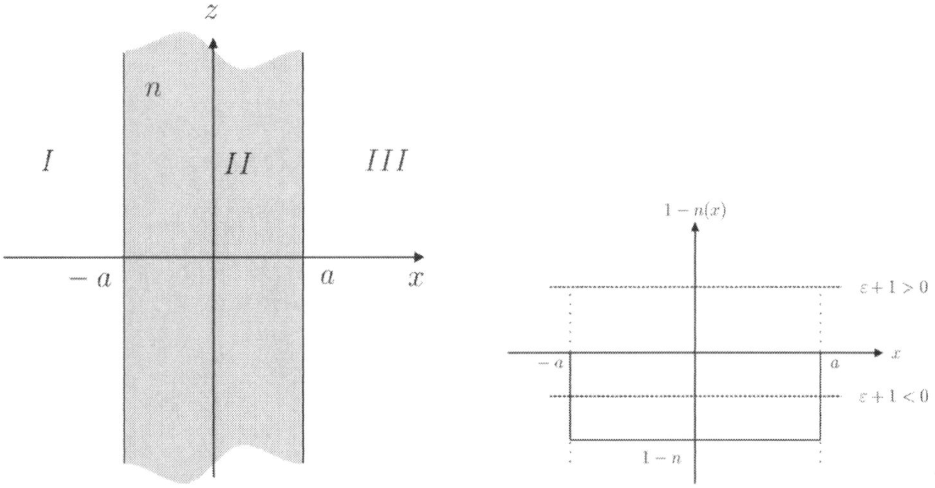

Figure 1. Left: Finite medium with constant refractive index n of width $2a$. Right: The system is analogous to a particle in a square well potential in Quantum Mechanics.

medium, equation (14) is valid within each region. However, the boundary conditions of the electromagnetic field state that the tangential components of **E** and **H** must be continuous at any interface separating two regions. This indicates that $E(x,z)$ and $H_z(x,z) \propto \partial E(x,z)/\partial x$, and thus the function ψ and its first derivative, must be continuous at $x = \pm a$. Then, for the internal and external fields we have respectively the equations

$$-\frac{1}{2k_0^2}\frac{d^2\psi_i}{dx^2} = (\varepsilon+1)\psi_i, \quad i = I, III \tag{19}$$

$$-\frac{1}{2k_0^2 n_0}\frac{d^2\psi_{II}}{dx^2} + (1-n)\psi_{II} = (\varepsilon+1)\psi_{II}. \tag{20}$$

The electric field amplitude has the form

$$\psi(x) = \begin{cases} I\,\mathrm{e}^{ikx} + A_1(k)\,\mathrm{e}^{-ikx} & x < -a \\ \\ A_2(k)\cos qx + B_2(k)\sin qx & -a \le x \le a \\ \\ A_3(k)\,\mathrm{e}^{-ikx} + B_3(k)\,\mathrm{e}^{ikx} & x > a \end{cases} \tag{21}$$

Journal of Physics: Conference Series **624** (2015) 012018 doi:10.1088/1742-6596/624/1/012018

with

$$k^2 = 2k_0^2(\varepsilon+1), \qquad q^2 = 2k_0^2 n(\varepsilon+n), \tag{22}$$

and A_1, A_2, B_2, A_3, B_3 – functions of the wave number k to be determined by applying the proper set of boundary conditions.

3.1. The reflection and transmission coefficients

For positive values of the parameter $\varepsilon+1$, we have real wave numbers $k = k_0\sqrt{2(\varepsilon+1)}$. The external field amplitude ψ_I and ψ_{III} remain finite as $|x| \to \infty$. In those regions $\varepsilon = -p_z = -\cos\theta$, meaning that ε ranges in the interval $[-1,0]$. Solution (21) describes the scattering of plane waves by the slab. If we assume that there exists a source of electromagnetic radiation at the left of the optical medium pumping waves into the interaction zone, then $I=1$ stand for the incident amplitude, A_1 and B_3 for the reflection and transmission ones, and the coefficient A_3 is set to zero as we expect no plane waves propagating to the left in region III. Imposing the continuity conditions to ψ and ψ' we find

$$A_1(k) = i\frac{q^2-k^2}{2\Delta(k)}\,\mathrm{e}^{-2ika}\sin 2qa, \qquad B_3(k) = \frac{kq}{\Delta(k)}\,\mathrm{e}^{-2ika}, \tag{23}$$

$$A_2(k) = \frac{k}{\Delta(k)}\,\mathrm{e}^{-ika}\left(q\cos qa - ik\sin qa\right), \quad B_2(k) = \frac{k}{\Delta(k)}\,\mathrm{e}^{-ika}\left(k\cos qa - iq\sin qa\right), \tag{24}$$

$$\Delta(k) = \left(k\cos qa - iq\sin qa\right)\left(q\cos qa - ik\sin qa\right). \tag{25}$$

By writing the coefficients A_1 and B_3 in polar form we can obtain information about the reflection and transmission coefficients and longitudinal shifts of the scattered wave packet [9–11,15,26–28]. Thus we write

$$A_1(k) = r(k)\,\mathrm{e}^{i(\phi(k)-2ka)}, \qquad B_3(k) = t(k)\,\mathrm{e}^{i(\phi(k)-2ka+\pi/2)}$$

with

$$r(k) = \frac{\left(k^2-q^2\right)\sin 2qa}{\left[\left(k^2+q^2\right)^2\sin^2 2qa + 4k^2q^2\cos^2 2qa\right]^{1/2}}, \quad t(k) = \frac{2kq}{\left[\left(k^2+q^2\right)^2\sin^2 2qa + 4k^2q^2\cos^2 2qa\right]^{1/2}}, \tag{26}$$

$$\phi(k) = -\arctan\frac{2kq\cos 2qa}{(k^2+q^2)\sin 2qa}. \tag{27}$$

The reflection and transmission coefficients are now

$$R = r^2(k) = \frac{2\left(\frac{k-q}{k+q}\right)^2(1-\cos 4qa)}{1+\left(\frac{k-q}{k+q}\right)^4 + 2\left(\frac{k-q}{k+q}\right)^2\cos 4qa}, \quad T = t^2(k) = \frac{\left(\frac{2k}{k+q}\right)^2\left(\frac{2q}{k+q}\right)^2}{1+\left(\frac{k-q}{k+q}\right)^4 + 2\left(\frac{k-q}{k+q}\right)^2\cos 4qa}. \tag{28}$$

Let θ_i, $i=1,2,3$ be the angle between the propagation direction and the optical axis in the regions I, II, III respectively. Then

$$k = k_0\sqrt{2(\varepsilon+1)} \approx k_0\theta_1, \qquad q = k_0\sqrt{2n(\varepsilon+n)} \approx k_0 n\theta_2.$$

This allows us to write

$$\frac{k-q}{k+q} = \frac{\theta_1 - n\theta_2}{\theta_1 + n\theta_2} = r_{12} = -r_{23}$$

$$\frac{2k}{k+q} = \frac{2\theta_1}{\theta_1+n\theta_2} = t_{12}, \qquad \frac{2q}{k+q} = \frac{2n\theta_2}{n\theta_2+\theta_3} = t_{23},$$

where r_{ij}, t_{ij} are the Fresnel reflection and transmission coefficients, in the paraxial approximation, from region i to region j respectively [34]. Therefore we may write the reflection and transmission coefficients as

$$R = \frac{2r_{12}^2(1-\cos 2\beta\theta_2)}{1+r_{12}^4-2r_{12}^2\cos 2\beta\theta_2}, \qquad T = \frac{t_{12}^2 t_{23}^2}{1+r_{12}^4-2r_{12}^2\cos 2\beta\theta_2}, \tag{29}$$

with $\beta = 2k_0 na$, which coincides with that reported for a dielectric film (see [34]). In Figure 2 we present the transmission coefficient for a homogeneous slab of refractive index $n = 1.3$. Observe the series of peaks typical of the behaviour of the transmission coefficient for particles in a short range interaction potential [21, 23]. In this case, however, the number of peaks does not only depend on the width and the refractive index of the slab, but also on the wavelength of the electromagnetic field. Figure 2 shows the transmission coefficient for (a) $k_0 a = 50$ and (b) $k_0 a = 10^5$. For a dielectric plate of $1cm$ these values correspond to wavelengths of $1256\mu m$ and $628nm$ respectively.

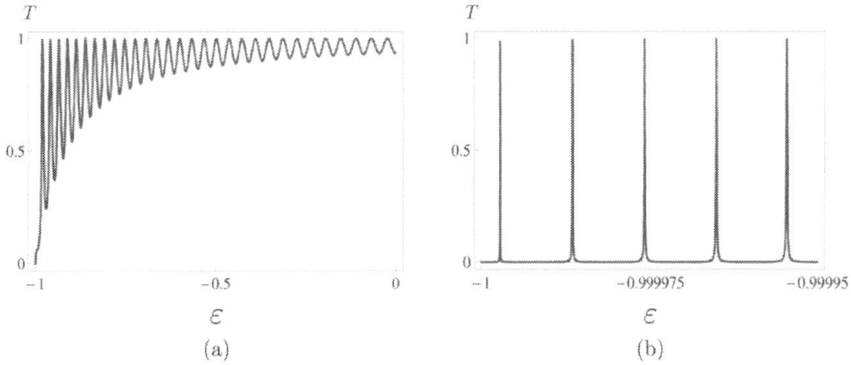

Figure 2. Transmission coefficient for a dielectric slab as a function of the eigenvalue $\varepsilon = -p_z$ for (a) $k_0 a = 50$ and (b) $k_0 a = 10^5$, in both cases $n = 1.3$.

3.2. The longitudinal shifts

The phases of the coefficients A_1 and B_3 give information about the longitudinal shift of a wave packet under scattering by the optical medium, this is equivalent to evaluate the reflection and transmission phase times for a particle in a rectangular well potential [9–11]. Due to the symmetry of the system this shift turns out to be the same under reflection and transmission. Consider the wave-packet scattering solution for the external electric field

$$E_I(x,z) = \int_0^\kappa dk f(k)\, e^{i(kx-k_0\varepsilon(k)z)} + \int_0^\kappa dk f(k) r(k)\, e^{i(-kx-\varepsilon(k)k_0 z-2ka+\phi(k))}, \tag{30}$$

$$E_{III}(x,z) = \int_0^\kappa dk f(k) t(k)\, e^{i(kx-\varepsilon(k)k_0 z-2ka+\phi(k)+\pi/2)}, \tag{31}$$

where $f(k)$ is the Fourier coefficient and κ is defined by $\varepsilon(\kappa) = 0$. At any point z, a particular component will contribute appreciable to the wave packet in the neighbourhood of the point x for which the phase of the integrand is stationary [9, 26–28]. This condition for the incident and the transmitted wave packet read, respectively, as

$$\frac{\partial}{\partial k}\left(kx - \frac{k^2}{2k_0}z + k_0 z\right) = 0, \quad \frac{\partial}{\partial k}\left(kx - 2ka - \frac{k^2}{2k_0}z + k_0 z + \phi(k) + \frac{\pi}{2}\right) = 0, \tag{32}$$

leading to the equations of motion for the corresponding Fourier components

$$x = \frac{k}{k_0} z, \quad x = 2a + \frac{k}{k_0} z + \frac{\partial \phi}{\partial k}. \tag{33}$$

The point z_{in} at which an incident plane wave of wave number k would reach the left wall of the optical medium is then $z_{in} = -\frac{k_0}{k} a$. In the same way the point z_t at which the reflected or transmitted wave leaves the optical medium is

$$z_t = \frac{k_0}{k} \left(-a + \frac{\partial \phi}{\partial k} \right). \tag{34}$$

Then the longitudinal shift of the reflected or transmitted wave is readily written

$$\delta z = z_t - z_{in} = \frac{k_0}{k} \frac{\partial \phi}{\partial k}. \tag{35}$$

Figure 3 shows the longitudinal shift of a Fourier component under reflection or transmission

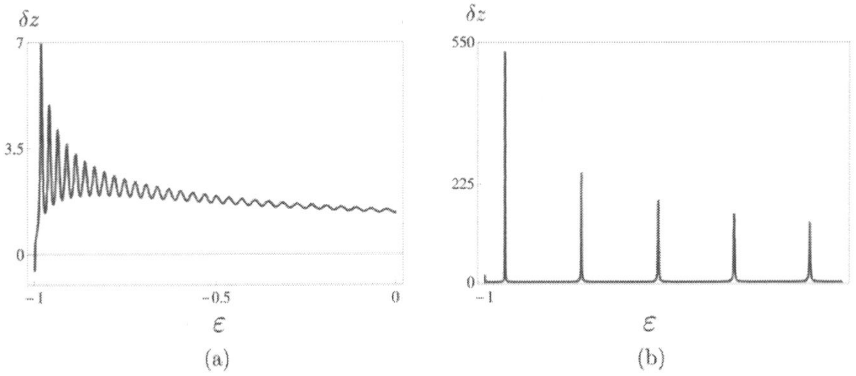

Figure 3. Reflection/transmission longitudinal shift of a Fourier component for a dielectric slab as a function of the eigenvalue $\varepsilon = -p_z$ for the product (a) $k_0 a = 50$, and (b) $k_0 a = 10^5$. In both cases $n = 1.3$.

as a function of ε for (a) $k_0 a = 50$ and for (b) $k_0 a = 10^5$. It is worthwhile to point out that the peaks of this shift coincide with the peaks of the corresponding transmission coefficient. This implies that the corresponding electromagnetic modes have maximum transmission probabilities but they travel a maximum longitudinal distance in the interaction zone before they are finally transmitted. Figure 4(a) shows the longitudinal shift as a function of $k_0 a$ for an eigenvalue $\varepsilon = -0.999$ and $n = 1.3$. As in the case of the transmission time in quantum mechanics, we can see that for some values of ε the transmission shift δz may take negative values. This negative shifts in optical systems have been already predicted for barriers as well as for rectangular wells [10, 11]. The analogy between the Schrödinger and the paraxial Helmholtz equations suggest that this phenomenon may be used to measure the negative phase time in the scattering of particles by a potential well [15]. Figure 4(b) shows the longitudinal shift surface as a function of ε and $k_0 a$ for $n = 1.3$.

4. Wave propagation in a dielectric homogeneous slab: bound and resonant states

4.1. The purely outgoing wave condition

It is also interesting to consider the solutions to the paraxial Helmholtz equation that fulfill the purely outgoing wave condition [21, 22]

$$\lim_{x \to \pm\infty} \left(\psi'(x) \mp ik\psi(x) \right) = 0, \quad k^2 = 2k_0^2(\varepsilon + 1), \tag{36}$$

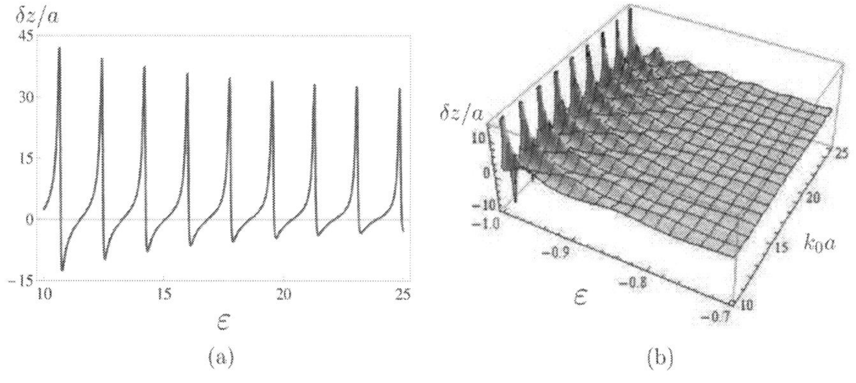

Figure 4. (a) Reflection/transmission longitudinal shift of a Fourier component for a dielectric slab as a function of k_0a with $\varepsilon = -0.999$ and $n = 1.3$. (b) Reflection/transmission longitudinal shift surface as a function of ε and k_0a with $n = 1.3$.

for ε a complex number. In this way $\varepsilon = \varepsilon_R - i\frac{\Gamma}{2}$ leading also to complex wave numbers $k = k_R + ik_I$. It is clear that the real and imaginary parts of k must obey

$$k_R^2 - k_I^2 = 2k_0^2(\varepsilon_R + 1), \quad k_R k_I = -k_0^2 \frac{\Gamma}{2}. \tag{37}$$

By applying the purely outgoing wave condition we conclude that the first terms in solution (21) for the external field must be negligible with respect to the second ones. The electric energy density $\rho_e(x,z) = \frac{\epsilon_0}{2}|E(x,z)|^2$ in regions I and III satisfies the continuity equation

$$\frac{\partial \rho_e}{\partial z} + \frac{\partial j_e}{\partial z} = 0, \quad j_e = \frac{i\epsilon_0}{4k_0}\left[\frac{\partial E^*}{\partial x}E - E^*\frac{\partial E}{\partial x}\right], \tag{38}$$

stating that the electric energy is preserved as the electromagnetic field propagates along the optical axis. The energy current density $j_e^{\mp}$ in regions I and III respectively can be written as

$$j_e^{\pm} = \pm v\rho_e, \qquad v = \frac{k_R}{k_0} \tag{39}$$

with v the flux velocity.

4.2. Bound and resonant modes

Suppose, in the first place, that ε is real. For $\varepsilon < -1$, $k^2 = 2k_0^2(\varepsilon + 1) < 0$ leading to purely imaginary values of k

$$k_{\pm} = \pm i\kappa, \quad \kappa = \sqrt{-2(\varepsilon + 1)}. \tag{40}$$

In this case $j_e^{\pm} = 0$, and the external electric field amplitudes are

$$\psi_I^{\pm}(x) = A_1^{\pm} e^{\pm k_0\kappa x}, \qquad \psi_{III}^{\pm} = B_3^{\pm} e^{\mp k_0 \kappa x}. \tag{41}$$

Observe that the solutions associated to k_+ are bounded in the external region and correspond precisely to the external vanishing fields associated to the bound modes propagating inside the slab. Therefore, even though the electric field does not disappear outside the dielectric medium, the electric energy density flux in the transversal direction is null, meaning that all the energy propagates in the direction of the optical axis. The modal fields in this case correspond to those described in (21) with the coefficients A_1, A_2, B_2, B_3 given by (23-25). The purely outgoing condition is fulfilled by demanding that $\Delta = 0$. Then, we may consider that $I \propto \Delta$ and then set $I = 0$. The eigenvalue ε is discretised, and thus, only a discrete number of modes will be able to propagate through the slab without losses. This case corresponds to the phenomenon of guided light in a waveguide by total internal reflection [35]. Since ε is related to the value of the $z-$component of the linear momentum $\mathbf{p}$, we conclude that, in order to have bound modes inside the slab, the angle between the beam and the optical axis must be discrete. The electric field modes corresponding to these eigenvalues ε_m are square integrable, showing that the transverse energy is bounded in a finite region around the optical axis. Different from the quantum mechanical case, in which the number of bound states only depends on the geometry of the potential well, in this case, the number of bound modes propagating along the slab depends on the profile of the optical medium, as well as the wavelength of the electromagnetic field.

The solutions associated to k_- are non physical since in this case the amplitude of the electric field increases exponentially as $|x| \to \infty$ (see (41)).

Consider now the case for $\varepsilon > -1$. In this case $k^2 = 2k_0^2(\varepsilon + 1) > 0$ and k is real

$$k_\pm = \pm k_0 \kappa, \qquad \kappa = \sqrt{2(\varepsilon + 1)}. \tag{42}$$

Now we have $k_I = 0$ and $k_R = k_\pm = \pm k_0 \kappa$. The energy current densities outside the interaction region are

$$j_e^\pm = \pm k_0 \kappa \rho_e, \quad \text{for} \quad k_+, \qquad j_e^\pm = \mp k_0 \kappa \rho_e, \quad \text{for} \quad k_-. \tag{43}$$

The corresponding solutions are not square integrable. There must be a net flux of electromagnetic radiation outwards the dielectric in the transversal direction. This case is discarded since we are considering no light sources or sinks inside the slab.

Finally, let us regard complex eigenvalues $\varepsilon = \varepsilon_R - i\frac{\Gamma}{2}$. Then $k = k_R + ik_I$ and

$$\varepsilon_R = \frac{k_R^2 - k_I^2}{2k_0^2}, \qquad -\frac{\Gamma}{2} = \frac{k_R k_I}{k_0^2}. \tag{44}$$

The energy density flux in this case is non trivial since $k_R \neq 0$. This may correspond to an incident wave which enters into resonance with the optical medium. The wave is in this way captured inside the interaction region traveling there through a finite distance until it is finally emitted. The result is a reflected wave propagating to the left in region I, and a transmitted wave propagating to the right in region III. The correct sign for k_R is thus positive. The expression for the electric energy density fulfills

$$\lim_{x \to \pm\infty} \rho_e \propto \mathrm{e}^{-k_0 \Gamma} \left| \psi(x) \right|^2. \tag{45}$$

From this expression it is clear that the energy density is damped as z grows whenever $\Gamma > 0$, and, as the sign of k_R is already positive, from (44) we conclude that $k_I < 0$. The external electric fields are then plane waves of increasing amplitudes. These modes are called resonances and correspond to wave numbers $k = k_R + ik_I$ in the fourth quadrant of the complex plane. The factor $1/k_0\Gamma$ is the mean distance traveled by the beam, along the optical axis, before its intensity decays by a factor of $1/e$ of its initial intensity. In this context Γ can be interpreted as a measure of the transitory confinement of the field mode inside the medium. In the limit $\Gamma \to 0$ we will have a quasi-stationary mode propagating inside the slab almost without losses. In this limit, it is possible to establish an analytic approach to calculate the resonances and we may write

$$\frac{\Gamma/2}{\Delta \varepsilon_R} \ll 1, \tag{46}$$

where $\Delta\varepsilon_R$ is the separation between two consecutive resonances. We will also assume that $\Delta\varepsilon_R < \varepsilon_R + n$, with ε_R any resonance. Additionally to (37) we have

$$q_R^2 - q_I^2 = 2k_0^2 n(n + \varepsilon_R), \quad q_R q_I = -k_0^2 n \frac{\Gamma}{2}, \tag{47}$$

that allows us to derive an expression for q_I^{-2} [21], namely

$$q_I^{-2} = \frac{\varepsilon_R + n}{k_0^2 n(\Gamma/2)^2}\left[1 \pm \sqrt{1 + \left(\frac{\Gamma/2}{\varepsilon_R + n}\right)^2}\right]. \tag{48}$$

Since condition (46) implies that $\Gamma/2 \ll \varepsilon_R + n$ this last expression leads us to (keeping only the positive root in order to get real values of q_I)

$$q_I \approx \frac{k_0^2 n\,(\Gamma/2)^2}{2\,(\varepsilon_R + n)}. \tag{49}$$

Combining this equation with (47) we obtain

$$q_R^2 \approx 2k_0^2 n\,(\varepsilon_R + n)\,, \quad q_I^2 a^2 \ll \theta^2, \quad \theta^2 = 2k_0 n(n-1)a^2. \tag{50}$$

Additionally, from (37) and (47) we may write approximate expressions for k_R and k_I in terms of q_R and q_I valid for $\theta \gg 1$

$$k_R \approx \sqrt{\frac{q_R^2 - 2k_0^2 n(n-1)}{n}}, \quad k_I = \frac{q_R q_I}{\sqrt{n\left(q_R^2 - 2k_0^2 n(n-1)\right)}}. \tag{51}$$

In order to fulfill the condition $\Delta = 0$ we take the left factor of (25) equal to zero. This leads to

$$q_I \approx -\frac{\sqrt{\frac{1}{n}\left(q_R^2 + \frac{\theta^2}{a^2}\right)}\cos q_R a}{\sin q_R a + q_R a \cos q_R a}, \quad \frac{a}{q_R}\left[(n-1)q_R^2 + \frac{\theta^2}{a^2}\right] \approx -n\tan q_R a - \cot q_R a. \tag{52}$$

The values of q_R that fulfill (52) in the limit $\theta \to \infty$ are given by

$$q_R \approx m\frac{\pi}{2a} \tag{53}$$

with m an integer number. Substituting (53) into (50) we obtain a set of discrete values for the resonance eigenvalue ε_R

$$\varepsilon_R \approx \frac{1}{2k_0^2 n}\left(\frac{m\pi}{2a}\right)^2 - n. \tag{54}$$

However, since $-1 < \varepsilon_R < 0$, the number m must be bounded by some limiting values. In fact

$$\frac{2a}{\pi}k_0\sqrt{2n(n-1)} < m < \frac{2a}{\pi}k_0\sqrt{2n^2}. \tag{55}$$

Thus, for each combination of the parameters (a, n, k_0), there is a finite number of resonances given by (54). The value of Γ is now determined by using (53) to obtain the values of q_I from (52) and substituting it into (51) to find $k_I \approx -\frac{1}{na}$. This result, together with (37) gives the value of Γ

$$\frac{\Gamma}{2} \approx \frac{1}{k_0 n a}\sqrt{2\,(\varepsilon_R + 1)}. \tag{56}$$

Similar results are obtained by setting to zero the second factor of (25).

5. Concluding remarks

We have described some propagation processes of electromagnetic waves in a homogeneous slab. In the scattering regime, the reflection and transmission coefficients were determined, which coincide with those previously reported for a dielectric film [34]. The corresponding longitudinal shifts were also obtained in the context of phase reflection and transmission time for a square well potential. It was shown that for some set of the parameters (a, n, k_0) the longitudinal shifts are negative, fact which is associated to negative group velocities of the wave packet [15]. The modal fields corresponding to guided light and resonances were presented as the eigenstates of the paraxial Helmholtz operator, corresponding to complex eigenvalues, satisfying the purely outgoing wave boundary condition. In some limiting cases this procedure allows to state analytic approaches to determine the bound modes and resonance eigenvalues. Light propagation through more general optical devices such as graded waveguides, optical fiber and multiple layered media can be studied in the same fashion. Some results on the matter are in progress.

References

[1] Evans J, Rosenquist M 1986 "$F = ma$" optics *Am. J. Phys.* **54** 876-883
[2] Wolf K B 2004 *Geometric optics of phase space* (Berlin: Srpinger-Verlag)
[3] Gloge D, Marcuse D 1969 Formal quantum theory of light rays *J. Opt. Soc. Am.* **59** 1629-1631
[4] Lax M, Louisell W H, McKnight W B 1975 From Maxwell to paraxial wave optics *Phys. Rev. A* **11** 1365-1370
[5] Deutsch I H, Garrison J C 1991 Paraxial quantum propagation *Phys. Rev. A* **43** 2498-2513
[6] Marte M A M, Stenholm S 1997 Paraxial light and atom optics: The optical Schrödinger equation and beyond *Phys. Rev. A* **36** 2940-2953
[7] Lamprecht C, Marte M A M, 1998 Comparing paraxial atom and light optics *Quantum Semiclass. Opt.* **10** 501-518
[8] Longhi S 2009 Quantum-optical analogies using photonic structures *Laser Photon Rev.* **3** 243-261
[9] Hartman T E 1963 Tunneling of a wave packet *J. Appl. Phys.* **33** 3427-3433
[10] Steinberg A M, Chiao R 1994 Tunneling delay times in one and two dimensions *Phys. Rev. A* **49** 3283-3295
[11] Li C-F, Wang Q 2000 Negative phase time for particles passing throug a potential well *Phys. Lett. A* **275** 287-291
[12] Ranfagni A, Mugnai D, Fabeni P, Pazzi G P 1990 Delay-time measurements in narrowed waveguides as a test of tunneling *Appl. Phys. Lett.* **58** 774-776
[13] Enders A, Nimtz G 1992 On superluminal barrier traversal *J. Phys. I France* **2** 1693-1698
[14] Chiao R Y, Steinberg A M 1998 Quantum optical studies of tunneling and other superluminal phenomena *Phys. Scripta* **T76** 61-66
[15] Vetter R-M, Haibel A, Nimtz G (2000) Negative time for scattering quantum wells: A microwave analogy experiment *Phys. Rev. E* **63** 046701 5 pages
[16] Stoler D 1981 Operator methods in physical optics *J. Opt. Soc. Am.* **71** 334-341
[17] van Enk S J, Nienhius G 1992 Eigenfunction description of laser beams and orbital angular momentum of light *Opt. Commun.* **94** 147-158
[18] Nienhius G, Allen L 1993 Paraxial wave optics and harmonic oscillatos *Phys. Rev. A* **48** 656-665
[19] Dartora C A et al 2011 Lagriangian-Hamiltonian formulation of paraxial optics and applications: Study of gauge symmetries and the optical spin Hall effect *Phys. Rev. A* **83** 012110 9 pages
[20] Causa F, Sarma J 2000 A versatile method for analyzing paraxial optical propagation in dielectric structures *J. Lightwave Tech.* **18** 1445-1452
[21] Fernández-García N, Rosas-Ortiz O 2008 Gamow-Siegert functions and Darboux-deformed short range potentials *Ann. Phys.* **323** 1397-1414
[22] Fernández-García N, Rosas-Ortiz O 2008 Optical potentials using resonance states in supersymmetric quantum mechanics *J. Phys. Conf. Ser.* **128** 012044 14 pages
[23] Rosas-Ortiz O, Fernández-García N, Cruz y Cruz S (2008) A primer on resonances in quantum mechanics *AIP CP* **1077** 31-57
[24] Fernández-García N, Rosas-Ortiz O 2011 Rectangular potentials in a semi-harmonic background: spectrum, resonances and dwell time *SIGMA* **7** 004 17 pages
[25] Rosas-Ortiz O, Cruz y Cruz S, Fernández-García N 2012 Time delay in reflection of particles by semi-harmonic wells *J. Phys. Conf. Ser.* **380** 012018 8 pages
[26] Rosas-Ortiz O, Cruz y Cruz S, Fernández-Gracía N (2012) Time delay in the reflection of particles by semiharmonic wells *J. Phys. Conf. Ser.* **380** 012018 8 pages

[27] Jaimes A 2012 Sobre el retraso temporal en la dispersión de un paquete de ondas por una barrera semiarmónica, *MSc. Thesis* Cinvestav (In Spanish)
[28] Rosas-Ortiz O, Cruz y Cruz S, Fernández-García N (2013) Negative delay for wave reflection from one-dimensional semiharmonic well in *Geometric Methods in Physics Trends in Mathematics* 2275-281
[29] Orozco Cortés L F, Fernández-García N 2014 Resonances in piecewise potentials and Supersymmetric Quantum Mechanics (SUSY-QM) for the construction of optical potentials *J. Phys. Conf. Ser.* **512** 012027 9 pages
[30] Chang D C, Kuester E K 1981 A hybrid method for paraxial beam propagation in multimode optical waveguides *IEEE Trans. Microwave Theor. Tech.* **MTT-29** 923-933
[31] Dolmatov V K, King J L, Oglesby J C 2012 Diffuse versus square well potentials in modelling $A@C_{60}$ atoms *J. Phys. B: At. Mol. Opt. Phys.* **45** 105102
[32] Levine B F 1993 Quantum-well infrared photodetectors *J. Appl. Phys.* **74** R1-R81
[33] Harrison P 2009 *Quantum wells, wires and dots* (Great Britain: Wiley)
[34] Born M and Wolf E 2002 *Principles of optics* (Cambridge: Cambridge University Press)
[35] Snyder A W, Love T D 1983 *Optical waveguide theory* (London: Chapman and Hall)

Hubbard operators in multiqubit systems

Marco Enríquez[1,2]
[1]Departamento de Física, Cinvestav, AP 14-740, 07000 México DF, Mexico
[2]SEPI-UPIITA, Instituto Politécnico Nacional, Av. IPN No. 2580, Col. La Laguna Ticomán, C.P. 07340 México DF, Mexico
E-mail: menriquezf@fis.cinvestav.mx

To Professor Bogdan Mielnik, the owner of Bar Quantum

Abstract. The properties of the Kronecker product of 2×2 matrices are reviewed in terms of Hubbard operators. This framework constitutes a shorthand notation to deal with the tensor algebra of operators acting on multiqubit states. As an application we derive some analytical expressions related to the geometric measure of entanglement of a certain class of multiqubit invariant permutation states. Our results can be straightaway extended for systems of larger dimensions.

1. Introduction

A system made up of several components is called multipartite [1, 2]. Of particular interest are the quantum systems integrated by several *qubits* because of their promising applications in quantum computing, quantum information, quantum cryptography, etc [3]. The Kronecker product (also known as tensor or direct product) plays an important role in the mathematical description of such systems. However, as the number of qubits increases some calculations involving the tensor product become tedious. For instance, consider the following operator acting on a Hilbert space of k qubits

$$\sigma_3^{\otimes k} = \underbrace{\sigma_3 \otimes \cdots \otimes \sigma_3}_{k\text{–times}}, \qquad \sigma_3 = \begin{pmatrix} 1 & 0 \\ 0 & -1 \end{pmatrix}. \tag{1}$$

As a matrix, $\sigma_3^{\otimes k}$ is clearly diagonal with entries equal to either 1 or -1. A problem arises when one wants to determine the sign of a given element if an explicit calculation should be performed. In a recent work [4] we show that the Hubbard operators (also referred as X-operators) can be used as an accurate tool to deal with the Kronecker product of arbitrary square matrices $A, B, C, \ldots$ In fact, this approach allows to find the element (i, j) of the operator $A \otimes B \otimes C \otimes \cdots$ by using simple relations of subscripts. The relevance of such simplification is clear by considering for instance that the multiple tensor product of operators lies on the basis of the quantum entanglement. Indeed, it is required for a given entanglement measure to be invariant under the action of local unitary operators. That is to say, the n-partite state $|\psi\rangle$ must have the same entanglement as the state $|\widetilde{\psi}\rangle = U_{\text{loc}}|\psi\rangle$, where the operator U_{loc} is the tensor product of n unitary operators. However, there is not a unique form to quantify

Journal of Physics: Conference Series **624** (2015) 012019 doi:10.1088/1742-6596/624/1/012019

entanglement, even in the simplest case of bipartite systems. For two qubits the concurrence introduced by Hill and Wootters is a widely used option [5,6]. This is the case for example, in the analysis of the entanglement of a pair of qubits in quantized radiation fields [7]. Nevertheless, the characterization of entanglement in multipartite systems is a tough problem in contemporary physics [8,9]. An interesting proposal is the *geometric measure of entanglement* [10,11]. This function is related to the maximum overlap of the state we are considering with the closest separable state. Being related to an optimization problem, the evaluation of such overlap is a non trivial problem. Nonetheless, it has been evaluated analytically for some specific states like the multiqubit permutation invariant states [12–14]. However, there is not a general way to compute such a quantity.

In this contribution we construct the appropriate representation of the Hubbard operators for multiqubit systems. The aim is to provide a suitable framework to deal with the involved tensor algebra in a simpler manner. Through this work we use some results concerning the Kronecker product in terms of the X-operators reported in Ref. [4]. The paper is structured as follows. In Section 2 Hubbard notation is introduced for one qubit and multiqubit systems. As an application of this formalism some results related to the geometric measure of entanglement for symmetric states are obtained in Section 3. We also compare our results with those reported in the literature. Finally, in Section 4 some concluding remarks are presented.

2. Formalism

The X-operators were introduced by Hubbard [15–17] and have applications in several branches of physics (see, e.g. [4, 18] and references quoted therein). These obey the multiplication rule

$$X^{i,j}X^{k,m} = \delta_{jk}X^{i,m}, \tag{2}$$

and satisfy the following properties

$$(X^{i,j})^\dagger = X^{j,i}, \quad \sum_k X^{k,k} = \mathbb{I}, \quad [X^{i,j}, X^{k,m}]_\pm = \delta_{jk}X^{i,m} \pm \delta_{mi}X^{k,j}. \tag{3}$$

where $[A, B]_\pm$ stands for either the commutator $(-)$ or the anticommutator $(+)$ of A and B. This representation allows a shorthand notation for cumbersome calculations involving several factors or large dimension matrix operators [4].

2.1. One qubit

To give a concrete realization of the Hubbard operators consider a system with just two levels. The excited $|e_0^2\rangle = |0\rangle$ and the ground states $|e_1^2\rangle = |1\rangle$ are orthogonal

$$\langle e_0^2|e_0^2\rangle = \langle e_1^2|e_1^2\rangle = 1, \qquad \langle e_0^2|e_1^2\rangle = \langle e_1^2|e_0^2\rangle = 0, \tag{4}$$

the supra-index makes reference to the dimension of the corresponding Hilbert space $\mathcal{H}_2 = \text{Span}\{|0\rangle, |1\rangle\}$. The basis vectors in their simplest representation are written

$$|e_1^2\rangle = \begin{pmatrix} 1 \\ 0 \end{pmatrix}, \qquad |e_2^2\rangle = \begin{pmatrix} 0 \\ 1 \end{pmatrix}. \tag{5}$$

Any vector $|\psi\rangle \in \mathcal{H}_2$ can be written as a normalized linear combination, that is

$$|\psi\rangle = c_0|e_0^2\rangle + c_1|e_1^2\rangle = c_0|0\rangle + c_1|1\rangle, \quad |c_0|^2 + |c_1|^2 = 1, \quad c_0, c_1 \in \mathbb{C}. \tag{6}$$

The simplest representation of the Hubbard operators can be written as the outer product of the basis vectors. In Dirac's notation

$$X_2^{i,j} := |e_i^2\rangle\langle e_j^2|, \quad i, j = 0, 1, \tag{7}$$

the subindex 2 indicates the dimension of the Hilbert space where $X_2^{i,j}$ acts on. Note that the matrix representation of this last operator has 1 in the entry $(i+1, j+1)$ and zero elsewhere. We want to stress that since the Hubbard operators are complete any linear operator $O : \mathcal{H}_2 \longrightarrow \mathcal{H}_2$ can be be written in X-representation

$$O = \sum_{i,j=0}^{1} o_{i,j} X_2^{i,j}, \quad o_{i,j} = \langle e_i^2|o|e_j^2\rangle. \tag{8}$$

The representation in terms of Hubbard operators can generalized to the case of any finite dimension Hilbert space. In particular we are interested in the case of several qubits.

2.2. Multiqubit systems

The four-dimensional Hilbert space $\mathcal{H}_4$ of a two qubit system is given by

$$\mathcal{H}_4 = \text{Span}\{|e_{i_1}^2\rangle \otimes |e_{i_0}^2\rangle, \quad i_0, i_1 = 0, 1\}. \tag{9}$$

The basis elements of $\mathcal{H}_4$ can be expressed using either a single index notation or the binary form

$$\begin{aligned} |e_0^4\rangle &= |e_0^2\rangle \otimes |e_1^2\rangle = |00\rangle, \qquad & |e_1^4\rangle &= |e_0^2\rangle \otimes |e_1^2\rangle = |01\rangle, \\ |e_2^4\rangle &= |e_1^2\rangle \otimes |e_0^2\rangle = |10\rangle, \qquad & |e_3^4\rangle &= |e_1^2\rangle \otimes |e_1^2\rangle = |11\rangle. \end{aligned} \tag{10}$$

In a more compact form we write

$$|e_i^4\rangle \equiv |i\rangle \equiv |i_1 i_0\rangle, \quad i = (i_1 i_0)_2, \tag{11}$$

where $(\alpha_1\alpha_0)_2$ indicates that the number α is expressed in its binary decomposition $\alpha = 2^1\alpha_1 + 2^0\alpha_0$, for $0 \leq \alpha \leq 3$. An arbitrary normalized vector $|\psi\rangle \in \mathcal{H}_4$ can be written as a linear combination of the basis elements

$$|\psi\rangle = \sum_{i=0}^{3} c_i|e_i^4\rangle = \sum_{i=0}^{3} c_i|i\rangle = \sum_{i_1,i_0=0}^{1} c_{i_1,i_0}|i_1 i_0\rangle, \quad c_i \equiv c_{i_1,i_0}, \quad i = (i_1 i_0)_2. \tag{12}$$

We can generalize this notation to the case of n qubits. The corresponding Hilbert space $\mathcal{H}_{2^n}$ is given as

$$\begin{aligned} \mathcal{H}_{2^n} &= \text{Span}\left\{|e_{i_{n-1}}^2\rangle \otimes \cdots \otimes |e_{i_0}^2\rangle, \quad i_\ell = 0, 1, \quad \ell = 0, \ldots, n-1\right\} \\ &= \text{Span}\left\{|e_i^{2^n}\rangle, \quad i = 0, \ldots, 2^n - 1\right\} = \text{Span}\left\{|i\rangle, \quad i = 0, \ldots, 2^n - 1\right\} \\ &= \text{Span}\left\{|i_{n-1} \cdots i_0\rangle, \quad i_\ell = 0, 1, \quad \ell = 0, \ldots, n-1\right\}, \end{aligned} \tag{13}$$

where

$$|e_i^{2^n}\rangle \equiv |i\rangle \equiv |i_{n-1} \cdots i_0\rangle, \quad i = (i_{n-1} \cdots i_0)_2. \tag{14}$$

Therefore, any element $|\psi\rangle \in \mathcal{H}_{2^n}$ is in general a linear combination of the form

$$|\psi\rangle = \sum_{i=0}^{2^n-1} c_i|e_i^{2^n}\rangle = \sum_{i=0}^{2^n-1} c_i|i\rangle = \sum_{i_{n-1},\ldots,i_0=0}^{1} c_{i_{n-1},\ldots,i_0}|i_{n-1} \cdots i_0\rangle,$$
$$c_i \equiv c_{i_{n-1},\ldots,i_0} \quad i = (i_{n-1} \cdots i_0)_2. \tag{15}$$

As an example consider the Dicke states [19]:

$$|S(n,k)\rangle = \binom{n}{k}^{-1/2} \sum_{\pi\in S_n} |\pi(\underbrace{0\cdots 0}_{k}\underbrace{1\cdots 1}_{n-k})\rangle, \quad \binom{n}{k} = \frac{n!}{k!(n-k)!}. \tag{16}$$

The sum is performed over all the distinct permutations of the qubits and S_n denotes the symmetric group of order n. For instance, let us write down the explicit form of a four qubit Dicke state with $k = 3$:

$$\begin{aligned} |S(4,3)\rangle &= \frac{1}{2}\left(|0001\rangle + |0010\rangle + |0100\rangle + |1000\rangle\right) \\ &= \frac{1}{2}\left(|e_1^4\rangle + |e_2^4\rangle + |e_4^4\rangle + |e_8^4\rangle\right) = \frac{1}{2}\left(|1\rangle + |2\rangle + |4\rangle + |8\rangle\right). \end{aligned} \tag{17}$$

where $1 = (0001)_2, 2 = (0010)_2, 4 = (0100)_2$, and $8 = (1000)_2$. On the other hand, any linear operator acting on $\mathcal{H}_{2^n}$ can be represented as a $2^n \times 2^n$ matrix. In this case the Hubbard operators constitute an useful tool to deal with the involved algebra. We start with the case of a two qubit system and then the general case will be discussed. The following proposition establishes that the tensor product is closed on the set of X-operators.

Proposition 1. *Let $X_2^{i_0,j_0}$ and $X_2^{i_1,j_1}$ be two Hubbard operators of order 2. Then the $\otimes$-product*

$$X_2^{i_1,j_1} \otimes X_2^{i_0,j_0} = X_4^{2i_1+i_0,2j_1+j_0}, \tag{18}$$

is a Hubbard operator of order 4.

The proof can be consulted in [4]. This last proposition indicates that to localize the row (column) where the non-vanishing element of the resulting operator is, one has to "count" i_1 (j_1) blocks of size 2 plus i_0 (j_0) rows (columns). We can use a single index notation by defining $i = 2i_1 + i_0$ and $j = 2j_1 + j_0$. Note that the column (row) indices of the factor operators are the binary coefficients of the column (row) index i (j) of the resultant Hubbard operator. This can be expressed as follows

$$X_2^{i_1,j_1} \otimes X_2^{i_0,j_0} = X_4^{i,j}, \quad i = (i_1 i_0)_2, \quad j = (j_1 j_0)_2. \tag{19}$$

Equation (19) is easily generalized by induction. Indeed,

$$X_2^{i_n,j_n} \otimes \cdots \otimes X_2^{i_0,j_0} = X_{2^{n+1}}^{i,j}, \quad i = (i_{n-1}\cdots i_0)_2, \quad j = (j_{n-1}\cdots j_0)_2. \tag{20}$$

As an example consider the following product

$$X_2^{1,0} \otimes X_2^{0,0} \otimes X_2^{1,1} = X_8^{5,1}, \quad 5 = (101)_2 \quad 1 = (001)_2.$$

In the following Theorem we show that the Kronecker product of two arbitrary 2×2 matrices of order two can be written in terms of the Hubbard operators.

Theorem 1. *The Kronecker product of the 2-square matrices $A = [a_{i_1,j_1}]$ and $B = [b_{i_0,j_0}]$ in terms of the X-operators of order 4 reads*

$$A \otimes B = \sum_{i,j=0}^{3} c_{i,j} X_4^{i,j}.$$

Proof. From the linearlity of $\otimes$ and Proposition 1 we have

$$A \otimes B = \sum_{\substack{i_0,j_0=0\\ i_1,j_1=0}}^{1} a_{i_1,j_1} b_{i_0,j_0} X_2^{i_1,j_1} \otimes X_2^{i_0,j_0} = \sum_{\substack{i_0,j_0=0\\ i_1,j_1=0}}^{1} a_{i_1,j_1} b_{i_0,j_0} X_4^{2i_1+i_0,2j_1+j_0}.$$

Now define $i = 2i_1 + i_0$ and $j = 2j_1 + j_0$, so that $i, j = 0, 1, 2, 3$. Then

$$A \otimes B = \sum_{i,j=0}^{3} a_{i_1,j_1} b_{i_0,j_0} X_4^{i,j}, \qquad i = (i_1 i_0)_2, \quad j = (j_1 j_0)_2. \tag{21}$$

Identifying $c_{i,j} := a_{i_1,j_1} b_{i_0,j_0}$ the proof is completed. □

The properties of the Kronecker product of two arbitrary matrices in terms of the Hubbard operators are reviewed in [4]. We can use the fact that the $\otimes$-product is associative as well as Theorem 1 to compute the tensor product powers of a 2×2 matrix with itself

$$A^{\otimes k} = \sum_{i,j=0}^{2^k-1} a_{i_{n-1},j_{n-1}} \cdots a_{i_0,j_0}\, X_{2^k}^{i,j}, \quad i = (i_{n-1} \cdots i_0)_2, \quad j = (j_{n-1} \cdots j_0)_2. \tag{22}$$

The most general case of the tensor product of any size matrices is reviewed in [4]. Equation (22) allows to determine the sign of a given entry of the operator (1). First consider

$$\sigma_3^{\otimes k} = \sum_{i=0}^{2^k-1} (-1)^{i_{k-1}+\ldots+i_0} X_{2^k}^{i,i}, \quad i = (i_{k-1} \ldots i_0)_2. \tag{23}$$

For instance, take $k = 4$ and ask for the sign of the entry (11,11). This corresponds to the coefficient of the Hubbard operator $X_{16}^{10,10}$. Note $10 = (1100)_2$, so the required entry is $(-1)^{1+1} = (-1)^2 = 1$. As another example consider the simplest case of the Hadamard matrices [3, 20]:

$$H = \frac{1}{\sqrt{2}} \sum_{i_0,j_0=1}^{1} (-1)^{i_0 j_0} X_2^{i_0,j_0} = \frac{1}{\sqrt{2}} \begin{pmatrix} 1 & 1 \\ 1 & -1 \end{pmatrix}. \tag{24}$$

This operator is unitary and its action on a vector (6) reads

$$H|\psi\rangle = \frac{1}{\sqrt{2}} \sum_{i_0,j_0=0}^{1} (-1)^{i_0 j_0} c_{j_0} |i_0\rangle = \frac{1}{\sqrt{2}} [(c_0 + c_1)|0\rangle + (c_0 - c_1)|1\rangle]. \tag{25}$$

In quantum computing algorithms it is of interest the calculation of tensor product powers of H. Using (22) we immediately get

$$H^{\otimes k} = \frac{1}{\sqrt{2^k}} \sum_{i,j=0}^{2^k-1} (-1)^{i \cdot j} X_{2^k}^{i,j}, \quad i \cdot j = \sum_{\ell=0}^{k-1} i_\ell j_\ell, \quad i = (i_{k-1} \cdots i_0)_2, \quad j = (j_{k-1} \cdots j_0)_2. \tag{26}$$

The action of the former operator on a vector $|e_x^{2^k}\rangle \equiv |x\rangle$ reads

$$H^{\otimes k}|x\rangle = \frac{1}{\sqrt{2^k}} \sum_{i=0}^{2^k-1} (-1)^{x \cdot i} |i\rangle. \tag{27}$$

The following section is devoted to review other applications of the Hubbard operators in quantum entanglement theory.

3. Application in multipartite entanglement

We now consider a multipartite system $\mathcal{S} = \mathcal{S}_1 + \cdots + \mathcal{S}_k$. The n_ℓ-dimensional Hilbert space describing the subsystem $\mathcal{S}_\ell$ is given as $\mathcal{H}_{n_\ell} = \text{Span}\{|e_s^{(\ell)}\rangle, \; s = 0, \ldots, n_\ell - 1\}$. A generic multipartite state $|\psi\rangle \in \mathcal{H}_{n_1} \otimes \cdots \otimes \mathcal{H}_{n_k}$, written

$$|\psi\rangle = \sum_{p_1,\ldots,p_n} c_{p_1,p_2,\ldots,p_n} |e_{p_1}^{(1)}\rangle \otimes |e_{p_2}^{(2)}\rangle \otimes \cdots \otimes |e_{p_n}^{(n)}\rangle, \tag{28}$$

is said to be *completely separable* if it can be expressed as a product state of the form

$$|\varphi\rangle = |\varphi_1\rangle \otimes |\varphi_2\rangle \otimes \cdots \otimes |\varphi_n\rangle, \qquad |\varphi_\ell\rangle = \sum_{p_\ell} a_{p_\ell} |e_{p_\ell}^{(\ell)}\rangle. \tag{29}$$

Otherwise (28) is referred as *entangled*. A natural form to quantify entanglement is by considering how "different" is a given state $|\psi\rangle$ from (29). In this sense the maximum overlap of an state $|\psi\rangle$ with the closest separable state is defined as follows [10]:

$$\Lambda_{\max}(\psi) := \max_{|\varphi\rangle} \{ |\langle\psi|\varphi\rangle| : |\varphi\rangle \in \mathcal{H} \text{ is complete separable} \}. \tag{30}$$

Observe that if $|\psi\rangle$ is completely separable, then $\Lambda_{\max}$ is one. Thus, the *geometric measure of entanglement* is defined in two different forms [14]:

$$E_{\sin^2}(\psi) := 1 - \Lambda_{\max}^2(\psi), \quad \text{and} \quad E_{\log}(\psi) := -2\log_2 \Lambda_{\max}(\psi). \tag{31}$$

The term "geometric" is justified by the fact that $\Lambda_{\max}$ is the angle between two vectors. Besides, notice that the amount of entanglement increases as $\Lambda_{\max}$ decreases. We want to point that evaluating (30) is in general a hard optimization problem, although for specific states analytic expressions have been derived by taking advantage of some symmetries. For instance, the maximum overlap with the closest separable state of the Dicke states (16) is given by [14]:

$$\Lambda_{\max}(n, k) = \sqrt{\frac{n!}{k!(n-k)!} \left(\frac{k}{n}\right)^{k/2} \left(\frac{n-k}{n}\right)^{(n-k)/2}}. \tag{32}$$

This result has been obtained by considering that the product state can be taken as the Kronecker product of the same real single-party state, that is

$$|\varphi\rangle = \bigotimes_{i=1}^{n} |\varphi_i(\theta)\rangle, \quad |\varphi_i(\theta)\rangle = \cos\theta|0\rangle + \sin\theta|1\rangle, \quad \theta \in [0, \pi/2]. \tag{33}$$

The overlap of $\langle S(n,k)|\phi\rangle$ can be bounded by using the Schwarz and MacLaurin inequalities, so that optimization on θ is accomplished [14]. However, employing the formalism of Hubbard operators the same result (32) can be obtained in a simpler manner. First, notice that the definition of $\Lambda_{\max}$ is equivalent to the following

$$\Lambda_{\max}(\psi) := \max_{U_{loc}} \{ |\langle\psi|U_{\text{loc}}|0\rangle| : U_{\text{loc}} = U_1 \otimes \cdots \otimes U_N, \;\; U_\ell \text{ unitary} \}. \tag{34}$$

In the case of Dicke states, the operator U_{loc} can be taken as

$$U_{\text{loc}}(\theta) = U(\theta)^{\otimes n}, \quad U(\theta) = \begin{pmatrix} \cos\theta & -\sin\theta \\ \sin\theta & \cos\theta \end{pmatrix}, \tag{35}$$

in terms of the X-operators this reads

$$U(\theta) = \sum_{i,j=0}^{1} u_{i,j}(\theta) X_2^{i,j}, \quad u_{i,j}(\theta) = \cos[\theta + \pi(j-i)/2]. \tag{36}$$

Using (22) we get

$$U_{\rm loc}(\theta) = \sum_{i,j=0}^{2^n-1} u_{i_{n-1}j_{n-1}}(\theta) \cdots u_{i_0 j_0}(\theta) X_{2^n}^{i,j}, \quad i = (i_{n-1} \dots i_0)_2, \ j = (j_{n-1} \dots j_0)_2. \tag{37}$$

Note

$$U_{\rm loc}|0\rangle = \sum_{i=0}^{2^n-1} u_{i_{n-1},0} \cdots u_{i_0,0}|i\rangle = \sum_{i_{n-1},\dots,i_0} u_{i_{n-1},0} \cdots u_{i_0,0}|i_{n-1} i_{n-1} \dots i_0\rangle, \tag{38}$$

where we have omitted the θ dependence for shortness. The overlap of $\Lambda_\theta(n,k) = \langle S(n,k)|U_{\rm loc}|0\rangle$ with the former state can be immediate computed

$$\Lambda_\theta(n,k) = \frac{1}{\sqrt{C_k^n}} \sum_{i_{n-1},\dots,i_0} \sum_{\pi \in S_n} u_{i_{n-1},0} \cdots u_{i_0,0} \langle \pi(\underbrace{0 \cdots 0}_{k} \underbrace{1 \cdots 1}_{n-k})|i_{n-1}, \dots i_0\rangle.$$

Remark that the sum is contracted to C_k^n terms, each one having n factors $u_{0,0}$ and $n-k$ factors $u_{1,0}$. Explicitly

$$\Lambda_\theta(n,k) = \sqrt{C_k^n}\; u_{0,0}^k(\theta) u_{1,0}^{n-k}(\theta) = \sqrt{C_k^n} \sin^k\theta \cos^{n-k}\theta, \tag{39}$$

so that, maximazing over θ the expression (32) is obtained. As an example consider the n-qubit states

$$|W_n\rangle = |S(n,1)\rangle, \quad |\widetilde{W}_n\rangle = |S(n,n-1)\rangle, \tag{40}$$

for which the maximum overlap with the closest separable state yields

$$\Lambda_{\max}(W_n) = \Lambda_{\max}(\widetilde{W}_n) = \left(\frac{n-1}{n}\right)^{(n-1)/2}. \tag{41}$$

For $n = 3$ it is known that the maximal state of the geometric measure of entanglement is the W-state with $\Lambda_{\max}(W_3) = 2/3$ [21]. Using the X-operators we can compute $\Lambda_{\max}$ for a more general kind of symmetric states, for instance a linear combination of q Dicke states

$$|S(n,\vec{k})\rangle = \sum_{\ell=1}^{q} \alpha_\ell |S(n,k_\ell)\rangle, \quad \alpha_\ell \geq 0, \quad \sum_\ell \alpha_\ell^2 = 1, \tag{42}$$

where $\vec{k} = (k_1, k_2, \dots, k_q)$ and $k_\ell \leq n$. For this kind of state we obtain

$$\Lambda_{\max}(n,\vec{k}) = \cos^n\theta \sum_{\ell=1}^{q} \alpha_\ell \sqrt{C_{k_\ell}^n}\; \tan^{k_\ell}\theta, \tag{43}$$

here θ is a solution of the following transcendental equation

$$\sum_{\ell=1}^{q} \frac{\alpha_\ell}{\sqrt{C_{k_\ell}^n}} [k_\ell + (k_\ell - n)\tan^2\theta] \tan^{k_\ell}\theta = 0. \tag{44}$$

From this last result we can compute $\Lambda_{\max}$ for the following three-qubit state

$$|W\widetilde{W}(s)\rangle = \sqrt{s}|W\rangle + \sqrt{1-s}|\widetilde{W}\rangle, \quad s \in [0,1], \tag{45}$$

with the parameters $q = 2$, $k_1 = 1$, $k_2 = 2$, $\alpha_1 = \sqrt{s}$ and $\alpha_2 = \sqrt{1-s}$ the result reported by Wei and Severini [14] is obtained

$$\Lambda_{\max}(W\widetilde{W}(s)) = \frac{1}{2\sqrt{3}}(\sqrt{s}\cos\theta + \sqrt{1-s}\sin\theta)\sin 2\theta, \tag{46}$$

where θ satisfies the equation

$$\sqrt{1-s}\tan^3\theta + 2\sqrt{s}\tan^2\theta - 2\sqrt{1-s}\tan\theta - \sqrt{s} = 0.$$

4. Conclusions

The tensor product algebra in multiqubit systems has been addressed in terms of Hubbard operators. In this notation complicated calculations are turned into simple relations of the binary coefficients of indices. The presented formalism can be easily generalized to qu$d$its, in such a case one should work in base d instead of the binary one. To show the usefulness of our results, we derived an analytic expression for the maximum overlap with respect the closest separable state of a linear superposition of Dicke states. This result includes some cases reported in the literature.

Acknowledgments

Thanks to Prof. O. Rosas-Ortiz for fruitful remarks. The autor is also in indebted to Prof. K. Życzkowski for valuable discussions on multipartite entanglement. The financial support of Conacyt is acknowledged.

References

[1] Mielnik B. and Rosas-Ortiz O., *Quantum Mechanical Laws*, in Fundamentals of Physics [Morán-López J.L. Ed.], Encyclopedia of Life Support Systems (EOLSS), Developed under the Auspices of the UNESCO, Eolss Publishers, Oxford, United Kingdom (2007/Rev. 2009)

[2] Audrestsch J., *Entangled Systems*, Wiley, 2005

[3] Nielsen M.A. and Chuang I.L., *Quantum Computation and Quantum Information*, Cambridge University Press, Cambridge (2000)

[4] Enríquez M. and Rosas-Ortiz O., The Kronecker product in terms of Hubbard operators and the Clebsch-Gordan decomposition of SU(2)×SU(2), *Annals of Physics* **339** (2013) 218

[5] Hill S. and Wootters W.K., Entanglement of a pair of quantum bits, *Phys. Rev. Lett.*, **78** (1997) 5022

[6] Wootters W.K., Entanglement of formation of an arbitrary state of two qubits, *Phys. Rev. Lett.*, **80** (1998) 2245

[7] Enríquez M., Quintana C. and Rosas-Ortiz O., Time-evolution of entangled bipartite atomic systems in quantized radiation fields, *J. Phys. Conf. Series* **512** (2014) 012022

[8] Horodecki R. *et al*, Quantum entanglement, *Rev. Mod. Phys.* **81** (2009) 865

[9] Enríquez M., Puchała Z. and Życzkowski K., Entanglement properties of generic multipartite states, preprint Jagiellonian University (2015)

[10] Wei T.- Ch. and Goldbart P.M., Geometric measure of entanglement and applications to bipartite and multipartite quantum states, *Phys. Rev.* **A 68** (2003) 042307

[11] Chen L., Xu A., and Zhu H., Computation of the geometric measure of entanglement for pure multiqubit states, *Phys. Rev. A* **82** (2010) 032301

[12] Hübener R., Kleinmann M., Wei T.-C. and Gühne O., Geometric measure of entanglement for symmetric states, *Phys. Rev. A* **80** (2009) 032324

[13] Hayashi M., Markham D., Murao M., Owari M., and Virmani S., The geometric measure of entanglement for a symmetric pure state with non-negative amplitudes, *J. Math. Phys.* **50** (2009) 122104

[14] Wei T.-C. and Severini S. Matrix permanent and quantum entanglement of permutation invariant states, *J. Math. Phys.* **51** (2010) 092203

[15] Hubbard J., Electron correlations in narrow energy bands, *Proc. R. Soc. (London)*, Ser. A **276** (1963) 238
[16] Hubbard J., Electron correlations in narrow energy bands II. The degenerate band case, *Proc. R. Soc. (London)*, Ser. A **277** (1964) 237
[17] Hubbard J., Electron correlations in narrow energy bands IV. The atomic representation, *Proc. R. Soc. (London)*, Ser. A **285** (1965) 542
[18] Enríquez M. and Rosas-Ortiz O., Some applications of Kronecker product in Hubbard representation, *J. Phys. Conf. Series* 538 (2014) 012007
[19] Dicke R. H., Coherence in spontaneous radiation processes, *Phys. Rev.* **93** (1954) 99
[20] Hadamard J., Resolution d'une question relative aux determinants, *Bull. Sci. Math.* **17** (1893) 240
[21] Aulbach M., Damian M. and Mio M., The maximally entangled symmetric state in terms of the geometric measure. *New Jour. of Phys.* **12** (2010) 073025

Remarks on the probability of finding a particle in the interaction zone of a scattering process

Alfonso Jaimes-Nájera
Physics Department, Cinvestav, A.P. 14-740, México DF 07000, Mexico
E-mail: ajaimes@fis.cinvestav.mx

Abstract. The probability of finding a particle in the interaction zone of a scattering process is analyzed in one dimension. The quantum state of the particle is described by a Gaussian wave packet that is narrow in the momentum distribution. We find the intervals of time that are required in order to get a maximum probability in the interaction zone as a lower bound of the duration of the scattering process. As an immediate application, times involved in the scattering associated with symmetric potentials (as the rectangular barriers and wells) are calculated.

1. Introduction

The intervals of time involved in a decaying or scattering process are relevant in contemporary quantum mechanics [1, 2]. Sound examples are the tunneling time [3–8], the delay produced in the crossover of particles in a given region of space [9–12], and the lifetimes associated to the resonance phenomenon [13, 14]. In this contribution we are interested in determining lower bounds to the duration of scattering processes in one dimension. Our motivation is mainly due to the fact that the reflection and transmission times associated to scattering depend on the profile of the scatterer, and the non-locality of the wave functions introduces a lot of difficulties in the determination of such times. That is, no matter what is the definition used to calculate these times, it is not clear when the scattering process has been concluded since the tails of the incoming and outgoing waves are formally different from zero in all the real line for any time. One option is to say that the scattering process is finished whenever the profile of the transmitted and reflected waves is well defined. However, the values of time required to fulfill such a condition might be extremely large, depending on the scattering potential we are interested in. Therefore, we propose to calculate the interval of time $(0, t_*)$ required to get a probability of finding the particle in the interaction zone that is maximum for a given scatterer. At that times the scattering process is still in progress, so that the transmitted and reflected times should not be shorter than t_*.

The organization of the paper is as follows. In Section 2 we revisit the generalities of the one-dimensional scattering of a Gaussian wave packet. Then, we calculate the probability of finding the particle in the interaction zone and make the assumption that the initial wave packet is very sharp in the momenta distribution. Conditions for the extremal probability are also found. In Section 3 we apply our approach to the case of symmetric scattering potentials. Final remarks are given in Section 4.

2. Scattering of Gaussian wave packets in one dimension

Consider a one dimensional scattering process defined by a function of compact support $V(x)$. That is, the potential $V(x)$ is different from zero in $(a, b) \in \mathbb{R}$ and null everywhere in $\mathbb{R}/(a, b)$, see e.g. [15]. Let us assume that the particle is injected from the right at $t = 0$ with a momentum k_0. Thus, we can represent the quantum state of the particle by a wave packet of Gaussian profile in the momentum space $\Lambda(k)$ that is centered at $k = k_0$ in the initial spatial coordinate $x = x_0$. Using dimensionless variables we write

$$\Psi(x,t) = \int_{-\infty}^{\infty} dk\, \Lambda(k)\psi_k(x)e^{i(kx_0-\omega t)}, \tag{1}$$

where $\psi_k(x)$ is solution of the stationary Schrödinger equation with eigenvalue $\omega = k^2$, and the coefficient of the superposition

$$\Lambda(k) = \frac{1}{\sqrt{\sqrt{\pi}\sigma\Delta}}\, e^{-(k-k_0)^2/2\sigma^2}, \tag{2}$$

is a normalized Gaussian distribution of width σ and normalization constant Δ. The center of this packet behaves like a classical particle of velocity equal to the packet's group velocity:

$$v_g(k_0) = \left.\frac{d\omega}{dk}\right|_{k=k_0} = 2k_0. \tag{3}$$

We are interested in conservative processes where the initial Gaussian profile in momentum space is preserved (neither sources nor sinks of probability are allowed). Then, to the right of the scatterer ($x > b$) one should get a superposition of incident $\Psi_{inc}(x, t)$ and reflected $\Psi_{\rm r}(x, t)$ wave packets of the form

$$\Psi_{\rm inc}(x,t) = \int_{-\infty}^{\infty} dk\, \Lambda(k)e^{-i(k(x-x_0)+\omega t)}, \tag{4}$$

and

$$\Psi_{\rm r}(x,t) = \int_{-\infty}^{\infty} dk\, \Lambda(k)S_{\rm R}(k)e^{i(k(x+x_0)-\omega t-\zeta)}, \tag{5}$$

where $S_{\rm R}(k)e^{-i\zeta(k)}$ is the reflection amplitude in polar form and $\zeta(k)$ is the reflection phase shift. To the left of the scatterer only a transmitted wave packet is expected

$$\Psi_{\rm t}(x,t) = \int_{-\infty}^{\infty} dk\, \Lambda(k)S_{\rm T}(k)e^{-i(k(x-x_0)+\omega t+\gamma)}, \tag{6}$$

with $S_{\rm T}(k)e^{-i\gamma(k)}$ the transmission amplitude in polar form and $\gamma(k)$ the transmission phase shift.

2.1. *Finding the particle in the interaction zone*

The idea of working with compact support functions $V(x)$ is to represent short range potentials in the scattering process. We define the support (a, b) of $V(x)$ as the interaction zone of the scattering. Our principal interest is to calculate the time at which the probability $P_{(a,b)}$ of finding the particle in the interaction zone is maximum. This information will be used to get intervals of time, bounded from below, for the complete scattering process. That is, if at $t = t_*$ the probability of finding the particle in (a, b) is maximum, then the scattering is still in progress.

As the initial wave packet $\Psi(x,t)$ is normalized, the probability of finding the particle in the interaction zone (a,b) is

$$P_{(a,b)}(t) = \int_a^b dx\, |\Psi(x,t)|^2 = 1 - \int_{-\infty}^a dx\, |\Psi(x,t)|^2 - \int_b^\infty dx\, |\Psi(x,t)|^2. \tag{7}$$

Considering the composition of the wave packet in the different regions, functions (4), (5) and (6), we get

$$P_{(a,b)}(t) = 1 - \int_{-\infty}^a dx\, |\Psi_{\rm t}(x,t)|^2 - \int_b^\infty dx\, \left[|\Psi_{\rm r}(x,t)|^2 + |\Psi_{\rm inc}(x,t)|^2\right] - 2\,{\rm Re}\left[\int_b^\infty dx\, \Psi^*_{\rm inc}(x,t)\,\Psi_{\rm r}(x,t)\right]. \tag{8}$$

To make the calculations we shall assume that the initial wave packet is (almost) monochromatic so that the width σ in (2) is arbitrarily narrow; that is, $\sigma << 1$. In this form, we hope small variations of the values of $S_{\rm R}$ and $S_{\rm T}$ in the vicinity of k_0. Thus, in order of removing the amplitudes from the integrals (5) and (6), we respectively make $S_{\rm R}(k) \approx S_{\rm R}(k_0) = S_{\rm R}$ and $S_{\rm T}(k) \approx S_{\rm T}(k_0) = S_{\rm T}$ in the interval $k_0 \pm \frac{1}{2}\sigma$, see e.g. [16]. Moreover, the sharpness of the initial wave packet also allows to expand the phase shifts around $k = k_0$,

$$\gamma(k) = \gamma_0 + \left.\frac{\partial\gamma}{\partial k}\right|_{k=k_0}(k-k_0) + \cdots, \qquad \zeta(k) = \zeta_0 + \left.\frac{\partial\zeta}{\partial k}\right|_{k=k_0}(k-k_0) + \cdots, \tag{9}$$

and to take into account the linear terms in k only. Therefore, the transmitted and reflected wave packets read as

$$\Psi_{\rm t}(x,t) = \frac{S_{\rm T}}{\sqrt{\Delta^2\sqrt{\pi}\sigma}} e^{i\left(k_0\frac{\partial\gamma}{\partial k}(k_0)-\gamma_0\right)} \int_{-\infty}^\infty dk\, e^{-(k-k_0)^2/2\sigma^2 - ik\left(\left(x-x_0+\frac{\partial\gamma}{\partial k}(k_0)\right)+kt\right)}, \tag{10}$$

$$\Psi_{\rm r}(x,t) = \frac{S_{\rm R}}{\sqrt{\Delta^2\sqrt{\pi}\sigma}} e^{i\left(k_0\frac{\partial\zeta}{\partial k}(k_0)-\zeta_0\right)} \int_{-\infty}^\infty dk\, e^{-(k-k_0)^2/2\sigma^2 + ik\left(\left(x+x_0-\frac{\partial\zeta}{\partial k}(k_0)\right)-kt\right)}, \tag{11}$$

where the explicit form of $A(k)$ given in (2) has been introduced. The probability (8) reads now as follows

$$\begin{aligned} P_{(a,b)}(t) = \frac{1}{2\Delta}\Bigg\{ & {\rm erf}\left[\frac{1}{\sigma_x}\left(2k_0t - (x_0-b)\right)\right] - (S_{\rm T})^2\,{\rm erf}\left[\frac{1}{\sigma_x}\left(2k_0t - \left(x_0 - b + \left.\frac{\partial\phi_{\rm t}}{\partial k}\right|_{k=k_0}\right)\right)\right] \\ & - (S_{\rm R})^2\,{\rm erf}\left[\frac{1}{\sigma_x}\left(2k_0t - \left(x_0 - b + \left.\frac{\partial\phi_{\rm r}}{\partial k}\right|_{k=k_0}\right)\right)\right] \\ & - S_{\rm R}\, e^{-c}\left[e^{i\varphi}\left({\rm erf}\left[\frac{d}{\sigma_x}\right]+1\right) + {\rm c.c.}\right] + 2(\Delta - 1)\Bigg\}, \end{aligned} \tag{12}$$

where ${\rm erf}(z)$ stands for the *error function* of z [17] and c.c. means complex conjugate; other expressions involved in (12) are the phase shifts

$$\phi_{\rm t} = k(b-a) - \gamma, \quad \phi_{\rm r} = 2kb - \zeta, \quad \varphi = k_0\left.\frac{\partial\zeta}{\partial k}\right|_{k=k_0} - \zeta_0, \tag{13}$$

the spatial shift

$$\sigma_x = \sqrt{\frac{1}{\sigma^2} + 4\sigma^2t^2}, \tag{14}$$

the interference argument

$$d = i\left(\frac{k_0}{\sigma^2} + 2\sigma^2 t\left(x_0 - b + \frac{1}{2}\frac{\partial\phi_{\rm r}}{\partial k}\bigg|_{k=k_0}\right)\right) - \frac{1}{2}\frac{\partial\phi_{\rm r}}{\partial k}\bigg|_{k=k_0} \tag{15}$$

and the constants

$$\begin{aligned} c &= \frac{k_0^2}{\sigma^2} + \sigma^2\left(x_0 - b + \frac{1}{2}\frac{\partial\phi_{\rm r}}{\partial k}\bigg|_{k=k_0}\right)^2, \\ \Delta &= 1 + e^{-c}S_{\rm R}\left(\cos\varphi - {\rm erfi}\left[\sigma\left(x_0 - b + \frac{1}{2}\frac{\partial\phi_{\rm r}}{\partial k}\bigg|_{k=k_0}\right)\right]\sin\varphi\right). \end{aligned} \tag{16}$$

In the sequel, it will be useful to rewrite the derivative with respect to the momentum k as a derivative with respect to the energy

$$\frac{\partial}{\partial E} = \frac{1}{2k}\frac{\partial}{\partial k}. \tag{17}$$

In this form the expressions

$$\tau_{\rm t} = \frac{\partial\phi_{\rm t}}{\partial E}\bigg|_{E=k_0^2}, \qquad \tau_{\rm r} = \frac{\partial\phi_{\rm r}}{\partial E}\bigg|_{E=k_0^2}, \tag{18}$$

correspond to the transmission and reflection phase times, respectively (for details see [10–12,18] and references quoted therein). Using the phase times (18) in Eqs. (12)–(16), the notation is notably simplified.

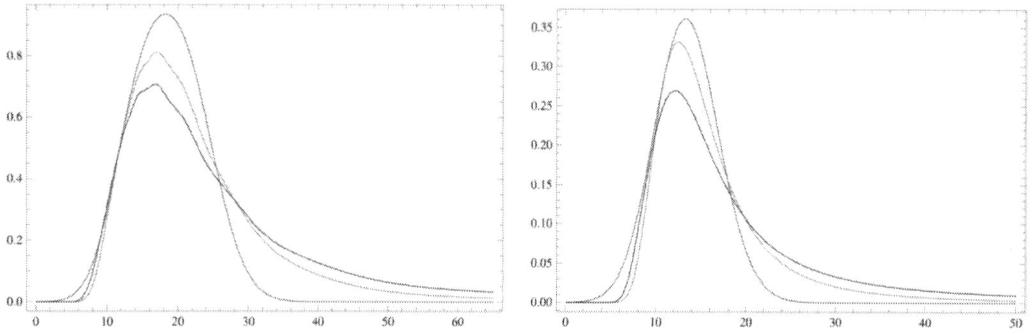

Figure 1. Probability in the interaction zone $P_{(a,b)}$ as a function of time for a rectangular barrier (left) and a rectangular well (right). In both cases $k_0 = 1.0724$, $x_0 = 30$, $b = -a = 3\pi/2$, and the time t_b spent by the particle in crossing the distance between x_0 and b is $t_b = (x_0 - b)/(2k_0) = 11.79$. The height $V_0 = 1$ of the barrier implies $S_{\rm T}^2 = 0.744$ and $\tau_{\rm t} = \tau_{\rm r} = 13.0671$, while the depth $V_0 = -1$ of the well leads to $S_{\rm T}^2 = 0.916$ and $\tau_{\rm t} = \tau_{\rm r} = 3.073$. The curves correspond to the width $\sigma = 0.55$ (black), $\sigma = 0.43$ (red), and $\sigma = 0.1$ (blue).

Our expression for the probability $P_{(a,b)}$ in (12) encodes the information of the scatterer in the amplitudes $S_{\rm R}$, $S_{\rm T}$, and in the phase (times) shifts $\phi_{\rm r}$ and $\phi_{\rm t}$ ($\tau_{\rm r}$ and $\tau_{\rm t}$). In Fig. 1 we have depicted the probability $P_{(a,b)}$ associated with a rectangular barrier (left) and a rectangular well (right) as a function of time. In both cases the values of $S_{\rm R}$, $S_{\rm T}$, $\tau_{\rm r}$ and $\tau_{\rm t}$ have been calculated for $(a,b) = (-3\pi/2, 3\pi/3)$. The well has a depth $V_0 = -1$ and the barrier a height $V_0 = 1$. The time scale goes from $t = 0$, the moment in which the particle crosses the point x_0, till an arbitrary value t at which the probability goes to zero. Initially, $P_{(a,b)}$ is negligible and reaches its maximum at a time $t = t_*$ that is smaller than $\min\{\tau_r + t_b, \tau_t + t_b\}$, with $t_b = (x_0 - b)/(2k_0)$ the time spent by the particle in crossing the distance between x_0 and b. As we can see, the value of the maximum increases as the initial width σ decreases. Thus, the sharpness of the momenta distribution $\Lambda(k)$ in the initial packet defines the height of $P_{(a,b)}$: the probability of finding the particle in the interaction zone (a, b) is as great as peaked is the distribution $\Lambda(k)$ in the value $k = k_0$. Another profile of $P_{(a,b)}$ that can be appreciated is that the time t_* at which this probability is maximum increases as the width σ of $\Lambda(k)$ decreases.

2.2. *Extreme values of the probability*

We are interested in calculating the time $t = t_*$ at which the probability (12) is maximum. As we have discussed in the previous section, such a time depends on the width σ of the initial wave packet: t_* is greater as σ is narrower. To get an idea of the influence of the sharpness of $\Lambda(k)$ on t_* let us consider the spatial width σ_x defined in (14). Given σ, the spatial distributions are broader as the time goes pass. In this context we realize that the approximation

$$\sigma_x = \frac{1}{\sigma}\sqrt{1 + 4\sigma^4 t^2} \approx \frac{1}{\sigma}\left(1 + 2\sigma^4 t^2\right) \tag{19}$$

holds whenever $t << \frac{1}{2\sigma}$. That is, we can identify an interval of time $(0, t_0)$ in which the spatial width σ_x is approximately constant $\sigma_x \approx \sigma^{-1}$; it will be enough to take $t_0 << \frac{1}{2\sigma}$. Note that this interval is larger as σ is narrower. Within this approximation, one can identify the extreme values of the probability (12) by calculating the zeros of its time-derivative:

$$\frac{dP_{(a,b)}(t)}{dt} = \frac{4k_0\sigma}{\sqrt{\pi}(1 + 4t^2\sigma^4)^{3/2}}\left\{(1 + 4\sigma^4 t_b t)\exp\left[-\frac{y_0^2}{1 + 4\sigma^4 t^2}\right]\right.$$
$$- (S_{\rm T})^2\left[1 + 4\sigma^4 t\,(t_b + \tau_{\rm t})\right]\exp\left[-\frac{y_{\rm t}^2}{1 + 4\sigma^4 t^2}\right] - (S_{\rm R})^2\left[1 + 4\sigma^4 t\,(t_b + \tau_{\rm r})\right]\exp\left[-\frac{y_{\rm r}^2}{1 + 4\sigma^4 t^2}\right]$$
$$\left. - 4\,S_{\rm R}\left(\sigma^4 t\,\tau_{\rm r}\cos\Theta + \sigma^2\left[t - \left(t_b + \frac{1}{2}\tau_{\rm r}\right)\right]\sin\Theta\right)\exp\left[-\frac{y_0(y_0 - 2\sigma k_0\tau_{\rm r})}{1 + 4\sigma^4 t^2}\right]\right\} = 0, \tag{20}$$

where

$$\Theta = \frac{1}{1 + 4\sigma^4 t^2}\left\{2k_0^2\tau_{\rm r}\left[1 + 4\sigma^4 t\left(t_b + \frac{1}{2}\tau_{\rm r}\right)\right]\right\} + \varphi, \tag{21}$$

$$y_0 = \sigma(2k_0 t - (x_0 - b)), \qquad y_{\rm t} = y_0 - 2\sigma k_0\,\tau_{\rm t}, \qquad y_{\rm r} = y_0 - 2\sigma k_0\,\tau_{\rm r}. \tag{22}$$

Thus, (20) is reduced to the expression

$$\exp\left[-(a_{\rm r}t - b_{\rm r})\sigma^2\right] - (S_{\rm R})^2\exp\left[(a_{\rm r}t - b_{\rm r})\sigma^2\right] - (S_{\rm T})^2\exp\left[(a_{\rm t}t - b_{\rm t})\sigma^2\right]$$
$$- 4\sigma^2 S_{\rm R}\left(t - \left(t_b + \frac{1}{2}\tau_{\rm r}\right)\right)\sin\phi_{\rm r} = 0, \tag{23}$$

with

$$\begin{aligned} a_{\rm r} = 4k_0^2\tau_{\rm r}, \quad b_{\rm r} = 2k_0\tau_{\rm r}\left(x_0 - b + k_0\tau_{\rm r}\right), \quad a_{\rm t} = 4k_0^2(2\tau_{\rm t} - \tau_{\rm r}), \\ b_{\rm t} = 4k_0\tau_{\rm t}\left(x_0 - b + k_0\tau_{\rm t}\right) - 2k_0\tau_{\rm r}\left(x_0 - b + k_0\tau_{\rm r}\right). \end{aligned} \tag{24}$$

Note that (23) is a transcendental equation of time. To solve it, remember that σ is arbitrarily narrow and $t \in (0, t_0)$, with $t_0 << \frac{1}{2\sigma}$. As the argument in the first three exponentials of (23) includes σ^2 as a global factor we look for the simultaneous solution of the inequalities

$$t << \frac{1}{a_{\rm r}}\left(\frac{1}{\sigma^2} + b_{\rm r}\right), \quad t << \frac{1}{a_{\rm t}}\left(\frac{1}{\sigma^2} + b_{\rm t}\right),$$

this last allows to approximate the exponentials by their first terms in a Taylor series. Therefore, if t is in $(0, t_0)$ and

$$b_{\rm r}, b_{\rm t} << \frac{1}{\sigma^2}$$

the above inequalities are fulfilled. Within this approximations the equation (23) is reduced to

$$1 - (a_{\rm r}t - b_{\rm r})(2k_0\sigma)^2 - (S_{extR})^2\left[1 + (a_{\rm r}t - b_{\rm r})(2k_0\sigma)^2\right] \\ - (S_{\rm T})^2\left[1 + (a_{\rm t}t - b_{\rm t})(2k_0\sigma)^2\right] - 4\sigma^2 S_{\rm R}\left(t - \left(t_b + \frac{1}{2}\tau_{\rm r}\right)\right)\sin\phi_{\rm r} = 0. \quad (25)$$

The root t_m of this last equation is given by the weighted average

$$t_m = \frac{(S_{\rm T})^2\,\tau_{\rm t}\left(t_b + \frac{1}{2}\tau_{\rm t}\right) + \left((S_{\rm R})^2\,\tau_{\rm r} + \frac{1}{2k_0^2}S_{\rm R}\sin\phi_{\rm r}\right)\left(t_b + \frac{1}{2}\tau_{\rm r}\right)}{(S_{\rm T})^2\,\tau_{\rm t} + (S_{\rm R})^2\,\tau_{\rm r} + \frac{1}{2k_0^2}S_{\rm R}\sin\phi_{\rm r}}. \quad (26)$$

The weight factor is the sum of three terms: $(S_{\rm T})^2\,\tau_{\rm t}$ and $(S_{\rm R})^2\,\tau_{\rm r}$ are the contribution of the transmission and reflection processes, and

$$\tau_i = -\frac{S_{\rm R}\sin\phi_{\rm r}}{2k_0^2} \quad (27)$$

gives account of the contribution due to the interference between the incident and the reflected wave packets. This last becomes important as k_0 goes to zero, i.e., for scattering at low energies. Therefore, due to the interference between the incident and the reflected wave packets, slower particles will spend larger intervals of time outside the interaction zone. The contribution to t_m associated to the interference is less important for faster particles (see the discussion on the matter in [18]).

3. Symmetric potentials

For symmetric potentials the transmission and reflection times coincide $\tau_{\rm t} = \tau_{\rm r} = \tau$, consequently (26) is reduced to its simplest expression

$$t_m = t_b + \frac{1}{2}\tau. \quad (28)$$

This extreme value gives the maximum of $P_{(a,b)}$ just at a half the time τ required to complete the scattering process (remember: the approach includes Gaussian wave packets that are peaked in the momentum k_0). That is, $t_m = t_*$. The probabilities depicted in Figure 1 correspond to symmetric barriers and wells for which $t_* \approx 11.79 + 6.53 = 18.32$ and $t_* \approx 11.79 + 1.53 = 13.32$, respectively.

Now, let us evaluate $P_{(a,b)}$ at the critical time defined in (28). We get

$$P_{(a,b)}\Big|_{t=t_m} = \frac{1}{\Delta}\text{erf}(z - ik_0/\sigma) + \frac{e^{-\frac{k_0^2}{\sigma^2}}}{\Delta}S_{\rm R}\left[\cos\varphi\left(\text{Re}\left[\text{erf}(z)\right] - 1\right) + \sin\varphi\,\text{Im}\left[\text{erf}(z)\right]\right], \quad (29)$$

with

$$z = \sigma k_0\tau + i\frac{k_0}{\sigma}.$$

The first term in (29) is a monotonically increasing function of τ, k_0, and σ with limit $1/\Delta$ at $t \to \infty$; this function corresponds to the individual contributions of the incident, transmitted and reflected wave packets. The second term, including S_R as a factor, is the contribution of the interference between the incident and reflected packets in the region $x > b$. This term oscillates with τ and goes to zero as τ increases (see Figure 2). Therefore, at $t = t_0$, the probability $P_{(a,b)}$ will oscillate around the values of $\text{erf}(\sigma k_0 \tau)$ from zero to $1/\Delta$ (within our approximation, $\sigma << 1$, one has $\Delta \approx 1$). We have depicted the behavior of $P_{(a,b)}$ as a function of τ in Figure 2. There, we can appreciate that the interference contributes mainly for small values of τ while the probability $P_{(a,b)}$ takes its smaller values. As τ depends on the analytic form of the function of compact support $V(x)$, among other variables, we realize that short times of scattering $\tau < 1$ imply small probabilities $P_{(a,b)}$ even for sharp momenta distributions in the wave packets. For the same packets and properly chosen scatterers $V(x)$, the scattering times τ are larger and the probabilities $P_{(a,b)}$ are bigger.

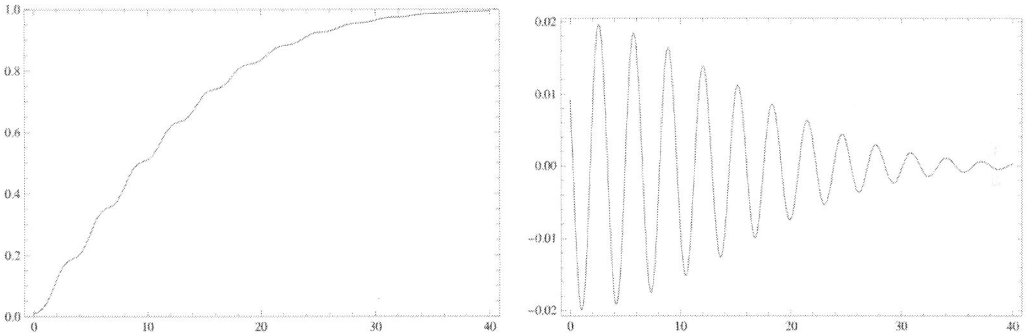

Figure 2. The probability $P_{(a,b)}$ defined in (29) as a function of the scattering time τ (left) and the contribution to $P_{(a,b)}$ due to the interference of the incident and reflected wave packets. In both cases $\sigma = 0.05$, $x_0 = 200$, $b = 1$, $S_R = 0.5$, $k_0 = 1$, $\varphi = 2.664$.

4. Concluding remarks

We have studied the probability of finding a particle, represented by a Gaussian wave packet, in the interaction zone of a scattering process that is characterized by a function of compact support $V(x)$. Our approach considers a series of approximations addressed to solve the transcendental equations that rule the times in which the probability is maximum. The main approximation is to consider (almost) monochromatic wave packets; that is, we assume that sharp momentum distributions can be used to construct the Gaussian wave packets. Within this approximation we have determined the extreme points of the probability we are interested in, and we have shown that these points can be used to determine the maximum of the probability. As an immediate application of our results, we have analyzed the case of symmetric potentials $V(x)$. Our results indicate that the interference between the incident and reflected waves plays a relevant role in the determination of the probabilities for scattering processes of short duration. That is, even for peaked distributions of momentum the analytic form of the potential $V(x)$ would produce small probabilities in the interaction zone. The times discussed in this work can be used to bound from below the duration of a scattering process, a condition that is crucial in the determination of dwell times and the related phenomena in quantum mechanics (see, e.g., [1, 2]).

Acknowledgments
The support of CONACyT and Cinvestav is acknowledged. The author appreciates the enlightening discussions with Prof. Oscar Rosas-Ortiz.

Journal of Physics: Conference Series **624** (2015) 012020 doi:10.1088/1742-6596/624/1/012020

References

[1] Muga J G, Sala Mayato R, and Egusquiza I L 2008 *Time in Quantum Mechanics* (Berlin: Springer)
[2] Muga J G, Ruschhaupt A and del Campo A 2010 *Time in Quantum Mechanics* vol 2 (Berlin: Springer)
[3] Wigner E P 1955 Lower limit for the energy derivative of the scattering phase shift *Phys. Rev.* **98** 145
[4] Smith F T 1960 Lifetime matrix in collision theory *Phys. Rev.* **118** 349
[5] Hartman T E 1962 Tunneling of a wave packet *J. Appl. Phys.* **33** 3427
[6] Büttiker M 1983 Larmor precession and the traversal time for tunneling *Phys. Rev.* E **27** 6178
[7] Golub R, Felber S, Gähler R and Gutsmiedl E 1990 A modest proposal concerning tunneling times *Phys. Lett.* A **148** 27-30
[8] Tsuchiya M, Matsusue T and Sakaki H 1987 Tunneling escape rate of electrons from quantum well in double-barrier heterostructures *Phys. Rev. Lett.* **59** 2356
[9] Steinberg A M, Kwiat P G and Chiao R Y 1993 Measurement of the single-photon tunneling time *Phys. Rev. Lett.* **71** 708
[10] Fernández-García N and Rosas-Ortiz O 2011 Rectangular potentials in a semi-harmonic background: spectrum, resonances and dwell time *SIGMA* **7** 044
[11] Rosas-Ortiz O, Cruz y Cruz S, and Fernández-García N 2012 Negative time delay for wave reflection from a one-dimensional semi-harmonic well *Proc. of the XXX Workshop on Geom. Meth. Phys.* ed P Kielanowski *et al* (Poland: Springer Basel) 275-281
[12] Rosas-Ortiz O, Cruz y Cruz S, and Fernández-García N 2012 Time delay in the reflection of particles by semi-harmonic wells *J. Phys.: Conf. Ser.* **380** 012018
[13] Rosas-Ortiz O, Fernández-García N and Cruz y Cruz S 2008 A primer on resonances in quantum mechanics *AIP Conf. Proc.* ed L M Montaño Zetina *et al* (Mexico city: American Institute of Physics) 31
[14] Fernández-García N and Rosas-Ortiz O 2008 Gamow-Siegert functions and Darboux-deformed short range potentials *Ann. Phys.* **323** 1397-1414
[15] Naylor A.W. and Sell G.R., *Linear Operator Theory in Engineering and Science*, Springer, New York, 1982
[16] Cohen-Tannoudji C, Diu B and Laloë F 1977 *Quantum Mechanics* vol 1 (New York: John Wiley) 83-84
[17] Abramowitz M and Stegun I A 1972 *Handbook of Mathematical Functions with Formulas, Graphs, and Mathematical Tables* (AMS-55) (Washington D.C.: National Bureau of Standars)
[18] Winful H G 2003 Delay time and the Hartman effect in quantum tunneling *Phys. Rev. Lett.* **91** 260401

Mielnik50 IOP Publishing
Journal of Physics: Conference Series **624** (2015) 012021 doi:10.1088/1742-6596/624/1/012021

Ternary generalization of Heisenberg's algebra

Richard Kerner
Laboratoire de Physique Théorique de la Matière Condensée (LPTMC), Univesité Pierre et Marie Curie - CNRS UMR 7600 Tour 23-13, 5-ème étage, Boîte Courrier 121, 4 Place Jussieu, 75005 Paris, FRANCE

E-mail: richard.kerner@upmc.fr

Dedicated to Bogdan Mielnik for the first 50 years of his scientific career

Abstract.
A concise study of ternary and cubic algebras with Z_3 grading is presented. We discuss some underlying ideas leading to the conclusion that the discrete symmetry group of permutations of three objects, S_3, and its abelian subgroup Z_3 may play an important role in quantum physics. We show then how most of important algebras with Z_2 grading can be generalized with ternary composition laws combined with a Z_3 grading.

We investigate in particular a ternary, Z_3-graded generalization of the Heisenberg algebra. It turns out that introducing a non-trivial cubic root of unity, $j = e^{\frac{2\pi i}{3}}$, one can define two types of creation operators instead of one, accompanying the usual annihilation operator. The two creation operators are non-hermitian, but they are mutually conjugate. Together, the three operators form a ternary algebra, and some of their cubic combinations generate the usual Heisenberg algebra.

An analogue of Hamiltonian operator is constructed by analogy with the usual harmonic oscillator, and some properties of its eigenfunctions are briefly discussed.

1. Introduction

More than fifty years ago Bogdan Mielnik has published several fundamental papers devoted to the analysis of physical and mathematical bases of quantum mechanics. The titles of these articles speak for themselves: the first two were named *"Geometry of quantum states"* ([1]) and *"Theory of Filters"* [2], and the third one was *"Generalized Quantum Mechanics"* [3].

In these articles, remarkable for the depth and clarity of developed argumentation as well as for their mathematical rigor, Bogdan Mielnik followed Bohr's interpretation of quantum mechanics rather than Einstein's point of view. Both approaches displayed in their famous discussions [4–6]. The essence of Bohr's approach was that neither a quantum object nor the classical object serving as measuring device can be totally separated from each other; any measurement describes the *interaction* between the quantum object and the observer. Einstein, on the other hand, in his opposition to Bohr, advocated a more classical idea of reality of quantum object independent of observer. In order to refute Bohr's arguments, Einstein (together with his collaborators, Boris Podolsky and Nathan Rosen) proposed an ideal experiment that would lead to a paradox if Bohr's interpretation of quantum mechanics were right. The paradox consisted in the possibility of a measurement performed on an atomic system to determine the issue of another measurement even in case of a space-like separation excluding any transfer of information. For example, in spite of the apparent randomness of photon emission by a single atom, according to Bohr, if a first photon emitted was found in a state with given polarization, the next photon will be found with a polarization perpendicular to the first one, as if the system "knew" in advance what to do next. Einstein thought that if such were the case, this would reveal the existence of *hidden parameters* if the following three principles were conserved: *1. Causality, 2. Relativity* (no physical influence can travel faster than the speed of light c) 3. *Locality* (the absence of action at a distance).

In 1964 John Bell [8, 9] formulated a theorem which gives a possibility to check the simultaneous validity of Einstein's point of view. The *Bell inequalities* introduced in these papers provide a criterion discriminating between quantum theory with hidden parameters, advocated by Einstein, and the probabilistic interpretation defended by Bohr. It took some time (more than twenty years) before a series of experiments carried out by A. Aspect and collaborators [10] has proven the existence of the *entangled states* of photons, thus confirming the non-local character of quantum physics, at least in some of its manifestations.

All that occurred long time after Bogdan Mielnik's first published papers; nevertheless he paid a great attention to mathematical and physical postulates of quantum theory and analyzed with an extremely acute vision the intricate framework of this fundamental branch of knowledge. Although he was active in other domains of physics, including General Relativity, his main and most original contributions concerned quantum theory and its possible extentions, including quaternionic and non-associative structures.

The models we shall display here represent a similar exploration effort, partly inspired by Bogdan Mielnik's work [3], with a novel algebraic approach based on *ternary* and $Z_3 - graded$ algebras. Before we proceed farther, let us expose the motivation for studying *ternary* and Z_3-*graded* generalizations of Heisenberg's algebra.

Quantum Mechanics provided an unexpected backing to ancient idea expressed by Plato in the seventh book of "Republic" known under the name of "Allegory of the Cave" [11–13]. Under closer scrutiny, it appears that continuous physical quantities of classical physics, described exclusively by real numbers, are the result of the averaging of measurements of quantum nature, described by complex probability amplitudes. Which leads to the conclusion that the underlying reality is of quantum and discrete nature, while the classical picture is due to the characteristic properties of our perception.

As a paradoxical conclusion, one is tempted to say that the so-called *complex numbers* deserve to be called "real", because they reflect the deepest properties of quantum physics, while the *real numbers* with which we describe our perceptible world appear to be the result of averaging over great number of microscopic events, and therefore deserve better the adjective "synthetic" or "imaginary".

Thus, the underlying physical realm is described in terms of states which are vectors, or rather *rays*, in a Hilbert space over complex numbers. One of the most striking properties of this space of states is the presence of universal discrete symmetry due to the representations of the simplest permutation group Z_2. All elementary particles we know belong to two distinct classes: fermions or bosons. The first ones ensure the stability of matter due to the Pauli exclusion principle [14], according to which two electrons cannot be in the same state characterized by identical quantum numbers. This principle not only explains the structure of atoms and therefore the entire content of the periodic table of elements, but it also guarantees their stability by preventing collapse, as suggested by Ehrenfest [15], and proved later by Dyson [16]. The relationship between the exclusion principle and particle's spin, known under the name of the "spin-and-statistic theorem", represents one of the deepest results in quantum field theory.

In purely algebraic terms Pauli's exclusion principle amounts to the anti-symmetry of wave functions describing two coexisting particle states. The easiest way to see how the principle works is to apply Dirac's formalism in which wave functions of particles in given state are obtained as products between the "bra" and "ket" vectors.

The wave function of a two-particle state of which one is in the state $\mid 1>$ and another in the state $\mid 2>$ (all other observables supposed to be the same for both states) is represented by a superposition

$$\mid \psi >= \Phi(1,2)\,(\mid 1 > \otimes \mid 2 >). \tag{1}$$

It is clear that if the wave function $\Phi(1,2)$ is anti-symmetric, i.e. if it satisfies $\Phi(1,2) = -\Phi(2,1)$, then $\Phi(1,1) = 0$ and such states have vanishing both their wave function and probability. It is easy to prove using the superposition principle, that this condition is not only sufficient, but also necessary.

After second quantization, when the states are obtained with creation and annihilation operators acting on the vacuum, the anti-symmetry is encoded in the anti-commutation relations

$$a^\dagger(1)a^\dagger(2) + a^\dagger(2)a^\dagger(1) = 0, \;\; a(1)a(2) + a(2)a(1) = 0 \tag{2}$$

The bottom line is that the Hilbert space of fermionic states is always divided in two sectors corresponding to the anti-commutation of creation of dichotomic spin state, admitting only two values which are labeled $+\frac{1}{2}$ and $-\frac{1}{2}$. The anti-commuting character of their operator algebra of observables is represented by the antisymmetric tensor $\epsilon_{\alpha\beta} = -\epsilon_{\beta\alpha}$, $\alpha, \beta = 1, 2$. The exclusion principle being universal, it is natural to require that it should be independent of the choice of a basis in the Hilbert space of states. Therefore, if the states undergo a linear transformation

$$\mid \psi^\alpha > \rightarrow \mid \psi_{\beta'} >= S^\alpha_{\beta'} \mid \psi_\alpha >, \tag{3}$$

the anti-symmetric form encoding the exclusion principle should remain the same as before:

$$\epsilon_{\alpha'\beta'} = S^\alpha_{\alpha'} S^\beta_{\beta'} \epsilon_{\alpha\beta}, \quad \text{with} \quad \epsilon_{1'2'} = -\epsilon_{2'1'} = 1, \quad \epsilon_{1'1'} = 0, \;\; \epsilon_{2'2'} = 0. \tag{4}$$

This invariance condition, akin to the invariance of the metric tensor $\eta_{\mu\nu}$ of the Minkowskian spacetime, defines the invariance group. It is easy to see that in this case the effect of (4) leads to the definition of the $SL(2, \mathbf{C})$ group. It is enough to check one of the four equations (4), e.g. choosing $\alpha' = 1$, $\beta' = 2$. We get then

$$\epsilon_{1'2'} = 1 = S^\alpha_{1'} S^\beta_{2'}\, \epsilon_{\alpha\beta} = S^1_{1'} S^2_{2'} - S^2_{1'} S^1_{2'} = \det S = 1. \tag{5}$$

The other three choices of index values in (4) are either redundant, or trivial, i.e. leading to the identity $0 = 0$. The conjugate matrices span an inequivalent representation of the $SL(2, \mathbf{C})$ group, labeled by dotted indeces; the antisymmetric 2-form $\epsilon_{\dot\alpha\dot\beta}$ leads to the same result when its invariance is required: $\epsilon_{\dot1\dot2} = -\epsilon_{\dot2\dot1} = 1$, $\epsilon_{\dot1\dot1} = 0$, $\epsilon_{\dot2\dot2} = 0$; $\epsilon_{\dot1'\dot2'} = \bar{S}^{\dot\alpha}_{\dot\alpha'} \bar{S}^{\dot\beta}_{\dot2'}\, \epsilon_{\dot\alpha\dot\beta}$ Spinors transforming with the $SL(2, \mathbf{C})$ group do not represent observable physical quantities; measurable quantities are formed by their quadratic and hermitian combinations, like e.g. a four-vector $\psi^\dagger\, \gamma^\mu \psi$, transforming under the vector representation of the Lorentz group. This can suggest that the very origin of the Lorentz transformations resides in the discrete properties of elementary particles, and not vice versa; more precisely, $SL(2, \mathbf{C})$ appears as the invariance group of the non-trivial action of the Z_2-group.

The mysterious properties of quarks, observable only by packs of three, or by quark-antiquark pairs, call for a non-trivial application of the Z_3-group, or even the entire S_3 group. Mathematical constructions acknowledging these properties via introduction of Z_3-graded algebras have been proposed in [19–21] and developed further in [23, 24]. A third-order generalization of Dirac equation was proposed, with waves that could propagate only as tensor products of three elementary solutions with complex wave vectors each.

In what follows, we shall present a Z_3-graded analogue of Heisenberg's algebra and corresponding generalization of quantum oscillator.

2. Cubic and ternary Z_3 graded algebras

Our goal being a ternary generalization of Heisenberg's algebra, let us start with recalling basic facts about ternary algebras [17, 18] The usual definition of an algebra involves a linear space $\mathcal{A}$ (over real or complex numbers) endowed with a *binary* constitutive relations

$$\mathcal{A} \times \mathcal{A} \rightarrow \mathcal{A} \tag{6}$$

In a finite dimensional case, dim $\mathcal{A}$ = N, in a chosen basis $\mathbf{e}_1, \mathbf{e}_2, ..., \mathbf{e}_N$, the constitutive relations (6) can be encoded in *structure constants* f^k_{ij} as follows:

$$\mathbf{e}_i \mathbf{e}_j = f^k_{ij}\, \mathbf{e}_k. \tag{7}$$

With the help of these structure constants all essential properties of a given algebra can be expressed, e.g. they will define a *Lie algebra* if they are antisymmetric and satisfy the Jacobi identity:

$$f^k_{ij} = -f^k_{ji}, \quad f^k_{im} f^m_{jl} + f^k_{jm} f^m_{li} + f^k_{lm} f^m_{ij} = 0, \tag{8}$$

whereas an abelian algebra will have its structure constants symmetric, $f^k_{ij} = f^k_{ji}$.

Usually, when we speak of algebras, we mean *binary algebras*, understanding that they are defined via *quadratic* constitutive relations (7). On such algebras the notion of Z_2 − *grading* can be naturally introduced. An algebra $\mathcal{A}$ is called a Z_2 − *graded* algebra if it is a direct sum of two parts, with symmetric (abelian) and anti-symmetric product respectively,

$$\mathcal{A} = \mathcal{A}_0 \oplus \mathcal{A}_1, \tag{9}$$

with *grade* of an element being 0 if it belongs to $\mathcal{A}_0$, and 1 if it belongs to $\mathcal{A}_1$. Under the multiplication in a Z_2-graded algebra the grades add up reproducing the composition law of the Z_2 permutation group: if the grade of an element A is a, and that of the element B is b, then the grade of their product will be $a + b$ *modulo* 2:

$$\text{grade}(AB) = \text{grade}(A) + \text{grade}(B), \quad \text{so that} \quad AB = (-1)^{ab} BA. \tag{10}$$

It is worthwhile to notice at this point that the above relationship can be written in an alternative form, with all the expressions on the left hand side as follows:

$$AB - (-1)^{\alpha\beta} BA = 0, \quad \text{or} \quad AB + (-1)^{(\alpha\beta+1)} BA = 0 \tag{11}$$

The equivalence between these two alternative definitions of commutation (anticommutation) relations inside a Z_2-graded algebra is no more possible if by analogy we want to impose *cubic* relations on algebras with Z_3-symmetry properties, in which the non-trivial cubic root of unity, $j = e^{\frac{2\pi i}{3}}$ plays the role similar to that of -1 in the binary relations displaying a Z_2-symmetry.

The Z_3 cyclic group is an abelian subgroup of the S_3 symmetry group of permutations of three objects. The S_3 groups contains *six* elements, including the group unity e (the identity permutation, leaving all objects in place: $(abc) \rightarrow (abc)$), the two cyclic permutations

$$(abc) \rightarrow (bca) \quad \text{and} \quad (abc) \rightarrow (cab),$$

and three odd permutations,

$$(abc) \rightarrow (cba), \quad (abc) \rightarrow (bac) \quad \text{and} \quad (abc) \rightarrow (acb).$$

There was a unique definition of *commutative* binary algebras given in two equivalent forms,

$$xy + (-1)yx = 0 \quad \text{or} \quad xy = yx. \tag{12}$$

In the case of cubic algebras [18] we have the following four generalizations of the notion of *commutative* algebras:

a) Generalizing the first form of the commutativity relation (12), which amounts to replacing the -1 generator of Z_2 by j-generator of Z_3 and binary products by products of three elements, we get

$$S: \quad x^\mu x^\nu x^\lambda + j \, x^\nu x^\lambda x^\mu + j^2 \, x^\lambda x^\mu x^\nu = 0, \tag{13}$$

where $j = e^{\frac{2\pi i}{3}}$ is a primitive third root of unity.

b) Another primitive third root, $j^2 = e^{\frac{4\pi i}{3}}$ can be used in place of the former one; this will define the conjugate algebra $\bar{S}$, satisfying the following cubic constitutive relations:

$$\bar{S}: \quad x^\mu x^\nu x^\lambda + j^2 \, x^\nu x^\lambda x^\mu + j \, x^\lambda x^\mu x^\nu = 0. \tag{14}$$

Clearly enough, both algebras are infinitely-dimensional and have the same structure. Each of them is a possible generalization of infinitely-dimensional algebra of usual commuting variables with a finite number of generators. In the usual Z_2-graded case such algebras are just polynomials in variables $x^1, \; x^2, ... x^N$; the algebras S and $\bar{S}$ defined above are also spanned by polynomials, but with different symmetry properties, and as a consequence, with different dimensions corresponding to a given power.

c) Then we can impose the following "weak" commutation, valid only for cyclic permutations of factors:

$$S_1: \quad x^\mu x^\nu x^\lambda = x^\nu x^\lambda x^\mu \neq x^\nu x^\mu x^\lambda, \tag{15}$$

d) Finally, we can impose the following "strong" commutation, valid for arbitrary (even or odd) permutations of three factors:

$$S_0: \quad x^\mu x^\nu x^\lambda = x^\nu x^\lambda x^\mu = x^\nu x^\mu x^\lambda \tag{16}$$

The four different associative algebras with cubic commutation relations can be represented in the following diagram, in which all arrows correspond to *surjective homomorphisms*. The commuting generators can be given the common grade 0.

$$\begin{array}{ccc} S & & \bar{S} \\ & \searrow \;\; \swarrow & \\ & S_1 & \\ & \downarrow & \\ & S_0 & \end{array}$$

Let us turn now to the Z_3 generalization of anti-commuting generators, which in the usual homogeneous case with Z_2-grading define Grassmann algebras. Here, too, we have four different choices:

a) The "strong" cubic anti-commutation,

$$\Lambda_0: \quad \Sigma_{\pi \in S_3} \, \theta^{\pi(A)} \theta^{\pi(B)} \theta^{\pi(C)} = 0, \tag{17}$$

i.e., the sum of *all* permutations of three factors, even and odd ones, must vanish.

b) The somewhat weaker "cyclic" anti-commutation relation,

$$\Lambda_1: \quad \theta^A \theta^B \theta^C + \theta^B \theta^C \theta^A + \theta^C \theta^A \theta^B = 0, \tag{18}$$

i.e., the sum of *cyclic* permutations of three elements must vanish. The same independent relation for the odd combination $\theta^C \theta^B \theta^A$ holds separately.

c) The j-skew-symmetric algebra:

$$\Lambda : \theta^A\theta^B\theta^C = j\,\theta^B\theta^C\theta^A. \tag{19}$$

and its conjugate algebra $\bar{\Lambda}$, isomorphic with Λ, which we distinguish by putting a bar on the generators and using dotted indices:

d) The j^2-skew-symmetric algebra:

$$\bar{\Lambda} : \quad \bar{\theta}^{\dot{A}}\bar{\theta}^{\dot{B}}\bar{\theta}^{\dot{C}} = j^2\bar{\theta}^{\dot{B}}\bar{\theta}^{\dot{C}}\bar{\theta}^{\dot{A}} \tag{20}$$

Both these algebras are finite dimensional. For j or j^2-skew-symmetric algebras with N generators the dimensions of their subspaces of given polynomial order are given by the following generating function:

$$H(t) = 1 + Nt + N^2t^2 + \frac{N(N-1)(N+1)}{3}, \tag{21}$$

where we include pure numbers (dimension 1), the N generators θ^A (or $\bar{\theta}^{\dot{B}}$), the N^2 independent quadratic combinations $\theta^A\theta^B$ and $N(N-1)(N+1)/3$ products of three generators $\theta^A\theta^B\theta^C$.

The above four cubic generalization of Grassmann algebra are represented in the following diagram, in which all the arrows are surjective homomorphisms.

$$\begin{array}{ccc} & \Lambda_0 & \\ & \downarrow & \\ & \Lambda_1 & \\ \swarrow & & \searrow \\ \Lambda & & \bar{\Lambda} \end{array}$$

3. Examples of Z_3-graded ternary algebras

3.1. The Z_3-graded analogue of Grassman algebra

Let us introduce N generators spanning a linear space over complex numbers, satisfying the following cubic relations [19, 20]:

$$\theta^A\theta^B\theta^C = j\,\theta^B\theta^C\theta^A = j^2\,\theta^C\theta^A\theta^B, \tag{22}$$

with $j = e^{2i\pi/3}$, the primitive root of 1. We have $1 + j + j^2 = 0$ and $\bar{j} = j^2$.

Let us denote the algebra spanned by the θ^A generators by $\mathcal{A}$ [19, 20].

We shall also introduce a similar set of *conjugate* generators, $\bar{\theta}^{\dot{A}}$, $\dot{A}, \dot{B}, ... = 1, 2, ..., N$, satisfying similar condition with j^2 replacing j:

$$\bar{\theta}^{\dot{A}}\bar{\theta}^{\dot{B}}\bar{\theta}^{\dot{C}} = j^2\,\bar{\theta}^{\dot{B}}\bar{\theta}^{\dot{C}}\bar{\theta}^{\dot{A}} = j\,\bar{\theta}^{\dot{C}}\bar{\theta}^{\dot{A}}\bar{\theta}^{\dot{B}}, \tag{23}$$

Let us denote this algebra by $\bar{\mathcal{A}}$.

We shall endow the algebra $\mathcal{A} \oplus \bar{\mathcal{A}}$ it with a natural Z_3 grading, considering the generators θ^A as grade 1 elements, their conjugates $\bar{\theta}^{\dot{A}}$ being of grade 2.

The grades add up modulo 3, so that the products $\theta^A\theta^B$ span a linear subspace of grade 2, and the cubic products $\theta^A\theta^B\theta^C$ being of grade 0. Similarly, all quadratic expressions in conjugate generators, $\bar{\theta}^{\dot{A}}\bar{\theta}^{\dot{B}}$ are of grade $2 + 2 = 4_{\text{mod}\,3} = 1$, whereas their cubic products are again of grade 0, like the cubic products od θ^A's. [23]

Combined with the associativity, these cubic relations impose finite dimension on the algebra generated by the Z_3 graded generators. As a matter of fact, cubic expressions are the highest order that does not vanish identically. The proof is immediate:

$$\theta^A\theta^B\theta^C\theta^D = j\,\theta^B\theta^C\theta^A\theta^D = j^2\,\theta^B\theta^A\theta^D\theta^C = j^3\,\theta^A\theta^D\theta^B\theta^C = j^4\,\theta^A\theta^B\theta^C\theta^D, \tag{24}$$

and because $j^4 = j \neq 1$, the only solution is $\theta^A\theta^B\theta^C\theta^D = 0$.

3.2. The Z_3 graded differential forms

Instead of the usual exterior differential operator satisfying $d^2 = 0$, we can postulate its Z_3-graded generalization satisfying

$$d^2 \neq 0, \quad d^3 f = 0$$

The first differential of a smooth function $f(x^i)$ is as usual

$$df = \partial_i f \, dx^i,$$

whereas the second differential is formally

$$d^2 f = (\partial_k \partial_i f) \, dx^k dx^i + (\partial_i f) \, d^2 x^i$$

We shall attribute the grade 1 to the 1-forms dx^i, $(i, j, k = 1, 2, ...N)$, and grade 2 to the forms d^2x^i, $(i, j, k = 1, 2, ...N)$; under associative multiplication of these forms the grades add up *modulo* 3

$$\text{grade}(\omega \, \theta) = \text{grade}(\omega) + \text{grade}(\theta) \; (\textit{modulo } 3).$$

The Z_3-graded differential operator d has the following property, compatible with grading we have chosen:

$$\begin{gathered} d(\omega \, \theta) = (d\omega) \, \theta + j^{\text{grade}_\omega} \, \omega \, d\theta. \\ d^2 f = (\partial_i \partial_k f) dx^i dx^k + (\partial_i f) \, d^2 x^i, \\ d^3 f = (\partial_m \partial_i \partial_k f) dx^m dx^i dx^k + (\partial_i \partial_k f) d^2 x^i dx^k \\ +j \, (\partial_i \partial_k f) dx^i d^2 x^k + (\partial_k \partial_i f) dx^k d^2 x^i + (\partial_i f) \, d^3 x^i. \end{gathered}$$

equivalent with

$$d^3 f = (\partial_m \partial_i \partial_k f) dx^m dx^i dx^k + (\partial_i \partial_k f) [d^2 x^k dx^i - j^2 \, dx^i d^2 x^k] + (\partial_i f) \, d^3 x^i.$$

Consequently, assuming that $d^3 x^k = 0$ and $d^3 f = 0$, to make the remaining terms vanish we must impose the following commutation relations on the products of forms:

$$dx^i dx^k dx^m = j \, dx^k dx^m dx^i, \qquad dx^i d^2 x^k = j \, d^2 x^k dx^i,$$

therefore

$$d^2 x^k dx^i = j^2 \, dx^i d^2 x^k.$$

As in the case of the abstract Z_3-graded Grassmann algebra, the fourth order expressions must vanish due to the associativity of the product:

$$dx^i dx^k dx^l dx^m = 0.$$

Consequently, we shall assume that also

$$d^2 x^i d^2 x^k = 0.$$

This completes the construction of algebra of Z_3-graded exterior forms.

3.3. Ternary Clifford algebra

Let us introduce the following three 3×3 matrices:

$$Q_1 = \begin{pmatrix} 0 & 1 & 0 \\ 0 & 0 & j \\ j^2 & 0 & 0 \end{pmatrix}, \quad Q_2 = \begin{pmatrix} 0 & j & 0 \\ 0 & 0 & 1 \\ j^2 & 0 & 0 \end{pmatrix}, \quad Q_3 = \begin{pmatrix} 0 & 1 & 0 \\ 0 & 0 & 1 \\ 1 & 0 & 0 \end{pmatrix}, \tag{25}$$

and their hermitian conjugates

$$Q_1^\dagger = \begin{pmatrix} 0 & 0 & j \\ 1 & 0 & 0 \\ 0 & j^2 & 0 \end{pmatrix}, \quad Q_2^\dagger = \begin{pmatrix} 0 & 0 & j \\ j^2 & 0 & 0 \\ 0 & 1 & 0 \end{pmatrix}, \quad Q_3^\dagger = \begin{pmatrix} 0 & 0 & 1 \\ 1 & 0 & 0 \\ 0 & 1 & 0 \end{pmatrix}. \tag{26}$$

These matrices can be allowed natural Z_3 grading,

$$\text{grade}(Q_k) = 1, \quad \text{grade}(Q_k^\dagger) = 2, \tag{27}$$

The above matrices span a very interesting ternary algebra. Out of three independent Z_3-graded ternary combinations, only one leads to a non-vanishing result. One can check without much effort that both j and j^2 skew ternary commutators do vanish:

$$\{Q_1, Q_2, Q_3\}_j = Q_1 Q_2 Q_3 + j Q_2 Q_3 Q_1 + j^2 Q_3 Q_1 Q_2 = 0,$$
$$\{Q_1, Q_2, Q_3\}_{j^2} = Q_1 Q_2 Q_3 + j^2 Q_2 Q_3 Q_1 + j Q_3 Q_1 Q_2 = 0,$$

and similarly for the odd permutation, $Q_2 Q_1 Q_3$. On the contrary, the totally symmetric combination does not vanish; it is proportional to the 3×3 identity matrix $\mathbf{1}$:

$$Q_a Q_b Q_c + Q_b Q_c Q_a + Q_c Q_a Q_b = \eta_{abc} \, \mathbf{1}, \quad a, b, ... = 1, 2, 3. \tag{28}$$

with η_{abc} given by the following non-zero components:

$$\eta_{111} = \eta_{222} = \eta_{333} = 1, \quad \eta_{123} = \eta_{231} = \eta_{312} = 1, \quad \eta_{213} = \eta_{321} = \eta_{132} = j^2, \tag{29}$$

all other components vanishing. The relation (28) may serve as the definition of *ternary Clifford algebra.*

Another set of three matrices is formed by the hermitian conjugates of Q_a, which we shall endow with dotted indices $\dot{a}, \dot{b}, ... = 1, 2, 3$: $Q_{\dot{a}} = Q_a^\dagger$ satisfying conjugate identities

$$Q_{\dot{a}} Q_{\dot{b}} Q_{\dot{c}} + Q_{\dot{b}} Q_{\dot{c}} Q_{\dot{a}} + Q_{\dot{c}} Q_{\dot{a}} Q_{\dot{b}} = \eta_{\dot{a}\dot{b}\dot{c}} \, \mathbf{1}, \quad \dot{a}, \dot{b}, ... = 1, 2, 3. \tag{30}$$

with $\eta_{\dot{a}\dot{b}\dot{c}} = \bar{\eta}_{abc}$.

It is obvious that any similarity transformation of the generators Q_a will keep the ternary anti-commutator (29) invariant. As a matter of fact, if we define $\tilde{Q}_b = P^{-1} Q_b P$, with P a non-singular 3×3 matrix, the new set of generators will satisfy the same ternary relations, because

$$\tilde{Q}_a \tilde{Q}_b \tilde{Q}_c = P^{-1} Q_a P P^{-1} Q_b P P^{-1} Q_c P = P^{-1} (Q_a Q_b Q_c) P,$$

and on the right-hand side we have the unit matrix which commutes with all other matrices, so that $P^{-1} \, \mathbf{1} \, P = \mathbf{1}$.

3.4. Ternary Z_3-graded commutator

In any associative algebra $\mathcal{A}$ one can introduce a new binary operation, the *commutator*, using the generator of the Z_2 group in form of multiplication by -1 :

$$X, Y \in A \rightarrow [X, Y] = XY + (-1)YX = XY - YX. \tag{31}$$

In the case of the Z_2 group the generator of its representation on complex numbers was equal to -1; note that $-1 + (-1)^2 = 0$ In the case of the Z_3 group, the generator of its complex representation can be chosen to be $j = e^{\frac{2\pi i}{3}}$, with $j + j^2 + j^3 = 0$

Consider the following cubic combination defined on an associative algebra $\mathcal{A}$:

$$X, Y, Z \in \mathcal{A}, \qquad \{X, Y, Z\} := XYZ + jYZX + j^2 ZXY.$$

One obviously has:

$$\{X, Y, Z\} = j\{Y, Z, X\} = j^2\{Z, X, Y\}, \quad \text{and consequently} \quad \{X, X, X\} = 0.$$

In the case when $\mathcal{A}$ is a *unital* algebra, i.e. it contains a unit element $\mathbf{1}$ such that $\mathbf{1}X = X\mathbf{1} = X$ for any $X \in \mathcal{A}$, a unique Lie algebra is naturally generated by the cubic commutator:

$$\begin{aligned}\{X, 1, Y\} &= X \cdot 1 \cdot Y + j\, 1 \cdot Y \cdot X + j^2\, Y \cdot X \cdot 1 \\ &= XY + j\, YX + j^2\, YX = XY + (j + j^2)YX = XY - YX = [X, Y].\end{aligned}$$

The following simple example of a Z_3 cubic algebra can be constructed with 2×2 complex matrices. Consider the Lie algebra spanned by three Pauli's matrices:

$$\sigma_1 = \begin{pmatrix} 0 & 1 \\ 1 & 0 \end{pmatrix}, \quad \sigma_2 = \begin{pmatrix} 0 & -i \\ i & 0 \end{pmatrix}, \quad \sigma_3 = \begin{pmatrix} 1 & 0 \\ 0 & -1 \end{pmatrix}$$

satisfying the well known commutation relations:

$$[\sigma_i, \sigma_j] = C^k_{ij}\, \sigma_k, \quad i, j, k = 1, 2, 3, \quad C^k_{ij} = 2i\epsilon_{kij}.$$

The enveloping algebra $\mathcal{A}_\sigma$ contains the unit matrix $\mathbf{1}$:

$$\sigma_i \sigma_j = i\, \epsilon_{ijk} + \delta_{ij}\mathbf{1}$$

Let us define the *cubic j-commutator* on the algebra $\mathcal{A}_\sigma$:

$$\{\sigma_i, \sigma_j, \sigma_k\} = \sigma_i\sigma_j\sigma_k + j\, \sigma_j\sigma_k\sigma_i + j^2\, \sigma_k\sigma_i\sigma_j, \quad j = e^{\frac{2\pi i}{3}}.$$

This cubic algebra contains three cubic subalgebras generated by two σ-matrices out of three: for example

$$\{\sigma_1, \sigma_2, \sigma_1\} = -2\, \sigma_2, \quad \{\sigma_2, \sigma_1\, \sigma_2\} = -2\, \sigma_1,$$

and similarly for the couples σ_2, σ_3 and σ_3, σ_1. We have also $\{\sigma_1, \sigma_2, \sigma_3\} = 0$.

4. Ternary Heisenberg algebra

4.1. The Z_2 Heisenberg algebra

Let us first remind the original construction of the Heisenberg algebra used in the analysis of *quantum harmonic oscillator.* The Hamiltonian of classical harmonic oscillator, expressed in reduced dimensionless variables reads:

$$H = \frac{p^2}{2} + \frac{x^2}{2}$$

after first quantization which attributes to canonical variables (p, x) corresponding quantum operators acting on the Hilbert space $L^2(R)$ of square-integrable functions

$$p \to \hat{p} = -i\hbar\frac{d}{dx}, \quad x \to \hat{x}$$

The quantum version of the Hamiltonian is the hermitian operator ($\hbar = 1$)

$$\hat{H} = \frac{1}{2}\left[-\frac{d^2}{dx^2} + x^2\right]$$

The classical Heisenberg algebra [25] is generated by the following two operators:

$$a = \frac{1}{\sqrt{2}}\left(\bar{x} + i\bar{p}\right), \quad a^{\dagger} = \frac{1}{\sqrt{2}}\left(\bar{x} - i\bar{p}\right),$$

(in terms of dimensionless operators $\bar{x}$ and $\bar{p}$ defined as follows:

$$\bar{x} = \frac{x}{\lambda}, \quad \bar{p} = \frac{\lambda p}{\hbar}.$$

with λ some unit of length) satisfying the well known commutation relations

$$[\bar{x},\, \bar{p}] = i. \quad [a,\, a^{\dagger}] = 1.$$

From now on we follow original Schrödinger's presentation [25], without the factors $1/\sqrt{2}$ in a and $a^{\dagger}$, and with Hamiltonian without the factor $1/2$. Now the quantum version of he Hamiltonian becomes a hermitian operator, which can be expressed by means of the operators a and $a^{\dagger}$:

$$\hat{H} = -\frac{d^2}{dx^2} + x^2 = aa^{\dagger} + a^{\dagger}a,$$

The eigenfunctions of this operator, corresponding to fixed values of energy E, are obtained by defining first the lowest energy state corresponding to the zero-eigenvalue state of the operator a, $a \mid 0 >= 0$. In coordinate representation the zero state is represented by a Gaussian function:

$$a \mid 0 >= 0 \;\to \left[\frac{d}{dx} + x\right]\Phi_0\left(x\right) = 0 \quad \to \Phi_0\left(x\right) = e^{-\frac{x^2}{2}}. \tag{32}$$

Positive eigenvalue functions of the Hamiltonian $\hat{H}$ are then obtained by acting on $f(x)$ with consecutive powers of $a^{\dagger}$ [25]. These eigenfunctions span the well-known discrete energy spectrum (for bound states with negative total energy). The hermitian operators $\hat{p}$ and $\hat{x}$ could be replaced in this construction by non-hermitian combinations $a = ip + x, \quad a^{\dagger} = -ip + x$ because their quadratic combination $H = a^{\dagger}a + aa^{\dagger}$ is hermitian and positive definite. Due to the commutation relations between $a^{\dagger}$ and a, it can be easily checked that the new function obtained by the action of $a^{\dagger}$ on the vacuum, which is

$$a^{\dagger} \mid 0 >=\mid 1 > \;\to\; \left[-\frac{d}{dx} + x\right]e^{-\frac{x^2}{2}} = 2x\,e^{-\frac{x^2}{2}} = \Phi_1, \tag{33}$$

is an eigenfunction of the operator H:

$$H \mid \Phi_1 >= 3 \mid \Phi_1 > \quad \text{because} \quad \left[-\frac{d^2}{x^2} + x^2\right]2xe^{-\frac{x^2}{2}} = 6x\,e^{-\frac{x^2}{2}}. \tag{34}$$

4.2. The Z_3 ternary Heisenberg algebra

In the Z_2-graded case the two independent combinations of the operators $\frac{d}{dx}$ and x are given by the following non-singular matrix:

$$\begin{pmatrix} 1 & 1 \\ -1 & 1 \end{pmatrix}.$$

By analogy, in the Z_3-graded case, the following transformation matrix should be used, producing three independent combinations of the operators $\frac{d}{dx}$, x and $\mathbf{1}$:

$$\begin{pmatrix} 1 & 1 & 1 \\ j & j^2 & 1 \\ j^2 & j & 1 \end{pmatrix}$$

A natural ternary generalization of Heisenberg's algebra should be spanned by three generators [26] instead of two. By analogy with the Z_2 case, the Z_3-generalization will be generated by the following three operators:

$$c_1 = \lambda \frac{d}{dx} + jx + j^2 1, \qquad c_2 = \lambda \frac{d}{dx} + j^2 x + j1, \qquad c_3 = \lambda \frac{d}{dx} + x + 1,$$

which span the following *binary* Lie algebra:

$$[c_1, c_2] = \lambda(j^2 - j)\, 1, \qquad [c_2, c_3] = \lambda(1 - j^2)\, 1, \qquad [c_3, c_1] = \lambda(j - 1)\, 1,$$

The following linear relations hold:

$$\frac{1}{3}(c_1 + c_2 + c_3) = \lambda \frac{d}{dx}, \qquad \frac{1}{3}\left(j\, c_1 + j^2\, c_2 + c_3\right) = \mathbf{1}, \qquad \frac{1}{3}\left(j^2\, c_1 + j c_2 + c_3\right) = x,$$

Here are the independent 3-commutators between the generators of our algebra:

$$\begin{gathered}
\{c_1, c_2, c_1\} = -3\lambda\, c_1, \quad \{c_2, c_1, c_2\} = 3\lambda\, c_2, \\
\{c_2, c_3, c_2\} = -3j\, \lambda\, c_2, \quad \{c_3, c_2, c_3\} = 3j \lambda\, c_3, \\
\{c_3, c_1, c_3\} = -3j^2\, \lambda\, c_3, \quad \{c_1, c_3, c_1\} = 3j^2\, \lambda\, c_1, \\
\{c_2, c_3, c_1\} = \lambda\, \left[(1 - j)c_1 + (j^2 - 1)c_2 + (j - j^2)c_3\right], \\
\{c_1, c_3, c_2\} = \lambda\, \left[(j^2 - j)c_1 + (j^2 - j)c_2 + (j^2 - j)c_3\right],
\end{gathered}$$

If we add the unit operator $\mathbf{1}$ to the three generators c_1, c_2 and c_3, the ordinary commutators between the generators can be interpreted as ternary commutators involving the unit operator 1 in the middle:

$$\begin{gathered}
[c_1, c_2] = \{c_1, 1, c_2\} = \lambda(j^2 - j)\, 1, \\
[c_2, c_3] = \{c_2, 1, c_3\} = \lambda(1 - j^2)\, 1, \\
[c_3, c_1] = \{c_3, 1, c_1\} = \lambda(j - 1)\, 1, \\
(c_1 c_3 c_2 + c_3 c_2 c_1 + c_2 c_1 c_3) - (c_2 c_3 c_1 + c_3 c_1 c_2 + c_1 c_2 c_3) = 3\lambda(j - j^2).
\end{gathered}$$

after renormalization $c_k \to \frac{1}{\sqrt{3}} c_k$ and setting $\lambda = -i$, the right-hand side of this cubic commutation relation will become equal to $\hbar \mathbf{1}$. Normalized operators are as follows

$$c_1 = \frac{1}{\sqrt{3}}\left[\frac{\sqrt{\hbar}}{i}\frac{d}{dx} + jx + j^2 1\right], \quad c_2 = \frac{1}{\sqrt{3}}\left[\frac{\sqrt{\hbar}}{i}\frac{d}{dx} + j^2 x + j1\right], \quad c_3 = \frac{1}{\sqrt{3}}\left[\frac{\sqrt{\hbar}}{i}\frac{d}{dx} + x + 1\right],$$

5. Ternary analogue of quantum oscillator

5.1. Cubic combinations of the generators

The Z_3-graded analogues of creation and annihilation operators a and $a^\dagger$ are the operators $c_3 = c_3^\dagger$, c_1 and $c_2 = c_1^\dagger$. Imposing hermiticity means that the factor λ in front of the derivation must be pure imaginary, so we shall set $\lambda = -i$, as in the usual quantization scheme.

The expression $aa^\dagger + a^\dagger a$ can be interpreted as a sum of all Z_2 permutations, or all S_2 permutations, because here all permutations are cyclic, which is not the case of Z_3 and S_3 groups. The sum of all cyclic permutations of three operators $c_1c_2c_3$, or even all (six) permutations does not lead to any simple expression. Moreover, in such combinations the complex representation of Z_3 does not appear.

However, the quadratic harmonic oscillator Hamiltonian $aa^\dagger + a^\dagger a$ can be represented in an alternative manner, in which the representation of Z_2 by multiplications by 1 and -1 does appear explicitly:

$$\hat{H} = \frac{1}{2}\left[(a+a^\dagger)^2 + (-1)(a-a^\dagger)^2\right] = aa^\dagger + a^\dagger a. \tag{35}$$

Now the Z_3 generalization becomes obvious: we should form the sum of *cubes* of similar expressions with three generators each, multiplied by j and j^2 in all possible combinations so as to ensure the hermiticity of the resulting expression, like in (35) above. Two expressions of this type can be formed with the generators c_1, c_2 and c_3,

$$\frac{1}{27}\left[(c_3+c_1+c_2)^3 + j(c_3+jc_1+j^2c_2)^3 + j^2(c_3+j^2c_1+jc_2)^3\right],$$

$$\text{and}\quad \frac{1}{27}\left[(c_3+c_1+c_2)^3 + j^2(c_3+jc_1+j^2c_2)^3 + j(c_3+j^2c_1+jc_2)^3\right].$$

Neither of these two combinations is hermitian, but their sum $\hat{K}_{Z_3}$ is (if we set λ pure imaginary, e.g. $-i$):

$$\hat{K}_{Z_3} = 2\lambda^3\frac{d^3}{dx^3} - x^3 - 1 = 2i\frac{d^3}{dx^3} - x^3 - 1. \tag{36}$$

The expression x^3+1 appearing as the "potential" part factorizes as follows:

$$x^3 + 1 = (x+1)(x+j)(x+j^2).$$

However, this operator does not seem to be a proper Z_3-graded generalization of quantum harmonic oscillator Hamiltonian. It is an odd function of momentum, and its physical dimension is a fractional power of energy, and its spectrum, although real, is not bound from below because it is not positive-definite like the ordinary quadratic Hamiltonian.

The eigenvalue equation

$$\hat{K}_{Z_3} f(x) = K f(x)$$

can be solved explicitly for the particular value of $K = -1$, which amounts to solve the following differential equation of third order:

$$2i\frac{d^3 f}{dx^3} - x^3 f(x) = 0. \tag{37}$$

This equation can be solved by expanding $f(x)$ into a power series

$$f(x) = \sum_{k=0}^{\infty} c_k\, x^k.$$

The solution is given in terms of the *generalized hypergeometric function* $F([\ ];p,q;\xi)$. In the convention used in "Maple" computing program, this symbol defines the following power series:

$$F([\ ];p,q;\xi) = \sum_{k=0}^{\infty} c_k x^k, \quad c_k = \frac{\Gamma(p)\Gamma(q)}{\Gamma(p+k)\Gamma(q+k)}. \tag{38}$$

The eigenfunction satisfying (37) is a linear combination of three independent solutions:

$$F([\ ];\frac{2}{3},\frac{5}{6};\frac{ix^6}{432}), \quad x\,F([\ ];\frac{5}{6},\frac{7}{6};\frac{ix^6}{432}), \quad \text{and} \quad x^2\,F([\ ];\frac{7}{3},\frac{4}{3};\frac{ix^6}{432}), \tag{39}$$

corresponding to three different initial conditions, as it should be for a third order differential equation. All these eigenfunctions are complex.

In the classical harmonic oscillator case the operators a and $a^\dagger$ were not hermitian; nevertheless their combination $aa^\dagger + a^\dagger a$ became hermitian. We could have as well abandoned the requirement of hermiticity, i.e. admit real value for the factor λ in the definition of our operators c_k, and still get real eigenvalues for the non-hermitian third order operator, in the spirit of C. Bender's work on the subject [27]. But the third-order differential operator is not a good candidate for Hamiltonian of any kind, because it is not positive definite and its spectrum can not display a minimal value. Moreover, in the operators c_k the momentum operator $\hat{p} = -id/dx$ can not be replaced by its real counterpart d/dx without destroying the hermiticity. Such a substitution will be nevertheless possible if we introduce quadratic combination of two third-order operators. This is why we propose to consider the following two third-order operators:

$$\hat{K}_1 = ip^3 + x^3 = -\frac{d^3}{dx^3} + x^3, \quad \hat{K}_2 = -ip^3 + x^3 = \frac{d^3}{dx^3} + x^3. \tag{40}$$

Then it is easy to see that one has

$$\hat{K}_1^\dagger = \hat{K}_2 \quad \text{and} \quad \frac{1}{2}\left[\hat{K}_1\hat{K}_2 + \hat{K}_2\hat{K}_1\right] = \hat{H}_{Z_3} = \frac{d^6}{dx^6} + x^6. \tag{41}$$

The zero-value eigenfunctions of the operator are of the same form as the eigenfunction (39), but with real argument $x^6/432$ instead of the imaginary one $ix^6/432$. The same operator can be represented as a cubic combination of second-order operators, one of which is the classical harmonic oscillator, the two remaining ones of the same form, but with complex conjugate potential parts:

$$\hat{H}_0 = -\frac{d^2}{dx^2} + x^2, \quad \hat{H}_1 = -\frac{d^2}{dx^2} + j\,x^2, \quad \hat{H}_2 = -\frac{d^2}{dx^2} + j^2\,x^2, \tag{42}$$

One easily checks that

$$\hat{H}_0\hat{H}_1\hat{H}_2 + \hat{H}_1\hat{H}_2\hat{H}_0 + \hat{H}_2\hat{H}_0\hat{H}_1 = -\frac{d^6}{dx^6} + x^6. \tag{43}$$

The odd combination of the same kind gives the identical result.

This suggests that what we are describing here is a kind of third power of ordinary harmonic oscillator with its complexified counterparts. Each of the two "exotic" operators, $\hat{H}_1 = i\hat{p} + jx$ and $\hat{H}_2 = \hat{H}_1^\dagger = i\hat{p} + j^2 x$ displaying a complex spectrum, with the energies multiplied by j and by j^2, respectively. This seems to indicate that the spectrum of the cubic combination $\hat{H}_{Z_3}$ should be proportional to the products of the corresponding three eigenvalues, E_0, $E_1 = jE_0$ and $E_2 = j^2 E_0$, leading to the real eigenvalue $E_0 \cdot E_1 \cdot E_2 = j \cdot j^2 \cdot E_0^3 = E_0^3$.

We shall show now that the classical Bohr-Sommerfeld quantization scheme applied to the Hamiltonian $\hat{H}_{Z_3}$ confirms this hypothesis.

5.2. Bohr-Sommerfeld quantization

In the case of a one-dimensional harmonic oscillator whose classical Hamiltonian is given by the quadratic expression

$$H(p,x) = \frac{p^2}{2m} + \frac{kx^2}{2},$$

the Bohr-Sommerfeld quantization rule imposes the condition of finite discrete energy levels via the following rule:

$$\int_{\text{period}} pdx = n\hbar \tag{44}$$

Expressing the momentum p as a function of the coordinate x by posing $H(p,x) = E = \text{Constant}$, one gets the explicit expression of this integral:

$$\int_{\text{period}} pdx = 2m\omega \int_{x_{min}}^{x_{max}} \sqrt{\frac{2E}{k} - x^2}\, dx = n\hbar, \tag{45}$$

where $\omega = \sqrt{k/m}$, and E is the energy. The integral is well known, and leads to the quantization rule $E = \omega\hbar\left(n + \frac{1}{2}\right)$.

Now let us apply a similar quantization condition to our "cubic Hamiltonian". We start with the expression which is supposed to describe something that is an analogue of a Hamiltonian function. This expression is obtained from ternary generalization of Heisenberg's algebra:

$$H_{(3)} = \frac{p^6}{6m^3} + \frac{k^3x^6}{6} \tag{46}$$

The dimension of this expression is the third power of energy, i.e., $\dim[H] = [\text{energy}]^3$.

We shall suppose that the expression (46) is constant in time. This gives the following equation:

$$H_{(6)} = \frac{p^6}{6m^3} + \frac{k^3x^6}{6} = E^3 = \text{Const.} \tag{47}$$

This constant enables us to get the following expression of p as a function of x:

$$p = 6^{\frac{1}{6}}\sqrt{mk}\left[\frac{6E^3}{k^3} - x^6\right]^{\frac{1}{6}}. \tag{48}$$

It is easy to see that the corresponding trajectory is closed, and the motion must be periodic.

Now let us impose the Bohr-Sommerfeld quantization condition (44), which is based on a quasi-classical approximation and gives very good agreement with observation in the large quantum number limit, as shown e.g., in [28]: which in this case yields the explicit integral that can be evaluated with a little effort, yielding the following result when we substitute for p its expression (48):

$$\int_{\text{period}} pdx = 2\int_{-x_{max}}^{+x_{max}} pdx = 2\sqrt{mk}\int_{-x_{max}}^{+x_{max}} \left[\frac{6E^3}{k^3} - x^6\right]^{\frac{1}{6}} dx. \tag{49}$$

where the limits of integration are defined by the condition $p = 0$, which yields $x_{max} = 6^{\frac{1}{6}}\sqrt{\frac{E}{k}}$.

The above integral can be expressed by means of Gamma-functions as follows:

$$2\sqrt{mk}\int_{-x_{max}}^{+x_{max}} \left[\frac{6E^3}{k^3} - x^6\right]^{\frac{1}{6}} dx = 4\cdot 6^{\frac{1}{3}}\sqrt{mk}\frac{\left[\Gamma(\frac{7}{6})\right]^2}{\Gamma(\frac{4}{3})}\frac{E}{k}, \tag{50}$$

The Bohr-Sommerfeld quantization rule supposes that this integral can take on discrete values $n\hbar$, which leads to the following discrete values of the energy E:

$$E_n = \left[\frac{\Gamma(\frac{4}{3})}{4 \cdot 6^{\frac{1}{3}} \left[\Gamma(\frac{7}{6})\right]^2} \right] n\hbar\omega, \tag{51}$$

with $\omega = \sqrt{k/m}$. Like in the usual harmonic oscillator, the energy levels are proportional to integers, $E_n \sim n\hbar\omega$? Consequently, the eigenvalues of $\hat{H}_{Z_3}$ follow the rule $\lambda_n \simeq n\omega^3\hbar^3$.

Acknowledgements
The author is greatly indebted to Karol Penson and Michel Dubois-Violette for numerous enlightening discussions and many useful remarks.

References

[1] Mielnik B. 1968 *Comm. Math. Phys.* **9** (1) pp.55-80.
[2] Mielnik B. 1969 *Comm. Math. Phys.* **15** (1) pp. 1-46
[3] Mielnik B. 1974 *Comm. Math. Phys.* **37** (3) pp.221-256.
[4] Bohr N. 1996 *Collected Works*, Vol. **6** p.21; Vol. **7**, p.263 Elsevier, Amsterdam
[5] Lurçat F. 2001 *"Niels Bohr et la physique quantique"*, Editions Seuil, Paris.
[6] Landé A. 1974 *American Journal of Physics* **42** p. 459-464
[7] Einstein A., Podolsky B., Rosen N. 1935 *Phys. Rev.* **47** 1935, p. 777-780
[8] Bell J.S, 1964 *Physics* 1 (1964), p.195
[9] Bell J.S, 1966 *Review of Modern Physics* **38** p.447.
[10] A. Aspect, P. Grangier, G. Roger, 1982 Physical Review Letters, **49** (2), p.91-94
[11] Plato: Republic, London: Wordsworth Editions, ISBN 1-85326-483-0
[12] Cohen S M. S. 2008 *Philosophy* **320**
[13] Hall D. 1980. *Interpreting Plato's Cave as an Allegory of the Human Condition. Apeiron* **14** (2), pp.74-86.
[14] Pauli W. 1940 *The Connection Between Spin and Statistics, Phys. Rev.* **58**, pp. 716-722
[15] Dyson F.J. 1967 *Journ. Math. Phys.* **8**, pp.1538-1545
[16] Dyson F.J., Lenard A 1968 *Stability of Matter, Parts I and II J. Math. Phys.*, **8**, pp.423-434; *J. Math. Phys.*, **9**, pp. 698-711
[17] Gelfand I M, Kapranov M M, Zelevinsky A V 1994 *Multidimensional Determinants, Discriminants and Resultants* (book), Birkhauser, Boston
[18] Vainerman L, Kerner R 1996 *Journal of Mathematical Physics* **37** (5) pp. 2553-2665
[19] Kerner R 1991 *Comptes Rendus Acad. Sci. Paris.* **10** 1237
[20] Kerner R 1992 *Journal of Mathematical Physics*, **33** (1) pp.403-4011
[21] Abramov V, Kerner R, Le Roy B 1997 *Journal of Mathematical Physics*, **38** (3) pp. 1650-1669
[22] Kerner R 1997 *Classical and Quantum Gravity* **14** (1A) pp. A203-A225
[23] Kerner R 2001 *Proceedings of the 23-rd ICGTMP colloquium, Dubna 2000* arXiv:math-ph/0011023
[24] Kerner R Suzuki O 2012 *Int. J. Geom. Methods Mod. Phys.* **09**, 1261007
[25] Schrödinger E 1940 *Proceedings of the Royal Irish Academy*, **46**(1940/1941), pp.9-16
[26] Nambu Y 1973. *Physical Review D* **7** (8) pp.2405-2412
[27] Bender C M, Boettcher S 1998 *Phys. Rev. Lett.* **80**, 5243
[28] Gorska K, Penson K.A., Horzela A, Duchamp G.H.E., Blasiak P and Solomon A.I. 2010 *J. Math. Phys.* **51**, 122102

AUTHOR INDEX